Dan Huang
96

EMULSION POLYMER TECHNOLOGY

PLASTICS ENGINEERING

Series Editor

Donald E. Hudgin
Princeton Polymer Laboratories
Plainsboro, New Jersey

1. Plastics Waste: Recovery of Economic Value, *Jacob Leidner*
2. Polyester Molding Compounds, *Robert Burns*
3. Carbon Black-Polymer Composites: The Physics of Electrically Conducting Composites, *Edited by Enid Keil Sichel*
4. The Strength and Stiffness of Polymers, *Edited by Anagnostis E. Zachariades and Roger S. Porter*
5. Selecting Thermoplastics for Engineering Applications, *Charles P. MacDermott*
6. Engineering with Rigid PVC: Processability and Applications, *Edited by I. Luis Gomez*
7. Computer-Aided Design of Polymers and Composites, *D. H. Kaelble*
8. Engineering Thermoplastics: Properties and Applications, *Edited by James M. Margolis*
9. Structural Foam: A Purchasing and Design Guide, *Bruce C. Wendle*
Plastics in Architecture: A Guide to Acrylic and Polycarbonate, *Ralph Montella*
11. Metal-Filled Polymers: Properties and Applications, *Edited by Swapan K. Bhattacharya*
12. Plastics Technology Handbook, *Manas Chanda and Salil K. Roy*
13. Reaction Injection Molding Machinery and Processes, *F. Melvin Sweeney*
14. Practical Thermoforming: Principles and Applications, *John Florian*
15. Injection and Compression Molding Fundamentals, *Edited by Avraam I. Isayev*
16. Polymer Mixing and Extrusion Technology, *Nicholas P. Cheremisinoff*
17. High Modulus Polymers: Approaches to Design and Development, *Edited by Anagnostis E. Zachariades and Roger S. Porter*
18. Corrosion-Resistant Plastic Composites in Chemical Plant Design, *John H. Mallinson*

19. Handbook of Elastomers: New Developments and Technology, *Edited by Anil K. Bhowmick and Howard L. Stephens*
20. Rubber Compounding: Principles, Materials, and Techniques, *Fred W. Barlow*
21. Thermoplastic Polymer Additives: Theory and Practice, *Edited by John T. Lutz, Jr.*
22. Emulsion Polymer Technology, *Robert D. Athey, Jr.*

Other Volumes in Preparation

EMULSION POLYMER TECHNOLOGY

ROBERT D. ATHEY, JR.
Athey Technologies
Los Gatos, California

Marcel Dekker, Inc. New York • Basel • Hong Kong

Athey, Robert D.
Emulsion polymer technology / Robert D. Athey, Jr.
p. cm. -- (Plastics engineering ; 22)
Includes bibliographical references and indexes.
ISBN 0-8247-7850-2
1. Emulsion polymerization. 2. Polymers. I. Title. II. Series:
Plastics engineering (Marcel Dekker, Inc.) ; 22.
QD281.P6A87 1991
668.4'2--dc20 90-28413
CIP

This book is printed on acid-free paper.

MARCEL DEKKER, INC.
270 Madison Avenue, New York, New York 10016

Current printing (last digit):
10 9 8 7 6 5 4 3 2 1

PRINTED IN THE UNITED STATES OF AMERICA

Preface

There is a real need for a book on emulsion polymers that speaks directly to the users (the mill engineers in paper mills, the chemists in paint or textiles plants, etc.). The need is not so much for erudition, but for simple explanation of operating mechanisms involved in the physics and chemistry of polymers and colloids as they apply to emulsion polymer manufacture and use. My own students (in the engineering extension division courses taught through the University of California at Berkeley) called my attention to the need for such a book.

Emulsion polymers or latexes are substantial commercial entities. However, the information available on these polymer systems is based on:

1. Classical reference books, at least 20 years old
2. An occasional chapter in a review book or an article in a trade magazine written for a specific industry (TAPPI, AATCC, various paint organizations, and so on)
3. Patents (often describing products that never make it to production scale)
4. Technical brochures describing products (again, specifically oriented to one industry)

This means that a paint formulator will often not know of a good textile polymer emulsion that would work well in a needed application *and* be cost-effective.

The key to finding the appropriate emulsion polymer system for a formulator is knowing how to compare compositions and properties. Our intention here is to clearly outline the various available materials, how they are made, and how their synthesis formulations and processes affect their fluid and dry film properties. We also provide some guidance on formulation with additives to make the emulsion into the desired end product.

We emphasize the similarities among the ink, paint, paper, textiles, adhesives, and rubber industries and note the differences where they are important. We look for connections between apparently unrelated facts that allow a better understanding of the emulsion polymer as both a colloidal system and a polymer system.

We do not intend this as a comprehensive review of all the literature, but as a representative sampling illustrating specific points. The information contained here is a combination of experience and the teachings of a variety of readings in colloid science, emulsion polymers, and their applications. We advise the reader that this is a brief reference, and further readings and experience will build expertise. The book may be used by the scientist/engineer in industrial practice or as a supplementary text for the advanced student in material science, polymer chemistry, or colloid science.

Recognizing that the text may include misinterpretations or be incomplete, we would appreciate readers' contributions of lore, references, reprints, and recommendations for additions, deletions, or modifications for future editions. A friend some years ago grumbled to me that a reprint I sent him meant the new edition of the book he'd just finished was out-of-date and needed a rewrite. I am not afraid of such a prospect. Current interpretation of observations is always subject to change as new evidence is gathered and we become more knowledgeable about how to conduct and interpret experiments. How can writing a general reference be any different?

We've experienced four levels of support from a variety of people over the years to help get this publication together. Thanks are offered to:

1. Barbara Hartmann, my three daughters, and Patricia Shaw for their moral support when needed
2. Fred Goetz, Ray Gates, Willem Van Essen, and Ted Herman for their professional support in setting my sights higher throughout my career
3. Professor Elizabeth Dyer, who asked me to give a talk on the first course I took on polymer science about emulsion polymerization soon after I started learning it at work (I called it a black art), and Prof. Steve Rosen, who wanted me to repeat it 15 years later (they got me thinking in the directions needed for this compilation)
4. Michael Murphy (Editor, *Metal Finishing Magazine*), Lothar Vincentz (Publisher, *Farbe Und Lack* and *European Coatings Journal*), and Maurits Dekker (Chairman, Marcel Dekker, Inc.) for their encouragement and patience with my polemics, style, and delays.

I also cannot forget the many coworkers who taught me what I needed when I needed it. The art of chemistry is really an apprenticeship, as it involves a lot of manipulative skills and on-site observation of the process and product performance. I often cite Zimmerman, Meincke, Witt, the TAPPI and FSCT organizations, and other coworkers, but there are many more. Recognize your contribution to this effort. I continue to learn simply because you opened my eyes.

Robert D. Athey, Jr.

Contents

Preface		*iii*
I	**Introduction**	**1**
1	The Common Ground	3
2	Colloid Science Applied to Emulsion Polymers	7
3	Polymer Concepts	19
4	Processing Emulsion Polymers	35
II	**The Monomers**	**59**
5	Vinyls	61
6	Styrenes	71
7	Acrylates and Methacrylates	79
8	Diene Monomers	87
9	Curing Monomers	93
10	Waterborne Condensation Polymers	103
11	By-Products in the Latex	109
III	**Analysis and Testing**	**115**
12	Class-I Tests for Emulsion Polymer Systems	119
13	Class-II Tests for Special Problems	131
14	Analyses	171
IV	**Additives for Postpolymerization Compounding**	**189**
15	Colloidal Stabilizers	191
16	Rheology Modifiers	201
17	Plasticizers, Cosolvents, and Coalescents	207

EMULSION POLYMER TECHNOLOGY

I

Introduction

1

The Common Ground

The emulsion polymer is a heterophase mixture of many ingredients. All the ingredients affect the colloidal or polymer properties (and sometimes both) and are of concern to the user who wants the adhesive to stick, the paint or coating to cure, or the dipped product to be free of holes and impermeable. The user may find this collection of information a good guide for thinking on the new better formulation to be designed, for solving the problem that crops up in production of the emulsion polymer, or in the production or use of a formulation based on the emulsion polymer.

We shall assume a basic understanding of the many concepts of the organic chemistry, electrochemistry, and colloid science that underlie the emulsion polymer theory. However, we will refer to these concepts occasionally to remind the reader of their importance, and to occasionally cull out a detail or two that will affect the formulation or performance. The necessity for this sort of examination is not immediately obvious, but we recognize how the emulsion polymer industry has grown over the past 50–60 years.

The original researchers in the industry were transplanted paint chemists, or adhesive chemists. Some were analytical chemists trying to follow the manufacture process using some of the techniques they knew from other fields. Some were physical chemists by training who tried optical, thermochemical, or electrochemical characterization techniques to make their contributions. Many were organic chemists who were brought in to do something related, and were taught emulsion polymerization so they could get new products on the market. I was initially hired to make unusual monomers, and had swamped the man making the polymers in a matter of about six months. I was then assigned to learn emulsion polymerization techniques, so I could evaluate my own monomers. In the ensuing years, I have probably worked for only one year in the area of the organic chemistry of my academic background and for more than 20 years in emulsion and polymer science. I have called myself a colloid scientist rather than an organic chemist for about 10 years.

So for many years, the makers of emulsion polymers in industrial labs were essentially formulators, using a variety of commercial products they knew as polymers, initiators, and colloidal stabilizers in a rather Edisonian approach to making a good latex. This is not bad, in any sense, as these people were smart enough to have a mental formulary library of things that worked; this formulary library grew every day. Many had good academic backgrounds that they kept updated by readings in colloid science. I have a great deal of respect for people like Russ Meincke of General Tire and Carl Zimmerman of International Latex, who could come up with new products ready for commercialization six months after they received the assignment. Their primary publications are patents, based on good analytical thinking. But these lights are hidden under bushels, compared to the more visible people in the field who publish from academic or research institute labs.

The field of industrial emulsion polymers is so broad and heterogeneous that no one discipline can be the basis for the understanding needed to make a workable product. Although my academic training was as an organic chemist, the physical and inorganic chemistry contributions to the process and products have extended my usable information base. As one cannot be a renaissance man, one must depend upon co-workers with special areas of expertise to help solve any problem that may come up. Indeed, I have called myself an experienced novice, as I visited various labs looking for a little help on my problem projects. The ideal research technique is probably to have a team of reviewers skilled in different fields look over the project results and reports to recommend new approaches for the experimentalist.

A contributing (and complicating) factor to the heterogeneity of the field of emulsion polymers is the variety of specialized applications. In some cases, the polymer is to be easily coagulated, whereas in others, the polymer should not coagulate at all. In some formulations, the polymer is used alone as a coating, in others with 10% pigment or filler, and in others with 600% pigment or filler, by weight on the dry polymer solids. We give examples in Table 1.1. Unfortunately, many of these differences in formulations are specific to different industrial applications, and the users in those differing industries do not talk to each other or read each others' technical or trade journals. It is interesting that they all use essentially the same components, albeit in different amounts. So the

Table 1.1 Industrial Formulations for Emulsion Polymer Products

Industry	Pigments and Fillers, Dry Wt.	Emulsion Polymer, Dry Wt.
Paint	up to 100	100
Paper	100	3–20[a]
Carpet	200–600	100

[a]May have other binders (starch or protein) with the latex.

mystery of the emulsion polymer deepens! This creates an opportunity for the consultant for cross-fertilization of ideas.

The wide variety of components used in each industry is an eye opener. Each industry uses what is essential to the end use, and no cognizance is given the other users in other industries, or what the other industries can teach about the usage. The typical components used by selected industries are shown in Table 1.2. Again, the mystery of the emulsion polymer deepens! Again, this creates a substantial opportunity for the consultant for cross-fertilization of ideas.

Given the difficulty of dealing with a complex set of formulation components like these, and given that a common formulation will have a dozen or more of these components, one could really be concerned with whether or not one should start at all. Indeed, some professor of engineering said many years ago that ". . . after a process or formulation has more than seven components, it is no longer capable of being rigorously modelled mathematically and controlled. Then

Table 1.2 Typical Components for Emulsion Polymer Formulations

Component	Paper Coatings	Carpet Backings	Paint
Surfactants	X	X	X
Dispersants	X		X
Thickeners	X	X	X
Defoamers	X	X	X
Biocides			X
Flame retardants		X	X
Plasticizers		X	X
Curatives	X	X	X
Reodorants		X	X
Antioxidants		X	X
UV stabilizers		X	X
Pigments		X	X
Buffer salts	X		X
pH Neutralizers	X	X	X

you are dealing with art of control, rather than engineering or science." That is perhaps an oversimplification in today's world of sophisticated statistical modeling techniques, powerful computers, and better information on what components do in formulations. However desirable process–product control modeling is, it is still not commonly done. It is a matter of economics, and only the better funded research and engineering departments of large companies in industry will even start such a statistical control program. The small ink or paint company, or the small mill (textile or paper), will more likely depend upon their local production trouble shooter to solve the problem or develop the product, from their mental formulary. Perhaps we can help in their effort with the chapters following.

We have organized the presentation here along several lines of logic. The first section deals briefly with the basic principles of colloid and polymer science, with some practical observations on latex preparation. The second section covers the common monomer systems used in making the latexes, with some guidelines as to the type of latex properties which can be expected. The third section recommends the analysis and testing schemes the producer or user may find useful in problem solving. The last section deals with additives for the formulated product, and some pitfalls in their use. We hope this integration of a common language approach to a complex industrial material set may be useful to the mill engineer, the lab person, and the formulator.

2

Colloid Science Applied to Emulsion Polymers

I. INTRODUCTION

Colloids are heterophase mixtures of submicroscopic particles. Colloid science is applied to the analysis, testing, use, and characterization of these mixtures, to achieve reproducibility and required properties in application. Examples of these mixtures are given in Table 2.1. The continuous phase is termed the dispersion medium, and the submicroscopic materials are termed the dispersed phase. The 3 × 3 matrix shown in Figure 2.1, can be filled with examples of colloids, some containing all three physical states, two dispersed in the third; no gas-in-gas colloids exist.

Heterophase combinations alone do not constitute a colloidal system. The key is the submicroscopic range of articles, between 500 nm and 50 μm. The lower limit borders on molecular size; at the upper limit individual particles can be seen in an optical microscope. The particle size is important to the stability of the colloid. When the fog or cloud particles are too big, it rains!

The basics of colloid science are complex and involve sophisticated physics, mathematics, and chemistry. Suffice it to say there are excellent treatments for these subjects. Among the best and most complete is Kruyt's two-volume set [1]. Our approach is to be qualitative in treatment, showing an occasional equation to point out the important implications for understanding and application to the emulsion polymer formulation, as latex or as compounded product. We prefer not to memorize an equation, but would rather remember the important implications from the equation's critical factors.

Table 2.1 Common Examples of Colloids

	Dispersed Phase		Dispersion Medium	
Colloid	Constituent	Phys. State	Constituent	Phys. State
Fog	Water	Liquid	Air	Gas
Smoke	Soot	Solid	Air	Gas
Foam rubber	Air	Gas	Rubber	Solid
Concrete	Sand, gravel	Solid	Cement	Solid
Cheese	Oil, fat	Liquid	Protein	Solid
Milk	Oil, fat	Liquid	Water	Liquid
Whipped cream	Air	Gas	Milk	Liquid
Ink	Carbon black	Solid	Solvent	Liquid

The discussion will start with the critical phenomena in colloid stability, settling and creaming, and electrophoretic mobility. Then the discussion will move to the typical emulsion polymer formulation for manufacture and a discussion of what components actually do. Other chapters will reemphasize some of these points or approach the discussion from another point of view.

II. CRITICAL FACTORS FOR COLLOID STABILITY

Instead of simply listing the factors affecting celluloid stability, we shall give some physical phenomenological descriptions based on performance of colloids and related to measurable properties of those colloids.

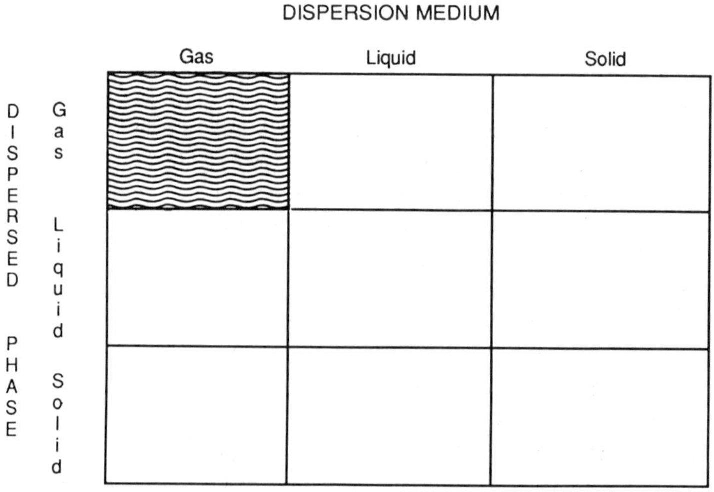

Figure 2.1 Colloid Example Matrix.

A. Stokes' Law of Settling and Creaming Rates

Based on the excellent discussion of Stokes' law by Becher [2], we show it as Eq. (2.1).

$$\frac{dx}{dt} = \frac{2a^2}{9\eta}(\rho_2 - \rho_1)g \tag{2.1}$$

where

η = viscosity of dispersion medium
ρ_1 = density of dispersion medium
ρ_2 = density of dispersed phase
$\frac{dx}{dt}$ = rate of settling
g = gravitational constant
a = particle radius for dispersed phase

This equation is a rate equation. No colloid is infinitely stable, especially not those with drastic density difference components. One should not be surprised to see settling or creaming in any formulation, and should be pleased with a low rate. The equation also tells us what to control to minimize the settling rate. Those control parameters are:

1. Viscosity of Dispersion Medium

In an emulsion polymer, the water phase viscosity may be high before adding the other ingredients to make the polymer. Unfortunately, the latex coagulated during the polymerization in trials made many years ago. But in the general case of the pigment dispersion, this control variable is useful. Indeed, a photo showing steel ball bearings and ping pong balls at the same liquid level in a CARBOPOL solution was supplied by BFG Chemical for a review article on the dispersion of pigments [3]. CARBOPOL is a highly crosslinked high molecular weight poly(acrylic acid), which is used in cosmetics, shampoos, hair setting gels, and similar products. Usually the formulation passes through all stages of processing before viscosity is finally adjusted to assure storage stability for the user.

2. Particle Size of Dispersed Phase

The smaller the particle can be made (or ground), the more stable the dispersion will be. That is the reason for the popularity of miroemulsions in the cosmetics industry. However, the emulsion polymer may need to be made in a particular particle size range for this successful application. This may not be a problem, as the emulsion polymer particle is typically 0.3 μm. Particle size is more important for pigments with larger, denser particles than the polymer particles.

The term for particle size in Eq. (2.1) is squared. If two or three particles

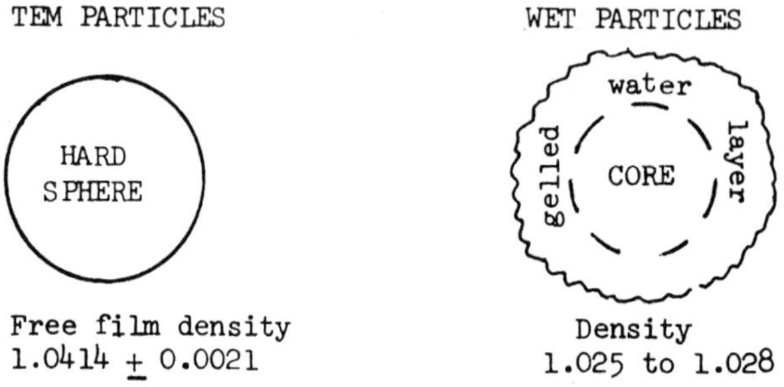

Figure 2.2 Latex Particles: dry (TEM) and wet [4].

flocculate, the radius may double, but the settling rate, based on the square of the radius, may quadruple.

3. Density Difference Between Phases

The larger the density difference, the more likely settling will be a problem. In emulsion polymers the density difference may be small, in which case the collision energy imparted by Brownian movement will be high enough to overcome the predicted settling rate. Lehigh University has some poly(styrene) emulsions over 30 years old which still appear to be stable. The styrene–butadiene latexes are commonly so close to the density of the water phase, that settling or creaming rarely occurs. However, some hydrocarbon latexes cream easily. The creaming of natural *cis*-poly(isoprene) latex for concentration was a common tool of the formulator in the early days of the rubber industry, in order to reduce shipping costs.

The use of protective colloids, or of a copolymerization imparting carboxylation to the surface of the polymer particle, may bring the effective particle density nearer to that of water. Surface carboxylation of the emulsion polymer makes it water swellable, with attendant change in density (and particle size) [4]. The swelling is crudely depicted in Figure 2.2. Salt or other soluble additives can be used also to adjust the water phase density.

B. Electrophoretic Mobility

In the main, colloidal particles are charged because of adsorbed ions. There is a technique (discussed in Chapter 13) to make the particles move in an electrical field. The motion of the particle depends on a set of variables, some of which are controlled by colloid physical properties. The equation governing speed of particle motion in an electrical field is shown in Eq. (2.2).

$$V_E = \frac{\mathsf{E} zE}{f\pi\eta} \qquad (2.2)$$

where

V_E = velocity of particle movement
E = dielectric of dispersion medium
E = electric field at particle
z = zeta potential (potential at the immobile aqueous layer surrounding the particle)
η = viscosity of dispersion medium
f = shape factor (between 4 and 6)

The important, or controlling, variables are discussed below.

1. Viscosity

The importance of this variable is obvious. The more viscous the dispersion medium, the slower the particle will move. In cases where the dispersion needs a high velocity (e.g., electrophoretic deposition of primer paint on auto bodies or frames), this term should be kept low.

2. Dielectric Constant for Dispersion Medium

Water has a high dielectric constant, which is reduce by additives. Organic solvents are efficient in reducing dielectric constant and useful as coagulants, as will be discussed in later sections. They seem to collapse the Helmholtz double layer, eliminating the electrostatic repulsion between particles.

Although not commonly thought of, salts also depress the dielectric constant. Debye showed the relation between salt concentration and dielectric constant; this is another control variable.

3. Electric Field at the Particle

The experimental setup has two electrodes in contact with the fluid. One can calculate the electric field strength based on cell geometry, electrode area, and voltage applied. A recent computer program models the electric field changes with position within the cell [6].

4. Zeta Potential

The change at the surface of the particle is usually quantized in experimentation. The charge is developed by the adsorption of surfactants or salts or by ionization of chargable groups from within the set of comonomers. The amount of any or all of these, and their adsorption-desorption equilibria, affect the perceived zeta potential.

C. Adsorption–Desorption Equilibria

1. Salts

a. Do Salts Adsorb? There has to be a reason why a simple salt such as sodium chloride adsorbs on a colloidal particle to help stabilize it. Since there

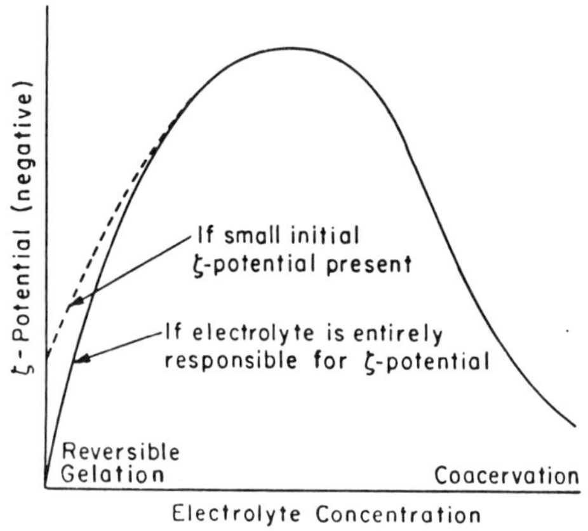

Figure 2.3 Salt addition effect on zeta potential [7].

is no surface activity for the simple salt, there must be some other driving force. The driving force is the simple concentration cell, as expressed by the Nernst equation for electromotive force, or the osmotic equivalent. An anthropomorphic argument describes ion perception and response to the concentration cell. One can even estimate the amount of salt that may be adsorbed on a surface for experimental verification at some time later.

For many years, the seminar we give on organic polymeric pigment dispersants has been quite popular. It has been requested by trade groups or industrial organizations at least yearly since first given in 1974 at the TAPPI Pigments Workshop, and twice published, as well [3]. One of the most frequent questions posed has been "Why does the poly(acrylic acid) or polyphosphate salt—not surface active—adsorb onto the pigment particle?" This question cannot be answered, but we do have a proposed mechanism [10].

Adsorption of nonsurface active species onto colloidal particles is a common phenomenon. Simple salts, such as sodium chloride or sodium sulfate, adsorb onto colloidal particles. Indeed, the early colloid scientists formulated their sols with a little extra salt to ensure stability. For instance, making a stable barium sulfate sol requires a slight excess over the stoichiometric amount of the sulfate to charge the sol surface. Blackley [7] used sodium sulfate addition to make a pseudo-gel latex into a fluid of reasonable viscosity, a graphic example of adsorbtion of a nonsurface active species. The plot of his results is shown as Figure 2.3. The adsorption of the peroxydisulfate initiator at the growing latex particle is critical to the success of emulsion polymerization, as it decomposes to form two sulfate radicals, which need to be close to the particle for efficient initiation.

There are several clues as to why a species migrates in a direction rather than simple random motion. There has to be a driving force. The clues must be examined to find precedent for a reasonable explanation of the move of a chemical species to adsorb on a colloid. A hypothesis is then constructed, and a test program is proposed.

b. Electrochemical Clue. The pH meter uses the Nernst equation in concentration cell mode to determine hydrogen ion concentration difference within the glass electrode and without. The form of the Nernst equation for concentration cells is shown in Eq. (2.3).

$$E = E_0 + 0.059RT\ (\log C_1 - \log C_2) \tag{2.3}$$

where

E = EMF measured in the cell
E_0 = EMF for the standard cell
C_1 = concentration in one half cell
C_2 = concentration in the other half cell
(EMF = electromotive force)

In the case of the pH electrode, there is a measurable hydrogen ion concentration on each side of the glass electrode. If one allows the cell to remain in the fluid being measured, there can be ion transport through the glass membrane, and the observed pH will drift. The movement of ions tries to equilibrate the pH.

c. The Adsorption Cell. In the case of species moving to adsorb onto a colloid, the preceding examples indicate the role of the concentration cell as a driving force. However, the cell for adsorption onto colloids is different. Let us take the simple case of the sodium chloride adsorbing onto an inert uncharged pigment particle. The salt concentration within the water is, for example, 0.5% NaCl, and the pigment concentration is 25% by volume. The chloride ion does not behave like a pigment dispersion, but like being in water alone. It responds to a driving force moving it to and through the pigment particle, as the salt concentration within the particle is zero. Indeed, we can set up the concentration cell as a Nernst equation, with c_1 as the salt molarity, and C_2 as zero; thus a substantial EMF moves that ion.

However, the ion stops at the particle surface. It cannot penetrate the pigment, in spite of the EMF. Therefore, the ion stays at the surface to impart a permanent charge for purposes of zeta potential measurement.

But the ion is not permanently attached. Ions adsorb and desorb as quickly as microequilibria allow them. They simply have an average residence time, and there are enough to make it appear as if there is always the same charge.

Given that the adsorption of the salt on the colloidal particles is a function of the volume fraction of the colloid, and the concentration of the salt in the rest of the volume, one should be able to analyze the amount of salt in the fluid phase, and determine how much is adsorbed. Indeed, this technique has been used to determine how much polyacrylate was adsorbed onto pigment [8]. However, the determination was not related to any driving-force argument, but is a simple answer to how much adsorbtion did occur.

The adsorption of polymers, such as polyphosphates or acrylic copolymers commonly used as pigment dispersants, is a different matter. The driving force is the same concentration cell, but the adsorption–desorption equilibria differ. Clayfield and Lumb [9] suggest that polymeric species adsorb randomly along their chain, rather than only at their ends, based on a computer model. We suggest that the chain segments adsorb and desorb fairly rapidly, but that the chain maintains contact at one or more points almost continuously, as it squirms about. The higher the molecular weight, the more contact points, and the longer between adsorption and desorption. Indeed, we suggested dispersant and pigment could be matched for best adsorption [3] based on the argument of Clayfield and Lumb.

d. Opposing Forces. A charged species is met by an opposing force, at the very least. The concentration of charge at the surface of the colloid is quite high, and this will electrostatically oppose the ion migration toward the colloid surface. This would be most significant for the smaller ions, which can pass quickly through the dispersion medium. However, the larger ions (polymeric species) are charge dense and slow to move.

e. Testing the Hypothesis. Adsorption tests of poly(acrylic acid) materials have been attempted by tagging the polymer with a copolymerizable species (like vinylanthracene or similar material) which is easily determined spectrophotometrically. Adsorptions of these copolymers are probably too high because they are more surface active. However, the electromotive or osmotic driving force compelling the unsubstituted species to adsorb may be quite large.

The test of this proposed hypothesis would be simple. The dispersion of some insoluble colloid (silica or titania) could be accomplished with some easily analyzed dispersant material. A polyphosphate would be useful, but a polymeric organic sulfate–sulfonate would also serve. Phosphorus and sulfur analyses are reasonably precise and accurate. The hypothesis predicts that the fraction adsorbed on the colloid surface would be equivalent to the volume fraction occupied by the colloid, and the supernatant dispersion medium would be reduced in dispersant by that same volume fraction. Too much dispersant would saturate the surface, and therefore the experiment should be repeated at the same dispersant concentration, preferably the lowest, to determine analytical precision. The dispersion should be allowed to settle to expose about 10–20% of the volume as supernatant; a small portion of the supernatant is then analyzed for the residual P or S content. An alternative to the P- or S-based dispersant would be radio-tagged carbon in poly(acrylic acid). The radio-tagged carbon is another easily analyzed species with good precision.

Here then is a challenge of the academics, or dispersant or pigment manufacturers. A test of the osmotic or electromotive concentration cell concept of the adsorption driving force based on volume fraction of the colloid, could be used to justify confirmation or rejection of the hypothesis.

2. Protective Colloids

a. Adsorption Examples. Broader examples of adsorption of nonsurface additives include the protective colloids. Poly(vinyl alcohol) is commonly used as a latex protective colloid in the synthesis of vinyl acetate copolymers. Hydroxyethyl cellulose, starch, and even dextrin may be used the same way. Only recently have researchers developed ways to convert these soluble polymers into surface-active moieties, by reaction with a fatty acid ketene.

The technology of noncharged protective colloids is old and commonly used. Their action after adsorption results in a highly viscous pseudo-gel layer at the surface of the colloid. Upon close approach of two particles, the two gel-like layers inhibit the two particles from moving closer because of the viscosity resistance in a direct line of approach. The steric stabilization species on particles is shown in Figure 2.4. I said once (in a lawsuit where this had to be explained to nontechnical people) that they slide past each other like mud-wrestlers. The more scientific explanation is that the particles are "sterically stabilized," that is, steric hindrance prohibits collision. That may well be, but that does not explain why they adsorb in the first place.

b. Osmometric Clue. Osmometry is a common molecular weight measurement technique for polymers. It is based on the principle that the colloid solution (dispersion) is on one side of the semipermeable membrane and only solvent on the other side. The concentration gradient causes the solvent to move through the semipermeable membrane in order to dilute the polymer solution. Some species small enough to pass through the pores of the semipermeable membrane migrate to the other side, where the concentration is zero. A driving force to equalize the concentration moves them through the pores. Again, the driving force is a concentration cell.

Thus, we can hypothesize that the noncharged protective colloid is driven to adsorb by a concentration cell-based driving force. This has to be entropically driven, as the concentrations do not develop the electric field ions do.

c. Testing the Hypothesis. Radio-carbon tagged material could be used effectively here to test the hypothetical prediction on volume concentration-controlled adsorption of non-surface-active nonionic polymers.

3. Surface-Active Agents

We discuss surface-active agents in Chapters 4 and 15 of this book. They are the easiest understood of the phenomena used to stabilize the colloidal particles in suspension. In the main, the surface-active agents have two segments: a

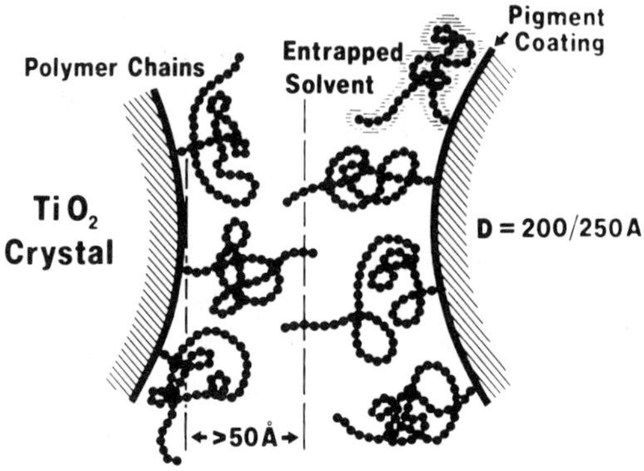

Figure 2.4 Steric Stabilization (Courtesy of Tioxide of Canada).

hydrophilic portion and a hydrophobic portion (see Chap. 15). Most of the hydrophilic portion must be ionized to be an effective developer of charge at the surface of the colloidal particle after adsorption. If the hydrophilic portion is nonionic, it must be large enough to extend a fair distance into the water phase to be an effective steric stabilizer.

The hydrophobe is particularly important for the success of a surfactant. Several studies have related hydrophobe size to effectiveness in stabilizing emulsions, or even to supporting emulsion polymerization without simultaneous coagulation. Suffice it to say eight to twelve carbon chains are the minimum for aqueous emulsions; the octadecyl-based natural materials (stearate, oleate) are useful at low cost.

Surface-active agents are surface active because they do not easily dissolve in water. They will align their hydrophobes on any surface (the stirrer, the vessel, the air–water interface, or any dispersed material) to separate from the water. If no surface is available, the surface-active agent will create micelles, wherein the surfactants aggregate to remove their hydrophobic tails from the water, and deposit their hydrophiles at the micelle–water interface. It reminds me, in concept, of the circling of musk oxen for defense. The onset of micelle formation, at low concentrations, is called the critical micelle concentration (CMC); it varies with hydrophobe and hydrophile.

One of the more interesting aspects of surfactant technology is the idea that they can cooperate or compete. Cooperation examples include the mixed micelles found in some polymerizations, a small amount of a dispersant or hydrotrope (see Chap. 15 for definitions) with a small amount of micelle former, at less than the CMC. Competitions are simply cases where adsorption occurs quickly for all species upon mixing. The adsorption–desorption equilibria take effect, with attendant changes in surface tension or viscosity or even colloidal stability with time.

IV. REPRISE

Colloids are submicroscopic heterophase dispersions or suspensions of one material in another, with neither being compatible (soluble) in the other. The incompatibility and the submicroscopic size are the critical qualifiers, as both are necessary for a mixture to be defined as colloidal.

The key points about colloids are their settling rates and electrophoretic mobility, as the physical characterization of these with the help of established equations allows definition of the controlling variables which are the key elements in making stable colloids. We introduce a proposal as to why stabilizing additives adsorb onto a colloidal particle's surface, and suggest test techniques to verify the driving forces suggested.

The latex or emulsion polymer is a special case of colloid science in general. The controlling factors of colloid science are the basis for making stable latex systems for paints, adhesives, and other applications. Selection of the materials chosen for use in the latex to provide stability can yield excellent field performance in application [5].

REFERENCES

1. H. R. Kruyt, *Colloid Science,* Elsevier, New York (1952).
2. P. Becher, p. 135, *Emulsions: Theory and Practice,* Reinhold, New York (1959).
3. R. D. Athey, Jr., *TAPPI, 58(9),* 66 (1975).
4. R. D. Athey, T. Provder, G. Poehlein, and J. Scolere, *Coll. and Polym. Sci., 255,* 1001 (1977).
5. R. D. Athey Jr., 1990 NACE National Meeting Presentation, No. 90472 (Las Vegas, 23 April 1990).
6. "ELECTROSTAT," Algor Interactive Systems Inc., 260 Alpha Dr., Pittsburgh, PA 15238 (415-967-2700).
7. D. C. Blackley, *High Polymer Latices,* Vol. 1, pp. 276–278, Palmerton Publishing Co., New York (1966).
8. D. E. Erickson, unpublished.
9. E. J. Clayfield and E. C. Lumb, *J. Coll. Interface Sci., 22,* 285 (1966).
10. R. D. Athey Jr., *Farbe and Lecke, 5/90,* 340.

3

Polymer Concepts

I. INTRODUCTION

This introductory material is included as a reminder to those who have the background in polymer science. The details are not included, and neophytes to the polymer field may need to refer to more basic references, such as Rudin [1], Rosen [2].

II. POLYMER STRUCTURE

The term *polymer* is a general descriptive term denoting a large molecule made from many smaller molecules termed *monomers*. Polymers low in molecular weight (2000–20000) are termed *oligomers* or *telomers*. The classifications of polymers are many and varied and generally relate to some feature, such as synthesis mechanism, thermal flow, or structural features.

The classifications based on synthesis mechanism go back to the days when small by-product molecules were taken off *condensation* polymers during synthesis, whereas in *addition* polymers monomers were added to the growing chain without by-products. Typical condensation polymers considered by industry are cellulose, nylons, and polyesters. Addition polymers include the acrylics, nitrile rubbers (NBRs), styrene–butadiene rubbers (SBRs), and vinyls found as emulsion polymers.

There are more rigorous kinetically derived definitions of the synthesis classifications now. The purists call the mechanism used for condensation polymers the *step-growth* mechanism, because of the way kinetics control the polymer growth and molecular weights. Step-growth polymers may add a monomer unit to any chain in the reaction volume, with no inhibition of relative reactivity of any molecule. The kinetically derived name for the addition synthesis mechanism is *chain-growth*. The chain-growth polymers grow by addition of monomer to a "live" chain. When growth is complete and terminated, there is no further growth of that chain. The synthesis-based definition system of classifications is not particularly useful to the polymer industry because emulsion polymers are usually addition or chain-growth polymers.

Another classification system of polymers is based upon their synthesis or polymerization process:

Bulk Polymerization: only monomer(s) and initiators are present, essentially a 100% reactive system.
Solution Polymerization monomer(s), initiators, and other components are dissolved in a solvent.
Dispersion Polymerization: monomer(s), initiator, and other components are slurried in a non-solvent, resulting in a large-particle-size unstable dispersion of polymer particles.
Emulsion Polymerization: Surface-active agents are added to a dispersion process, yielding a colloidally stable small-particle-size polymer system.

Many books have been written on the basis of these processes, contrasting them, but we intend to concentrate on the emulsion process and the polymers made therefrom, without comparison to the other systems.

The thermal-flow-derived definitions of polymer classifications are useful to polymer industry. *Thermoplastics* are polymers that turn to a flowing fluid (albeit, a high-viscosity liquid) upon heating. This mass can distort, move from or to a bond, or get tacky upon heating. The polymers are generally linear and modestly branched and can dissolve in an appropriately chosen solvent. *Thermoset* polymers, on the other hand, are *cross-linked* and cannot be appreciably made to flow upon heating. Strictly speaking, one can see evidence for flow in a cross-linked acrylic, as in hot-press-bonded cured acrylic-coated fibers to paper. However, a cross-linked thermoset cannot be passed through an extruder, where it would disintegrate to granules. Cross-linking renders the polymer insoluble *in all solvents*. We will make use of that later on in Chapters 17 on solvents and plasticizers.

There are several structurally derived polymer classifications. The configurational isomers are really stereoisomers and are named according to the way the monomers array themselves in the chain backbone. Emulsion polymers are generally random in configuration, and in the terminology of the configurational isomers, random (*atactic*). If some organized stereoisomerization occurs, the terms to describe the isomers are *isotactic* and *syndiotactic*. These are not particularly important in the emulsion polymer industry, although some fibers can take these forms.

The structural characterization of copolymers is of importance, however. A *copolymer* is a polymer made of two or more different types of monomers, in any percentage. This distinction of "in any percentage" is a fine point often ignored by polymer producers if the main monomer exceeds 95% of the monomer mix. Hence, the plastic window in an airplane may or may not be 100% poly(methyl methacrylate), but it is usually about 95% methyl methacrylate.

The copolymer is assumed to be a *random* copolymer, unless otherwise specified. The most ordered copolymer is the *alternating* copolymer, having exactly the structure of ABABABABABABA. Another highly ordered structure, so-called *block* copolymers, is present in many adhesive formulations. The properties of the adhesive depend on the monomer composition in the blocks. A typical textbook representation of a block copolymer may be AAAABBBBAAAAABBB, which would be an oversimplification. One is more likely to find A blocks of 5000–10,000 molecular weight (more appropriately called a segment weight), with a short random block of A and B, and B blocks of 5000 molecular weight.

Another variety of copolymer is the *graft* copolymer which is made in two steps; one making the "backbone" of one monomer or set of monomers, and the subsequent step adding different monomers to that backbone ("grafting") to yield a multiphase product. When this is done as an emulsion process, the original emulsion particle can be considered the "seed" for the subsequent polymerization procedures. The seed emulsion is charged into the reactor, additional monomers added, and the polymerization is restarted with new initiator. The final product particle has a "core" based on the original "seed" polymer backbone, and a "shell" around that core based on the subsequent grafting step monomer composition. However, the main portion of the emulsion polymers are simply random, or as random as copolymerization kinetics and other factors (to be discussed later) will allow them to be.

There are many classes of polymer based on the form in which they are supplied. The polymer may be supplied as the bulk dry solid, in slabs for natural rubber, pellets for extruder-fed thermoplastics, or drums for hot melts. The polymer may be in the form of undiluted bulk liquid polymer *oligomers,* or in solution with some solvent. Water-based systems may be solutions, dispersions, or latexes. The latter two are complex mixtures containing not only polymer, but a number of additives and from production by-products [3].

Just how the polymer systems are supplied governs the test selection and methods. The systems known as emulsion polymers have been used before their nature was known, and handling, testing, analysis, and usage have evolved to be fairly complex. Many forms of the polymers were made in emulsion to be isolated for compounding and processing into the form preferred by the consumer, such as powder, pellets, sheet, and others. However, our main concern will be with those industries that deal with latex.

III. ADDITION POLYMERIZATION

A. Polymerization Reactions

The addition polymers used for most of the emulsion polymers are generally based on a very simple set of polymerization reactions. A molecule with an active center approaches a monomer molecule, the latter having a set of bonds that may allow the activated molecule to add to one side and transfer the active center *unchanged* to the other side of the monomer molecule. In most cases, the monomer molecule has a carbon-carbon double bond.

Strictly speaking, there are five different mechanisms for addition polymerizations in bulk or in solution: free radical, anionic, cationic, coordination, and group transfer. Details of the kinetics of the first four of these mechanisms may be found in standard references [4,5]. Only one of these five mechanisms, the free-radical mechanism, can be invoked in aqueous media wherein most emulsion polymers are made. Equation (3.1) shows the addition of a radical species to a monomer to generate another radical species. The asterisk is the standard way to

$$R^{*} + R'CH{=}CHR'' \rightarrow RR'CH{-}CHR''^{*} \qquad (3.1)$$

denote a "live" free radical. This reaction is sometimes given in shorter form as $P^* + M \rightarrow P^*$ or $M^* + M \rightarrow M^*$. Unfortunately, the process is not that simple. Several reactions are needed to start the addition process, and there are many termination reactions to consider in order to inactive the polymer molecules (dead polymers, no longer carrying the radical-activated end to continue addition).

B. Initiation Reactions

The initiation process requires an initiator radical, which may be simply the product of a homolytic cleavage of a peroxide or azo compound to yield two similar radicals. For instance, di-*tert*-butyl peroxide cleaves thermally to form two tertiary butoxy radicals, whereas azobisisobutyronitrile (AIBN) liberates a molecule of nitrogen and two isobutyronitrile radicals, as shown in Eqs. (3.2) and (3.3), respectively. The presumption is that as soon as this initiator radical species adds to a monomer, the combination acts as if it is a polymer radical no matter how many monomer units have been added.

$$\begin{array}{ccccccccccc} & CH_3 & & & CH_3 & & & & CH_3 & \\ & | & & & | & & & & | & \\ CH_3- & C & -O-O- & & C & -CH_3 & \rightarrow & 2\,CH_3- & C & -O^{*} \\ & | & & & | & & & & | & \\ & CH_3 & & & CH_3 & & & & CH_3 & \end{array} \qquad (3.2)$$

$$\begin{array}{ccccccccc} & CH_3 & & & CH_3 & & & & CH_3 \\ & | & & & | & & & & | \\ CH_3- & C & -N=N- & & C & -CH_3 & \rightarrow N_2 + 2\ & CH_3- & C^* \\ & | & & & | & & & & | \\ & CN & & & CN & & & & CN \end{array} \qquad (3.3)$$

These two organic initiators are assumed to be oil-phase (i.e., monomer phase) initiators The more common initiator is peroxydisulfate (termed incorrectly persulfate in the emulsion polymer industry), which is a water-soluble species. There is argument that the initiation by peroxydisulfate is most effective at the interface of the polymer–monomer-filled micelle with the water phase. Since the salts in a formulation migrate to that interface in some considerable concentration, it should not be surprising that the initiation takes place where monomer and initiator have access to each other.

No initiator is 100% effective. Indeed, the figures are usually on the order of 40–60% of the decomposing initiator molecules actually starting the polymerization. The other decomposing initiator molecules react differently and form by-products in substantial quantity. AIBN undergoes so-called cage recombination to form tetramethylsuccinonitrile (TMSN), a potentially toxic material. The term cage recombination indicates that the AIBN molecule is surrounded by a cage of solvent molecules, and the radicals must migrate from the cage to be available to add to a monomer before they find one another to form the inactive TMSN. Other by-products of the AIBN have been found, as well [6].

Peroxydisulfate decomposes to form two sulfate radicals and cage recombination merely reforms the peroxydisulfate, as with most peroxide recombinations. But if the sulfate radical plucks a hydrogen atom from a water molecule, a hydroxy radical splits off and the inactive sulfate anion remains. Indeed, this formation of an acid species is the origin of an acid-direction pH drift during polymerization.

Because the initiator *adds* to the monomer molecule, it becomes an integral part of the polymer structure, most importantly, an *end group*. Detailed discussion of this has appeared in a review of telechelic polymers [7], telechelic meaning "having end-group functionality." The free-radical initiation schemes that produce such end groups are based on two types of initiation systems: those using a molecule capable of homolytic clevage to form free radicals while having some functional group on another segment of the radical, and those using multiple molecules whose reaction generates a free radical having some functional group on another segment of the radical. Hydrogen peroxide or the peroxydisulfate molecules are examples of the former types, while the latter may be exemplified by a reaction pair such as thioglycolic acid and some peroxide which generates the mercaptide radical containing a carboxy group on the other end of the molecule [8].

Some redox initiation systems use multiple molecules to generate the free radicals. Examples include peroxidisulfate–bisulfite reaction pairs, or systems containing peroxide(cumene hydroperoxide, for instance), iron (typically ferrous sulfate), chelant (EDTA or phosphate anions) and a reducing agent (sodium formaldehyde sulfoxalate). These are useful to reduce the temperature at which the polymerization is run. The peroxidisulfate–bisulfate can be used to promote an ambient-temperature polymerization of acrylic–methacrylic monomers, eliminating the need for a constant temperature bath at 45–65°C required for peroxidisulfate alone. Some monomers benefit from lower polymerizations temperatures. The *cis–trans* ratio of 1,4-additions of butadiene shifts to more *cis* at lower temperatures, and some common butadiene copolymer products have been made at 5°C with iron–peroxide-type initiators.

The peroxydisulfate-derived radical produces a sulfate ester end group. John VanderHoff [9] has demonstrated the existence and importance of the sulfate ester groups on the surface of poly(styrene) made by emulsion polymerization. Aqueous solutions of methyl methacrylate become latexes upon peroxydisulfate initiation and poly(methyl methacrylate) molecules increase to a size where they are no longer water soluble, but gather into micelles and emulsify monomer to continue as an emulsion polymerization [10]. This gave me the clue needed to identify the possible results of the peroxydisulfate-derived species in a latex for a presentation on by-products to a group interested in coatings for food packaging at Rutgers University. The species should be all sulfate esters (or their hydroxy hydrolyzates), such as:

1. Water-soluble oligomers, having a high ratio of sulfate groups to monomers added;
2. Surface active species—only modestly soluble—adsorbed to any surface available, be it on latex, pigment, vessel, or stirrer;
3. High polymer species with the sulfate group(s) exposed to the aqueous phase to aid in latex-particle stabilization by contributing to the Helmholtz double layer; and,
4. High polymer species with the sulfate group(s) buried beneath the latex particle surface, thus not contributing to stabilization.

In addition, the water phase will contain acid sulfate salts based on decomposed peroxdisulfates that did not find a monomer molecule, *and* the yet undecomposed peroxydisulfate molecules. I offer this hypothesis as the challenge to the researcher to find all these species in the latex, and to make use of them to stabilize the colloid or perform other functions.

Interestingly, the sulfate materials are not always desirable, as they can promote or catalyze corrosion of the metal carefully coated to prevent corrosion. However, in other systems, the acid sulfate desirably may act as an acid catalyst for subsequent cure of the polymer, if formulated properly. There is trade off in using a particular catalyst, and advantages and disadvantages must be carefully considered.

C. Termination Reactions

The termination reactions for radical polymerizations are transfer, addition, and disproportionation. Details of these reactions' kinetics may be obtained in standard references, such as Flory [4] or Burnett [5]. Our concern is for their structural implications. The first set of terminations are the transfer reactions. We denote them as terminations because they cause the polymer growth to stop, although the radical reaction continues to grow *another* polymer molecule. There are molecular weight implications from transfer reactions, which are discussed by Mayo [11]. The simplest form of the Mayo equation is shown in Eq. (3.4)

$$\overline{X}_n^{-1} = \overline{X}_{n_0}^{-1} + C_S\frac{[S]}{[M]} \tag{3.4}$$

where

$\overline{X}_n^{-1}$ = Number average degree of polymerization
$\overline{X}_{n_0}^{-1}$ = Number average degree of polymerization without additives
C_s = Chain transfer constant for S
$[S]$ = Concentration of S
$[M]$ = Concentration of monomer

The concentration terms are for monomer (M) and solvent (S), while the terms with negative exponents are the number average degree of polymerization. The multiplier for the concentration ratio is the *chain-transfer constant*, which is the ratio of rates of polymerization and of transfer reactions. It is essentially a quantization of the instances of transfer per monomer addition to a growing polymer.

A chemical is usually added to the emulsion polymerization to achieve the desired molecular weight. These chemicals are called *chain-transfer agents* or *modifiers*. Typically, a long-chain mercaptan is used in emulsion polymerizations, but to some vinyl acetate polymerizations a small amount of isopropanol is added. Uraneck [12] has given reasons for injections of long-chain tertiary mercaptans to styrene–butadiene polymerizations. These mercaptans occasionally speed up the peroxydisulfate-initiated styrene–butadiene emulsion polymerizations as well.

Strictly speaking, many transfer reactions take place. There is a transfer reaction to every species present in the emulsion polymerization. Thus, we can speak of an extended form of the Mayo equation having terms for transfer to monomer(s), polymer, initiator, or any other species that happens to be present. A change from an anionic surfactant in an emulsion polymerization of styrene and butadiene to a quaternary amine cationic surfactant caused a substantial reduction of molecular weight of the polymer. Since the quaternary ammonium salts are in equilibrium with tertiary amine, and since tertiary amines can have chain transfer constants as high as those for mercaptans, we surmised the chain transfer to surfactant was responsible for the substantial molecular weight reduction seen. The addition of hydroxyethyl cellulose (HEC) to vinyl–acrylic emulsion polymerizations grafted the HEC by chain transfer to the polymer particles' surface, allowing use of titanate chelate esters to act as rheology modifiers [13]. Any copper, brass, or bronze in a reaction vessel will kill a free-radical reaction, as they are preferred transfer sites. An attempt to make a liquid polybutadiene with a large amount of mercaptan in the emulsion polymer formulation failed as the transfer reaction to polymer was so successful in the later stages of the reaction that the product was an insoluble nonfriable solid mass upon isolation.

The last example is the exception in the molecular weight reducing activity earlier noted for transfer reactions. Many emulsion polymerizations need gel-free polymer for subsequent processing. The transfer to polymer cannot be avoided as the reaction progresses (see Eq. 2.3) as the monomer concentration is decreasing with increasing polymer concentration. Hence, the technique used is to *short-stop* the polymerization with some effective radical scavenger. The reaction may be terminated by hydroxylamine or its *N,N*-diethyl derivative, or hydroquinone or its monomethyl ether. These "short-stop" materials are very important to certain segments of the rubber industry, for products such as hoses, tires, wire insulation, and others needing bulk thermoplasticity. The last example from the preceding paragraph led to an experiment using a short-stopping agent, and the liquid polybutadiene was successfully isolated at less than 100% conversion. For butadiene, short-stopping must occur at 60–65% to assure completely gel-free polymer (or copolymer).

One important implication of the Mayo equation is the fact that telechelic polymers cannot be made by chain transfer to a modifier having functional groups. The Mayo equation can be converted to a probability of functionality equation by dividing by the left-hand term. The equation is changed to Eq. (3.5) where the first term describes the probability of functionalities from all nontransfer reactions, and the second chain-transfer term describes the probability of functionality derived from chain transfer. You can see that if the chain transfer is the only source of functionality, you can only approach difunctionality as an asymptote [14].

$$1 = \frac{\overline{X}_n}{X_{n_0}} + \overline{X}_n\, C_S \frac{[S]}{[M]} \tag{3.5}$$

Addition terminations or disproportionation terminations occur when two radicals meet and kill each other. The difference between these two mechanisms is simply that the first yields one molecule, the second two, that is, the effect on molecular weight differs substantially. These are two competing reactions; addition is preferred at lower temperatures and disproportionation at higher temperatures. The implications are structurally important as well. The disproportionation yields a molecule having an unsaturated carbon-carbon double bond as an end group, and another molecule with a saturated end carbon-carbon single bond. That unsaturated end group may be the origin of thermally initiated unzipping decompositions, yielding free monomer at high temperature. This *unzipping* is most prevalent for α-methylstyrene or methyl methacrylate.

Another implication of the addition termination is that it is possible, with appropriate choice of polymerization components, to make difunctional telechelic polymers. Hence, cure through chain extension can be a formulation option. One formulates to make telechelic polymers through choices that promote addition terminations, decrease chain transfers to polymer and monomer, and use a symmetrical initiator with functional groups.

One interesting aspect of the current theory of emulsion polymerization kinetics is that there is only one growing radical per latex particle, or there are none because an entering radical killed the one that was there. The termination rate must be fairly substantial to occur so quickly. This allows the nonradicalized latex particle to absorb monomer and swell to have a high concentration of monomer to polymerize to high molecular weight when a radical does enter.

D. Copolymerization

Few homopolymers are useful, and the practice of copolymerization to modify the properties of a product is a main stock-in-trade for the polymer chemist. The nominal composition of a copolymer is not necessarily how every molecule is composed, as there may be a range of compositions in the mixture. To understand why, we must examine the copolymerization reaction.

First of all, each monomer has its own energy level with respect to reaction with an approaching radical. The radical approaching the monomer also has its own energy level. In addition, there is an energy barrier for those two to have crossed before they react. This is *activation energy* of the reaction. We are all familiar with the hump-shaped plot describing the conversion of monomer to growing radical. In the case of copolymerization, the monomers compete to add to the growing radicals. In the case of two monomer copolymerizations, we have:

1. M1 and M2, each at their own energy levels;
2. M1* and M2*, the growing chain ends (each having M1 or M2 as the last monomer species added to the growing polymer chain), each at their own energy levels; and,
3. the activation energy for *each* of the *four* additions.

The energy curves and the monomer concentrations govern the rates of additions of monomer to the growing polymer. They can provide very useful information. Indeed, in the days of small "garage-shop" vinyl acetate polymer manufacturers, the runaway reaction exotherm was often brought to a screeching stop with the addition of a small amount of styrene monomer, because the high activation energy barrier of the growing polymer radical in a styrene end group prevented addition to a vinyl acetate monomer molecule.

The reactions of copolymerization are described by the following shorthand form of the chemical equations:

$$M_1^* + M_1 \xrightarrow{k_{11}} M_1^* \tag{3.6a}$$

$$M_1^* + M_2 \xrightarrow{k_{12}} M_2^* \tag{3.6b}$$

$$M_2^* + M_1 \xrightarrow{k_{21}} M_1^* \tag{3.6c}$$

$$M_2^* + M_2 \xrightarrow{k_{22}} M_2^* \tag{3.6d}$$

These rates are contrasted by use of the *reactivity ratios* defined as the ratio of reaction rates.

$$r_1 = \frac{k_{11}}{k_{12}} \tag{3.7a}$$

$$r_2 = \frac{k_{22}}{k_{21}} \tag{3.7b}$$

These are simple measures of the number of additions of monomer like radicalized monomer unit terminating the polymer, in comparison to the unlike monomer reaction with that radical. For more details on these kinetics distinctions, see Rudin [1], Burnette [5] Rosen [2] or Flory [4].

However, this simple explanation is not sufficient for the complications one experiences in the practice of the art. There are kinetic rate equations describing copolymerizations, but they describe only the reaction in bulk systems. In solution polymerizations, the reactivity ratios shift, suggesting activation or deactivation of radicals by solvents or some other interaction [15].

In addition, the rate equations (including concentration terms for the monomers) describe only the initial reactions. However, since one monomer is normally depleted more quickly than another, the copolymer composition drifts during the polymerization. We saw an instance where a carboxylic monomer copolymerized with an ester monomer to low molecular weights (large amount of chain transfer agent) yielded 85–95% water-soluble species, whereas at higher molecular weights, water-soluble yield was 100%. The acid comonomer was depleted more rapidly from the mixture and at the very end of the reaction for low molecular weight materials an ester homopolymer was obtained. This can be avoided by carefully metering the addition of monomer(s) most quickly depleted throughout the reaction. This yields a larger quantity of the desired copolymer as compared to those copolymers of different composition.

Seiner has published a description of the compositional shift during the polymerization process in his computer modeling of copolymerization of up to six monomers [16]. He will give a FORTRAN listing of the computer program on request without charge. Provder and Kuo [17], and Mirabella with co-workers [18] have described analyses that separated copolymers showing compositional variations. Recent work describing a hydroxy-functional poly(vinyl chloride) (PVC) for high solids coatings gives details of these oligomers with at least two hydroxy groups per molecule [19].

Another problem may be encountered in emulsion copolymerizations. Most monomers have a degree of water solubility, as they have some polarity. Only the hydrocarbon monomers have water solubilities small enough not to detract from the monomer concentrations for copolymerization terms. For the polar monomers, the concentration terms should include a water–oil phase distribution coefficient to reflect the reduced monomer concentrations on the droplet, especially in later stages of the emulsion polymerizations.

This compositional drift may have advantages. In the best formulations of the carboxylic acid monomer styrene–butadiene copolymerizations little acid comonomer was incorporated in the first 80% conversion of the monomers. However, thereafter the percentage of the acid comonomer incorporated increased rapidly until most was used up at 95% + conversion. It was found that the acid comonomer was mainly on the surface of the latex particles, adding to the Helmholtz double layer and increasing the emulsion's stability. These carboxylic monomers are soluble in the water phase, although not completely.

The compositional drift does not *always* occur. For monomer pairs with r1 and r2 less than one, a *constant-composition copolymer* (CCC) can be found at one (and only one) monomer concentration ratio [20]. This ratio is sometimes called the azeotropic composition, although this terminology is less than clear. I prefer the CCC designation. One may calculate the CCC from the reactivity ratios, although an error in the calculation may occur due to a calculation or analysis error in the published data.

However, it is an interesting guideline for starting a polymerization process wherein the polymer composition would be *known* from the start to the finish. One must also recognize that most of the reactivity ratio data come from bulk polymerizations. Application to emulsion processing may be less than accurate for copolymer composition. We calculated some interesting copolymer compositions from the Alfrey-Price Q and e data in the search for water soluble elastomers [21]. Some of the polymers reported therein were commercially available. Indeed, we have a more extensive listing which could be the source of many commercial copolymers [25].

These reactivity ratios have another purpose for the user as well. They can provide a guide to methods to make an otherwise unobtainable copolymer with appropriate the process modifications. For instance, the DuPont company produced a series of hot-melt polymers called the Elvax series, having mostly ethylene in the copolymer believed to be from a bulk polymerization. However emulsion polymerization gave the Elvace product line (recently sold to Reichold) with 80% or more vinyl acetate in the copolymer. The emulsion polymerization is run at high ethylene pressure until the vinyl acetate is all used up, and the remaining ethylene is recycled. Reactivity ratios would predict little if any copolymerized ethylene, but pressurization forces an extremely high concentration of ethylene and copolymerization. An alternative route to the ethylene-vinyl acetate (EVA) emulsion polymer was found (and patented for General Tire) by Meincke, wherein he used a low-temperature polymerization process promoting emulsification of ethylene, and eliminating the need for a high pressure reactor.

IV. EFFECT OF PHYSICAL AND CHEMICAL STRUCTURE ON PROPERTIES

A. Strength

The test methods for polymer strength is not always straightforward. The pure or ideal strength tests are tensile, shear, and compression tests, but the actual techniques contain some elements of each of these. Even more complex are other tests, such as impact, tear, or flexural tests. I often ask students to analyze the forces in a complex test to see what of the three pure stresses contributes to impact or reverse impact tests of a coating on a substrate.

In all instances, it helps to think of the polymer as a random coil of spaghetti subjected to those forces, in order to construct a mental model of the response of the polymer to these stresses. One can then envision the contributions from cross-links and chain entanglements. One can even envision the response in an acrylonidrile–butadiene–styrene copolymer (ABS) made of styrene and acrylonitrile grafts onto a poly(butadiene), with the brittle breaks of the plastic portion and the elastomeric breaks (soaking up energy) in the softer poly(butadiene) sections.

The entanglement contribution to strength is considerable. The higher the molecular weight, the more entangled the structures become, and the stronger the polymer becomes upon testing. This was clearly demonstrated in a series of adhesives compounded with soluble cellulosics, where table sugar was the weakest and pearl starches the strongest, while all other modified starches fell between the two extremes [22]. Thermoplastic elastomers (such as the segmented polyesters developed by W. Witseipe of DuPont) are based on a rubbery continuous phase tied by covalent bonds to a hard plastic phase which acts as reinforcement. These segments do not separate into rubbery and plastic phases unless the segment molecular weights of the block copolymer are above 5000–7000.

Although the emulsion polymers do not have the stereochemistry to do so, one may rationalize the contributions of crystallites to strength in the same way that hard plastic segments in thermoplastic elastomers contribute. In crystallizing polymers, the crystallites are tied to the amorphous (rubbery, elastomeric) regions by the molecular segments crossing the boundary with convalent bonds. Such molecules, having some segments in amorphous regions and other segments in crystallite regions, are called tie molecules. The only emulsion polymer to exhibit crystallinity (to my knowledge) is the natural *Hevea* rubber latex, which has 95% or more cis configuration. The crystallinity only is only apparent in lightly cured rubbers at high elongation, wherein the molecules can rearrange to become aligned well enough to crystallize and reinforce the rubber. Although I know of past research that had stereoregular emulsion polymers as an objective, I know of no all *cis*-poly(butadienes) made by emulsion polymerization.

Reinforcement by additives is another important contribution to the development of strength. Carbon black increases the strength of the rubber in tires, and silica the strength of silicone rubbers. The key point is that each of these develops a bonding to the rubber that provides a hard strong barrier for a stress. Other methods for making hard particles act as reinforcement include the addition of surface modifiers to the hard particles. There are several interpretations of what happens when a reactive silane or titanate is added to a particle surface. The additive makes the surface match the polymer in solubility parameter; cross-links or covalently by bonds to the polymer; or allows wetting of the particle by the polymer. In any case, the model that works for your formulation is the one you should use. Such modified pigments aid in latex-coating formulation strength development [23].

Crosslinking adds to strength, as well. A measure of how effective cross-linking is can be seen by comparing a tire (usually cured with 3–5 parts sulfur to 100 parts of rubber) to a bowling ball (which may be made from 20 to 30 parts sulfer per 100 parts of rubber), where the only difference is in how much sulfur was added to the curing recipe leading to shorter chain segments between cure sites in the bowling ball. The cross-links may be incorporated in the emulsion polymer during the polymerization by chain transfer or by adding cross-linking monomers such as divinylbenzene or ethylene glycol dimethacrylate. Indeed, we found two acrylic latexes having almost identical physical properties in nonwoven fabric bonding, but one latex had significantly less nitrogen incorporated. The nitrogen was indicative of *N*-methylolacrylamide, a curing monomer, so the difference was likely to have been the use of something like ethylene glycol dimethacrylate during the polymerization. Monomers (such as *N*-methylolacrylamide) that cross-link on drying and cure may also be added to the polymerization. Chapter 9 contains a partial listing of such monomers, including the cross-linking mechanism. Cross-linking monomers are available from specialty suppliers.

So far the presumption has been that the polymer is anisotropic, that is, that the strength is the same in all directions of testing. This is not always the case. Cast acrylic sheet is stretched in two directions simultaneously to make it stronger by orienting the molecules in those directions. Fibers are also stretched along their long axis to increase their strength, and several forms of poly(propylene) are stretched to make plastic armor because of the amazing strength-to-weight ratio achieved. However, all these stretching techniques are only effective on bulk polymer materials, while the emulsion polymers are most commonly used as coatings or adhesives with little or no stretching after drying. Stretching these bulk polymers has a weakening effect on the directions normal to the stretching direction.

Thus it appears that to design strength into a emulsion polymer, one must use the molecular weight and crosslinking monomers as the approaches controlled during the emulsion polymer synthesis. Other compounding techniques take advantage of the interactions of the polymer with a solid material, most likely treated to yield the bonding needed for reinforcement.

B. Flexibility and Elasticity

Rubbery or elastomeric behavior differs from plastic or elastic behavior in one other respect, and that is in the energy absorbtion by the polymer during a stress. Because of its plasticity, the polymer acts like a Hookean spring with essentially constant elongation over all the range of stress, whereas the elastomer elongates more at the early stages of stress than at the later stages. Since the work energy is the area under the stress–strain curve, the plot shows more work going into the stretching of the plastic initially.

This effect is a temperature dependence, as all polymers behave as elastomers above their glass transition temperatures (T_g). Materials that we perceive as plastics have a T_g above room temperature, whereas those we perceive as rubbery have a T_g below ambient. For textiles applications, the "hand feel" or "hand" of a fabric treated with a polymer emulsion is a function of whether the T_g of the polymer in the emulsion is above or below the ambient temperature. Saturated paper stiffness correlates well to T_g for equal amounts of polymer emulsion used in saturation, as long as the T_g is above ambient [24], but if the polymer T_g is below ambient, no difference can be detected. No correlation to initial modulus was found, either.

The rubber behavior when the T_g is below ambient means that the polymer segments are sufficiently mobile to rearrange themselves with small amounts of stress applied. An example could be a plate of spaghetti. When cooked and entangled, the spaghetti are quite flexible as individual segments slither past each other. However, upon drying, they behave more like the plastic whose segments are frozen into a particular position. One may use the same example to see how cross-linking can immobilize even a rubber material. Stapling the wet spaghetti segments together at the crossover points between segments would inhibit the free slithering while maintaining flexibility over all the segments taken as a group. Adding more staples further inhibits individual segment motions.

REFERENCES

1. A. Rudin, *The Elements of Polymer Science and Engineering,* Academic Press, New York (1982).
2. S. L. Rosen *Fundamental Principles of Polymeric Materials,* Wiley-Interscience, New York (1982).
3. R. D. Athey Jr., "Colloid Chemistry—The Key and the Challenge to Aqueous Coatings' Formulation and Properties," *Pigment Resin Technol.,* 10 (Sept. 1977).
4. P. J. Flory, *Principles of Polymer Chemistry,* Cornell Univ. Press, Ithaca, New York (1953).
5. G. M. Burnette, *Mechanisms of Polymer Reactions,* Interscience, New York (1954).
6. W. H. Starnes et al., *ACS Polym Div. Prepr., 25(2),* 75 (1984) and references therein.
7. R. D. Athey Jr., *Prog. Org. Ctgs., 7(3),* 289 (1979).
8. R. V. Lindsey and E. L. Jenner, *JACS, 83,* 1911 (1963).
9. J. W. Vanderhoff, 1976 Lehigh Emulsion Polymer Symposium, Lehigh Univ., Bethlehem, PA.
10. R. M. Fitch, Dept. of Chem., Univ. of CT, Storrs, Ct. personal communication.
11. F. R. Mayo, *JACS, 65,* 2324 (1943).
12. C. A. Uraneck, *Rubber Chem. Technol., 49,* 610 (1976).
13. J. Hall, 1986 FSCT National Meeting presentation, Atlanta, GA.
14. R. D. Athey Jr., W. A. Mosher, and N. E. Weston, *J Polym. Sci.—Polym. Chem. Ed., 15,* 1523 (1977).
15. D. R. Campbell, *J. Appl. Polym. Sci., 19,* 1283 (1975).
16. J. Seiner, *J. Polym. Sci., 3A,* 2401 (1965) and subsequent publications.
17. T. Provder and C.-Y. Kuo, *ACS Org. Ctgs. Plast. Chem. Div. Prepr., 36,* 7 (1976).
18. F. M. Mirabella et al., *J. Polym. Sci.—Polym. Chem. Ed., 14,* 581 (1976).
19. G. D. Shields and W. P. Mayer, *J. Ctgs. Technol., 58(742),* 39 (Nov. 1986).
20. F. W. Billmeyer, *Textbook in Polymer Science,* 2nd ed., Wiley-Interscience, p. 333 New York (1962).
21. R. D. Athey Jr., *Makromol. Chem., 179,* 2323 (1978).
22. R. D. Athey Jr. and Sam Hill, unpublished work, 1968.
23. R. D. Athey Jr., *J. Water Borne Coatings* (Feb. and May 1986).
24. R. D. Athey Jr. and M. M. Conrad, unpublished work, 1976.
25. R. D. Athey Jr., J. K. Satryan and S. Chilcoat, *Europ. Coatings J., 1990(1/2),* 25 (Feb. 1990).

4

Processing Emulsion Polymers

I. INTRODUCTION

The processing of emulsion polymers varies from the laboratory to the manufacturing production plant. The laboratory techniques are of particular interest, as they will be a guide to the manufacturing operations, but only a guide. The manufacturing process must be modified from that used in the lab, for several reasons as shall be discussed later.

Since there are three classes of emulsion polymerizations, from an operating standpoint, we shall deal with each as a separate entity. The basis for considering them as different classes lies in the pressure needed to accomplish the polymerization, and the differences in polymerization vessels in operation and materials handling.

There are two tools for process modification. The first is the formulation, wherein the adjustments are made by chemists in types or amounts of components. One day may add or delete a particular component from the formulation, or one may add or subtract from the amount of the formulation component. The second form of adjustment is in the processing details controlled by engineers. Guidelines on engineering control have been suggested by Fan and Shastry [1] and Penlidis and co-workers [2]. The design engineers may raise or reduce the temperature or the rate of stirring, or they may modify vessel geometry. In either case (formulation or design change), the changes must be documented and all parties to the product formulation, processing, marketing, and sales must be informed. There have been cases of customer complaints that the manufacturing engineers had made production efficiency changes that changed the properties of the emulsion polymer, and it was no longer useful in the customer's application.

II. LABORATORY OPERATIONS

A. Polymerizations

1. Ambient Pressure Systems

These are the most common systems for emulsion polymerizations. They are frequently taught in laboratory courses for training polymer chemists and engineers. Many supplier companies use these techniques to assess their products as potential offerings to the polymer industry, though they do not make polymers themselves.

The lab operations are most simply performed in the round-bottom standard-taper ground-glass-jointed flasks used for organic chemical syntheses. Slightly more sophisticated labs use so-called resin flasks, which consist of two parts with a flat ground-glass flange and a mechanical holder (spring or wing nut tightened) to hold the two parts together. The upper portion of the resin flask has standard taper joints, like the round-bottom flasks, so the added items of equipment needed may be put into use. Figure 4.1 shows a typical resin flask setup.

Stirring the emulsion polymerization promotes heat transfer and moves additives to where they are needed. Although one may use magnetic stirrers (oval bars are recommended for round-bottom flasks), mechanical stirrer blades are preferred for viscosities above 100 centipoise (cP). The mechanical stirrer is usually a Teflon blade on a glass rod, though glass or stainless steel blade can be used. The glass rod is passed through a ground-glass bushing with a silicone or glycerine lubricant. There is some wear of the moving glass-to-glass contact, but this is not a serious problem, though lubricant leaching into the formulation can be. The top of the glass rod is attached to a stirrer motor axle by a rubber-tubing bushing, and a hose clamp ensures there is no slippage. Electric stirrer motors are preferred, as air motors are difficult to keep clean enough for reproducible operation, and their low torque impedes operation in high viscosity fluids. There are metal bushings that fit into the ground-glass joints on the top of the resin flasks. These metal bushings have stirrer systems that can be attached to the stirrer motor axle by an Allen bolt. However, these systems are oil lubricated, and the oil and metal-wear particles can contaminate the emulsion polymer. This is not desirable but closely approximates the processing conditions in a production facility.

Other ground-glass fittings for the top of the resin flask include:

1. Thermometer holder;

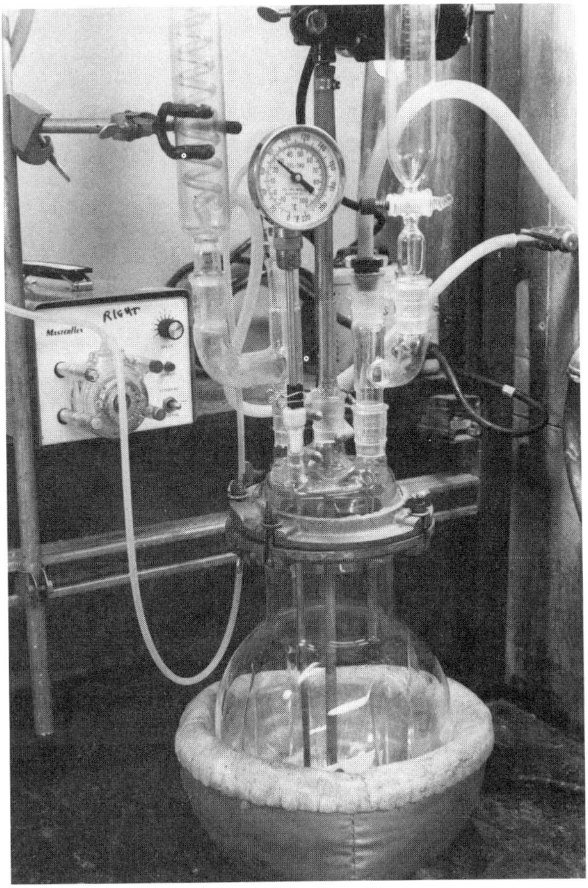

Figure 4.1 Typical reflux polymerization lab set-up. The resin flask has a detachable top for easy cleaning, and inlets for stirrer, fluid addition, inert gas blanket addition, reflux condenser, and thermometer. (Courtesy of Gencorp Polymer Products.)

2. Reflux condenser;
3. Addition funnel(s);
4. Gas injector; and,
5. Sampling probe.

The thermometer must immersed in the liquid surface; most glass thermometers have a line, marking the section that should be immersed. Stainless steel thermometers are more rugged (in case of being hit by a stirrer blade), but their scale is unreliable. A gas sweep or injection bubbler is usually installed to deoxygenate the water; it can be removed before the process is started. The monomer or a mixture of monomer, initiator, and additives for colloidal stabilization are added through a funnel. It may not be desirable to mix initiator and monomer in the funnel. The sampling probe may be a long needle on a hypodermic syringe or the gas bubbler hooked up to a capture container and a vacuum line.

The lower half or two thirds of the resin flask are usually heated in a water bath, though an oil bath or an electric heating mantle with a good Variac-style controller may also be used. The water bath may be heated by an immersion heater, a steam line, or a hot plate. Controllers are desirable, such as those supplied by Instruments for Industry and Research (Cheltenham, PA). Water bath may be stirred by a magnetic or mechanical (even air driven) stirrer.

The heating baths are usually maintained permanently in a setup on a lab bench with all the heating and stirring equipment in place. A series of one to eight such setups can be operated by one or two people once daily. The batches are first prepared for formulation on the afternoon before the actual runs. The water (distilled or deionized) is boiled, as this is the simplest method of deaeration. The colloidal stabilizers and other water-soluble (noninitiator or monomer) ingredients are added, and the mixture is placed in the polymerization vessel. The lid is closed, if a resin kettle is used, and stirrer, condenser, and other fittings put in place. A nonoxygenated gas (nitrogen, or deoxygenated air) may be swept slowly over the surface to displace oxygen, or it may be bubbled through the water to deoxygenate, if the water was not previously boiled. There is some fluid loss by evaporation, which is replaced. The nonoxygenated air may be made in several ways, as simple as a filtration with an apparatus from Balston, or by mixing natural gas into a closed volume of air and igniting the mixture to generate carbon oxides and water from the oxygen.

In large experiments, the water batch may be made in a slightly larger quantity than the sum of the volume of all the previous experiments. In other words, if you are making four 1-L batches of water (wherein only initiator of the monomers or monomer ratios will vary), prepare a master batch or 4.1 L or more of the water phase, and place 1L each into the polymerization kettles. Masterbatching has the advantage of eliminating an uncontrolled variable in the polymerizations, as all the water phases are identical. Table 4.1 shows a typical master batch. All components should be soluble in the water phase. Weighing batch components is preferred, unless volumetric convenience and knowledge of density give a reasonable assurance of accuracy.

The next morning, the heat is turned on for the water baths and the remaining ingredients are collected. At this point the variable water-soluble components may be added to the resin kettle. Label the resin kettle at this point, to keep track of the formulation variations. An example of such a master batch with formulation variations is shown in Table 4.2.

Table 4.1 A Typical Masterbatch Formulation[a]

Material	For 1 Batch	For 4.1 Batches
Water, distilled, mL	1000	4100
Sodium lauryl sulfate, 30% solution, mL	30	123
Methyl cellulose, g	5	20.5
Sodium bicarbonate, g	3	12.3
Sodium EDTA[b], g	0.5	2.05
Total, g	1038.5 g	4256.85 g

[a]One may use 1038 mL of this batch per flask, as such an error is small with respect to batch size.
[b]Ethylene diamine tetraacetic acid.

The monomers (usually acrylics, styrenes, or vinyl acetate) are weighed in, and the stirring and condenser are started up. When the mixture reaches the desired temperature (a few degrees below that of the water bath), the initiator may be changed (in water, if possible) and the time noted as the nominal start of polymerization. Further additions through funnel ports are recorded as well as the rates of addition. The temperature of the bath and the polymerizing fluid should be recorded at reasonable intervals, as well as observations on batch clarity, turbidity, color, reflux intensity, coagulations, or whatever else is of interest or unusual. The ideal technique for record keeping is a table, showing batch number, experimenter, components, date(s), temperature(s), sample(s) taken, time, and remarks.

Sampling may be by removing the thermometer and looking at the fluid clinging to it. This procedure may even supply a 100-mg sample for solids determination of viscous material. For larger samples a syringe or other device may be needed; a one-gram sample is preferred for percent solids determination.

Table 4.2 A Typical Masterbatch Formulation with Variations

Material	Batch 1	Batch 2	Batch 3	All batches
Water, distilled, mL	1000	1000	1000	4100
Sodium lauryl sulfate, 30% solution, mL	30	30	30	123
Methyl cellulose, g	5	5	5	20.5
Sodium bicarbonate, g	3	3	3	12.3
Sodium EDTA[a],	0.5	0.5	0.5	2.05
Subtotal, g	1038.5	1038.5	1038.5	4256.85
Sodium sulfate, g	0.5	1.5	2.5	

[a]Ethylene diamine tetraacetic acid.

Percent solids is, by the way, a linear measure of conversion. For every polymerization, you may calculate the nonvolatile percent solids at zero conversion and at 100% conversion. A plot of these on quadrille paper gives a calibration curve for estimates of conversion from samples taken through or after the polymerization.

Watch the polymerization! You will note the time at which the polymerization starts by an exotherm that often exceeds the heating bath temperature. The reflux will become more vigorous, foaming may be a problem, and flocculation may occur. These are indications that modifications are needed. The color may be a clue; for example, a monomer emulsion may turn blue just before the reaction control is lost. That is, the white opaque, large emulsion particles have suddenly been converted to a small fast-polymerizing particles, giving off a large amount of heat. This is related to particle size. The emulsion is likely to be foaming out of all openings very shortly, and even an ice bath may not help control the exotherm. In a vinyl acetate-based polymerization, 1 mL of styrene may be able to control the exotherm (See Copolymerization chapter.)

After the polymerization reaction is complete, one may add the stabilizers or adjust the pH, or one may pour the latex (through a filter) into a 1500 mL beaker or quart jar for such formulation. There are standard disposable paint filters (cheese cloth on a stiff paper frame) suitable for the filtration of low viscosity products, but a preferred technique employs a circular stainless-steel screen (about 100 mesh) formed into a cone to fit into a ring stand. It can be used for filtration of high viscosity liquids, and is easily washed after recovering any coagulum for drying and weighing (record the weight!). Any sticky adhering polymer may be burned off with a Meker burner in a hood. The quart jar is preferred as the container for the polymerized product, as it is capped to give good storage stability and safe transport for analysis or application testing. *Label the jar well!* Nothing is as useless as an unlabelled sample.

This sequence of events, from heating of the water bath to completion of the polymerization and recovering the latex, usually takes four to six hours. The time when the polymerization is in progress may be spent in preparing for the next day's polymerizations, and in characterizing the latexes from the preceding day's work (determinations of pH, viscosity, total solids content, etc.). With practice, one may do four to six polymerizations a day, for four days a week, and have time for reports, meetings, and administrivia on the fifth day.

2. Modest Pressure Polymerizations

The monomers that are gases at room temperature, but may be condensed at some reasonable low temperature, generate only modest pressure in containment vessels. The monomers most frequently thought of in this context are butadiene and vinyl chloride. Both condense at some temperature above 0°F (− 17°C), and may be liquid under pressure at room temperature.

Figure 4.2 Typical 28 oz. emulsion bottle with guard. (Courtesy of Gencorp Polymer Products.)

Many years ago, a simple inexpensive technique was developed for polymerizations involving these monomers, based on soda-bottle technology. The 28 oz. soda bottle was the most common form used, but citrate (pint) bottles and even Coke bottles have been used in recorded literature. The typical bottle with crown and a bottle guard is shown in Figure 4.2. All that is required is a bit of care in handling the gaseous monomer and some small items not used the preceding section, along with a large thermostatted water bath (ca. 50 gal) called a thermostat for short. This bath has a rotating creel or rack within to hold and mix the polymerization bottles. This technique and equipment are also useful for other low boiling monomers, such as isoprene or methyl acrylate, with a little pressurization of the bottle with nitrogen. Both vinyl chloride and butadiene monomers are suspected carcinogens, and there is substantial new literature on their handling from suppliers of the monomers.

The polymerization bottle itself may be bought from standard sources at a premium, because only perfect bottles are acceptable. The bottle should be carefully washed and labeled before the day's experiment. Bottles with even slight imperfections should be discarded. Invert the bottles and let them dry overnight for filling the next morning.

The master batch of waterborne materials is made up late in afternoon before the day of the experiment, using the techniques described in the preceding section. Volumetric fill of bottles from a large (500 mL), buret, for the master batch or the liquid monomers (styrenes, acrylics, etc.) is most convenient. Weigh in the solid monomers from a weighing dish through a funnel, and wash the funnel with the master batch addition. A syringe is a convenient device for small amounts of liquids, such as the mercaptan chain transfer agents. After all but the gaseous components are added, one transfers the bottles to the capping station. One may sweep with nitrogen to deaerate the bottle at this point (particularly necessary for acrylic ester copolymers).

Capping is not a difficulty and is easily learned. The caps are new soda bottle crowns, without labels, and some have a hole (about 1/16th in.) in the center for additions of components with a syringe during polymerization or sampling or for adding a gaseous monomer after closure. Gaskets (usually neoprene or nitrile rubber, with a Teflon face on some) are inserted into the inverted crown caps. The crown closure press is a lever about 18 in. long that lowers the concave face over the crown to clamp it tightly onto the bottle. Use the left hand to hold the crown and gasket on the bottle top, and the right hand to slowly make the contact, so that the left hand can guide the bottle and crown to the center of the concave press face. An uncentered bottle will likely shatter, so be careful. Safety precautions for capping and filling the bottle with gaseous monomer include a full face shield, gloves, and an apron.

The gaseous monomer is commonly added by weight in a hood just after closure. A small pressure bomb is filled by weight with a 2-5 g excess of monomer. The bomb is fitted with a needle to penetrate the rubber exposed by the bottle crown, and a valve for the transfer. Weigh the bottle before and after the addition, and bleed out any exceed monomer, if necessary. With an earlier technique the monomer was liquified by pressure and an excess piped into the preweighed bottle. The bottle was weighed (in a hood), and the excess monomer boiled off, sweeping any air simultaneously. With a gram or two left before attaining the desired weight, the bottle was moved to the crowning press and capped as above. With skill, the weight of gaseous monomer added was within a gram of that desired, a permissible surplus.

The polymerization bottles should now be enclosed in safety-shield containers. These are stainless steel cans, perforated to let the thermostat water come through to the glass. These cans consist of three parts: a base that covers 50–75% of the bottle with lugs reaching out radially from the open upper lip; a cap of large enough diameter to encircle the base, with "J" slots at the lip of the opening for the base lugs to lock into; and, a large conical spring fitting in the base to apply upward pressure to keep the lugs in the J slots.

The protective shields are essential to prevent injury caused by explosions. During two years of handling these bottles, I experienced three or four explosions of unshielded bottles but none occurred during the following six years working with shielded bottles.

The shielded bottles are put into the thermostat, and the time of initiator addition is recorded. The thermostat resembles a big steel box with a counter-weighted lid. It contains at least 50 gal of water, covering a rack of spring clips that hold the safety shields, a thermostat temperature sensor, and steam lines along the bottom for heating. Some thermostats are designed for cooling for operations below ambient temperatures. These may use a glycol–water mixture as the heat transfer fluid instead of water alone. A switch is installed, a temperature indicator with three positions (stop, jog, and run) for controlling the rotary motion of the rack within the water. You load the rack (still wearing protective gear) using the jog switch to move the rack so that the empty clips can be reach to be inserted into the shielded bottles. Mount the bottles on the rack in such a way that they are distributed evenly around the axis of rotation, so that their weight is balanced to even the wear on the axle bushings. When the rack is loaded, push the run button for continuous rotation.

A disadvantage of this technique is the gentle (though very efficient) agitation of the rotating bottles in the thermostat. Some thermostats are designed to rotate the bottles end-over-end, whereas others rotate the bottles over their shorter dimensions. In either case, there is no shear region. The reactors described in the preceding section or in following sections have points of high shear at stirrer blade tips or other points in the process. High shear regions are likely to cause mechanical coagulation in marginally stable latexes, and flocculation will occur. The bottle reactions may give no indication that this mechanical coagulation problem may exist, and it may first appear as a major problem in scale-up or pilot operations.

Sampling, for total solids determinations, is done by tapping the bottle at the exposed rubber gasket with a syringe needle. The syringe is a standard glass 2 mL diabetic or tuberculin Luer-Lok syringe, especially fitted with a gas-tight valve connected to a standard 1-in. needle and a bracket that surrounds the syringe cylinder, holding the plunger to a maximum 1–1.5 mL sample. The syringe is shown in Figure 4.3.

Safety gloves and full face mask are again needed for sampling. The technique of sampling is as follows:

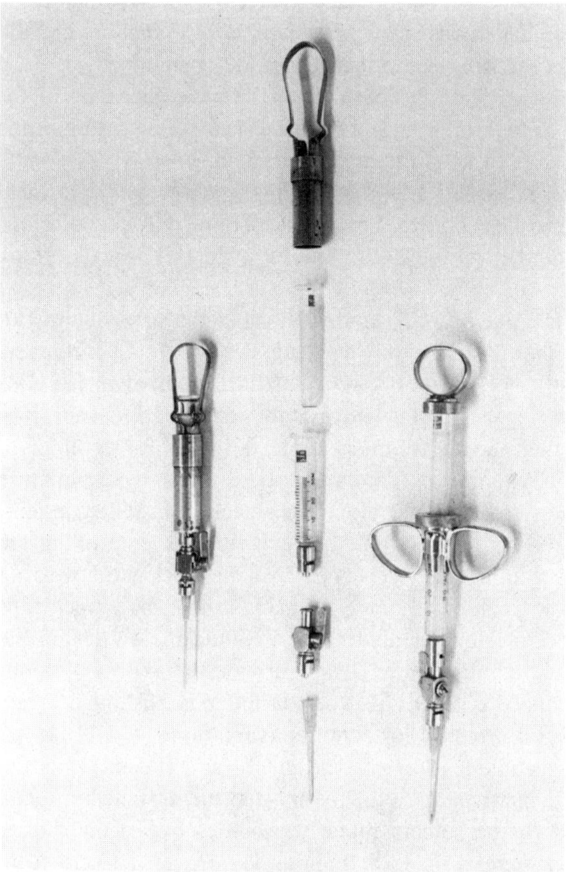

Figure 4.3 Typical syringes for bottle polymerizations. The syringe on the left is for withdrawing samples from pressurized emulsion polymerizations. The center shows the five disassembled parts of the syringe on the left. The parts, from top to bottom, are, plunger retainer, plunger, cylinder body, gas valve, and needle. The syringe on the right is for injection into the pressurized bottle. (Courtesy of Gencorp Polymer Products.)

1. Stop the rotation of the thermostat rack, and move it to the point where the first bottle may be taken out;
2. Remove the bottle and hold it in the left arm with base at the elbow and the crown exposed between the fingers of the hand;
3. Take the closed syringe in the right hand, insert the needle through the rubber gasket, and rotate the bottle so the crown is downward and the needle is covered with latex;

4. With the palm of the right hand covering the syringe plunger, open the valve to let in the sample;
5. When the syringe plunger has reached its maximum extension, close the valve, remove needle from gasket, and lay the syringe down;
6. Replace the shielded bottle in the thermostat rack, and turn on the rotation again;
7. Record the weight of the full syringe;
8. Point the needle at the preweighed aluminum weighing dish containing 1 mL of methanolic MEHQ (0.5%);
9. Open the valve slowly to spray the pressurized latex into the methanol coagulant–short-stop solution;
10. Close the valve when no more sample is forthcoming;
11. Record the weight of the empty syringe;
12. Dry sample under heat lamp or on a hot plate; and,
13. Record weight of weighing dish and dried sample.

Sampling for percent conversion estimates (using the type of calibration curve described in the preceding section) should be done every 2–3 hr, though this depends on the expected rate of the polymerization. Styrene–butadiene copolymers may be sampled every 3–4 hr during a 12–16-hr run, while butadiene and acrylics may need sampling every 90 min in 6–8-hr run. Try to get three or four points during the 10–80% conversion segment of the polymerization, as the rate is generally linear, and may be easily estimated with a % conversion vs. time plot.

There are times when injections will be needed. In cold polymerizations (usually about 5°C) initiator must be injected after the bottles have equilibrated with the thermostat. In polymerizations wherein the conversion must be stopped (about 60–65% conversion for butadiene copolymers where it is desired to have no gel content, i.e., no chain transfer to polymer), a *short-stop* solution is injected to trap all radicals or kill all the initiator. For the injection a prefilled syringe of the sort used for sampling suffices if 1–1.5 mL of the reagent is enough (shown also in Figure 4.3 with the finger rings). In other instances, a special one-way syringe with two ball valves is used. One valve allows exertion of pressure on the syringe to open the needle to the inside of the bottle, while the other allows suction to fill the syringe body from a tube extending into a fluid reservoir. The practice here is to stop the thermostat rotation, pull out the desired shielded bottle and set it on its base, insert the needle into the top of the bottle through the gasket, pull the plunger to fill the needle volume, and push down the syringe to transfer that volume into the bottle. When the injection is completed, pull out the needle and replace the shielded bottle in the thermostat. Again, full face mask, gloves, and apron are required safety protection.

At high conversions, the pressure is reduced, and in some cases, a vacuum may be created. A pressure sensor on a needle is a good indicator of completion of overnight runs where solids determination was not continuous. Occasionally a bottle may be contaminated with thermostat water, if the cap was not tight, and a vacuum developed. For some polymerizations of small amounts of butadiene or vinyl chloride and acrylic monomers the appropriate pressure is reached by nitrogen pressure (70–100 psig).

When the run is completed, remove all the shielded bottles from the thermostat, and let them cool in a hood to ambient temperature; check the pressure. Insert an open needle through the gasket of each bottle for venting or pressurizing, and remove the bottle from the shielding. Remove the cap with a bottle opener but *not without checking pressure*. Strong foaming by latex and monomer may occur if the venting needle is not open. Listen for the sound of vacuum loss or gas escaping! Short-stopped runs will be under butadiene pressure, which must be released slowly in small increments. Cooling the bottle in ice water controls the release of residual butadiene. Again, wear the protective mask, gloves, and apron.

The bottles may be sampled for final total solids content, and closed with a small cork for storage. Flocs may be removed by pouring the contents through the conical screen on a ringstand, but occasionally large floc will remain in the bottle and inhibit pouring. Large flocs are removed by an 18-in. brass or bronze rod with a half-in. hook on the end. The bottles must be properly and clearly labeled. Duplicate bottles can be combined in a 1500 mL beaker, and mixed for sampling and additions. A half-gallon jar with a well-fitted lid is suitable for long-term storage or shipment.

The bronze or brass hook and a bronze metal hammer for breaking up ice or dry ice, are the only copper-bearing metal that should be in a lab *or production* facility for emulsion polymers. These brass/bronze items are preferred as these metals do not emit sparks upon dropping on other metals. No iron or steel tools are allowed! Copper tubes, piping, screws, nuts, or bolts should not be present, since they interfere with emulsion polymerizations. Copper-bearing metals are notorious contaminants, inhibiting the start of polymerizations and causing discoloration later.

A skilled operator can make duplicates of a 750 mL formulation for each bottle, and run five to eight per day. These polymerizations procede well, unless there are problems with initiator formulation or concentration or with coagulation. The emptied bottles are discarded; if rinsed they can be recycled for uses other than polymerizations.

3. High Pressure Systems

High pressure systems are required for gaseous monomers such as ethylene. They operate above 200 psig, the maximum pressure applicable to the glass bottles described in the preceding discussion. Preferred are metal reactors, such

Figure 4.4 Typical pressure reactor for pilot scale reactions. The metal reaction vessel has the sight glass on the side, though larger vessels have it on top. The gas cylinder on the top is used to load butadiene or vinyl chloride. Other top ports let in the thermometer, stirrer, and another addition/sampling port. The pipes entering the side are heating oil inlet/outlet pipes. (Courtesy of Gencorp Polymer Products.)

as available from Parr or Bench-Scale Reactor suppliers. They are essentially steel cylinders with glass, phenolic, or epoxy linings or made completely of stainless steel. The bottom section may be welded on and may have a drain valve for emptying. A typical setup is shown in Figure 4.4.

The top section is usually separate and flanged with a gasket for sealing; and a number of bolts at the outer edge of the flange hold the top and main body together. The top may have been pierced around the periphery to hold thermometer well, injection port, sampling port, pressure gauge, or other devices. The top must be equipped with a safety device called a rupture disk which is easily broken and through which excess stress can be diverted. Some reactors also have a so-called sight glass, which allows observation of the process in progress, provided light source and eye fit closely together.

The center of the top is fitted with a metal stirrer bushing, passing the axle through from the stirrer motor above to a stirrer of various possible designs within.

The stirrer may be bladed like a propeller, or it may be of a paddle, anchor, or turbine shape, depending upon the shear needed. Additional shear may be provided within by baffles in the sides of the reactor. The best designs allow simple bolt connection and disconnection of baffles and stirrer blades.

The reactor may be heated or cooled by an external wrap of tubing or a jacket, or by tubing within the reactor. Some reactors have electrical tape heaters as well. Heating and cooling should be connected to a thermostatic control with a recording device to indicate the temperatures of the jacket and tubing fluids as well as of the internal reactor contents.

The volume of these reactors ranges from a few liters to 5 gal. Because of the difficulty of clean-up and setup, only one reaction a day is usually run. Many companies use these reactors only as a last resort for experimental work, employing formulations assumed to be equivalent. These pressure reactors are also the starting point for scale-up of lower pressure reactions on their way to production.

B. Stripping and Concentration

The latexes made in the laboratory will often have residual monomer that needs to be removed. Most customers require the lowest possible concentrations of vinyl chloride or acrylonitrile in their copolymers for protection against possible carcinogenicity. Styrene and vinylcyclohexene impart undesirable tastes and odors to food packaging. The technique to remove residual monomer is called *stripping*.

Latexes are frequently made a certain total solids content, but shipped or used at *higher* total solids content by removing the water, that is, concentrating the latex if for no other reason than to reduce shipping weight. A distillation or vacuum distillation process is easily set up. One may calculate the amount of water to be removed, based on the total solids, the target total solids, and the weight of the latex sample. As a practice, one should collect 80–90% of the theoretical amount of water, and check the total solids content, as there are losses of water due to inefficient condensation in the best of systems. There are collection traps that isolate the distillate in a volumetrically scaled column, so distillate is easily measured.

Since the undesirable monomers are hydrophobic, steam distillation in a variety of forms is used to separate the organic liquids from the latex. In the lab a simple hot water bath may be used to raise the temperature; a modified Dean-Stark trap is inserted to collect the organic fluids at ambient temperature. More sophisticated are vacuum systems that collect the entire distillate in a scaled vessel, as noted in the preceding paragraph. Thus stripping and concentration are combined in one operation, a common practice.

Both of the above operations may be carried out in a standard round-bottom single-neck laboratory flask. It may be easier to use the vacuum rotary evaporators (e.g., Buchi) for the agitation and increased evaporation area exposed. If vacuum is used, a cold trap (dry ice in acetone) is inserted between the vacuum source and the condenser, to protect the vacuum source from water or organic vapors that might pass through the condenser. Flanged flasks, such as thinner spring-clamped flat-topped resin flasks or rubber-flanged vacuum freeze dry flasks (Virtis) may be preferred for rotary evaporators. They facilitate the handling of the latex, cleaning and floc isolation. Some flocculation occurs in stripping or concentration processes. A rotary evaporator is shown in Figure 4.5.

The stripping operation is usually done to a specification (e.g., to less than 0.1% styrene), and analysis for residual monomer may be needed unless the process is well enough characterized to ensure the required low level. Occasionally, an additional stripping-concentration step is required, if so indicated by the analysis. Add more water and reprocess. Chapter 13 will deal with residual monomer analysis.

Stripping can be done where concentration is not needed. The "add-water" approach can be one way, but many facilities use steam injection in their standard round-bottom flask stripping process. Balancing a steam injection and vacuum system takes practice but can be done. Given the fluctuations in laboratory steam lines and vacuum lines, this sort of setup must be carefully watched. The steam stripping process may dilute the latex to a point where a concentration step is needed, and therefore the total solids content should be monitored. A typical lab stripper is shown in Figure 4.6.

The mixture of residual monomer and other organics isolated from the stripping process presents a disposal problem to the laboratory. Some by-products are contaminants of the recovered monomers, such as the benzene from benzoyl peroxide decomposition, tetramethylsuccinonitrile from AIBN cage recombination, monomer Diels-Alder addition products (vinylcyclohexene in butadiene based latexes), or other organic materials. Chapter 11 will deal with by-product forming reactions. Some of these contaminants present toxicity hazards, and others are known to reduce the rate of polymerization. Hence, these recovered organics must be disposed of in the standard hazardous waste product streams.

The stripping and concentration processes generate considerable foam on occasion. Foaming into the condenser system, or worse yet, into the vacuum

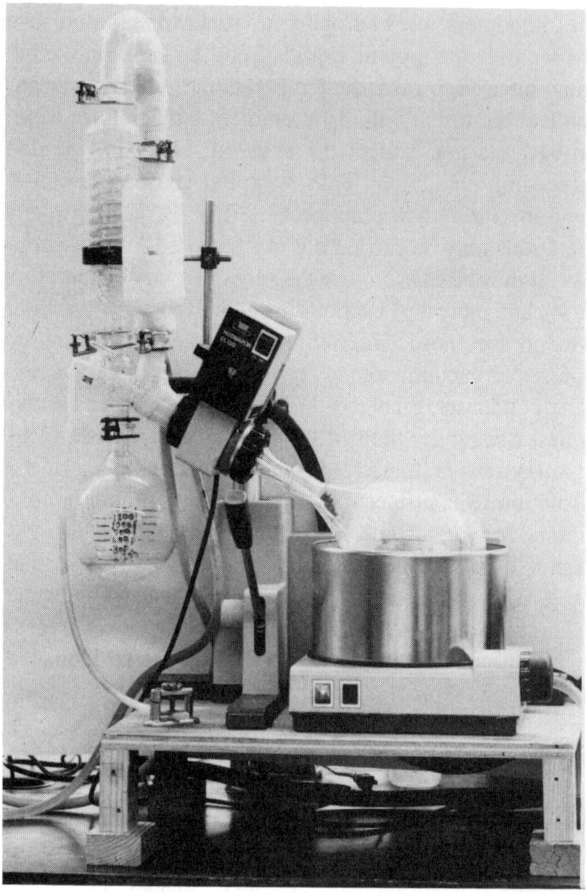

Figure 4.5 Typical emulsion polymer lab concentration set-up. This vacuum rotary evaporator arrangement makes the concentration of emulsion polymers quite a bit less troublesome than the static vacuum distilation apparatus used 25–30 years ago. Note the hose clamp in the foreground, used to control foaming in the early concentration stages by bleeding in air. The VIRTIS freeze-drying flasks may also be used instead of this round bottom flask, as they are easier to clean. (Courtesy of Gencorp Polymer Products.)

system can be troublesome. The process must be watched closely to balance foaming minimization with maximum output of distillate by adjustment of heat, agitation, and vacuum. The vacuum is adjusted with a needle-valve bleeder to admit a small amount of air, as in Figure 4.1. Agitation will sometimes mechanically reduce foaming, and a magnetic stirrer on a hot plate may be useful.

Defoamers and antifoams should be added at the rate of 0.1 mL per injection;

Figure 4.6 Typical steam stripping lab set-up. Without the steam injection, this would be the static concentration apparatus of years ago. This is a simple resin flask with heating mantle, thermometer, stirrer, steam injection, and vapor outlet to condensor. Note the vacuum line has connection to the collector flask, through an Ehrlenmeyer vacuum flask. The latter is for trapping fluids by immersion in coolant (usually dry ice/acetone) to prevent condensation in the vacuum lines or pump. (Courtesy of Gencorp Polymer Products.)

they might cause problems, but they are sometimes the only solution to the foaming problem.

C. Compounding

The finished emulsion polymer may have had several additions along the route of processing, such as a defoamer or antifoam agent before stripping and concentration or a surfactant to improve mechanical stability in subsequent steps. However, some additives are needed to finish the latex, so it may be used in applications evaluation, or for stabilization. These materials can be added to the formulated latex in a large beaker or wide mouth jar with stirring.

Choice of stirrer is important as the shear stress may result in coagulation. Use a large propeller or flat blade stirrer at low speed to move the latex around without causing air entrapment. Dilute additives with water, if at all possible, to minimize chemical coagulation or chemical shock. Follow the solutions or dispersions on the surface so you can watch the incorporation at the vortex, and look for signs of coagulation on the axle of the stirrer. Use a plastic 10–25 mL syringe with the Luer tip cut off for addition of high viscosity materials, such as thickener solutions. Stir about 30 min after the last addition, to ensure complete mixing.

Some high viscosity formulations are efficient at transmitting the stirrer system torque throughout the container. The container that suddenly starts rotating with the formulation is a threat for spillage. Chain clamps and a rubber base will help hold the container in place. On wet surfaces a so-called "turtle" arrangement is used; a series of small suction cups on both sides of a rubber disk about 3 in. in diameter, that will help hold the bottom of the container.

III. PILOT PLANT AND PRODUCTION

All the vessels are essentially variations on the high pressure vessels used for lab operations, that is, jacketed metal containers of stainless or clad steel. Black iron vessels have been used for the GRS (Government Rubber Service) in Akron for very slow, very large volume polymerizations. Because of the corrosion of the vessel and the deposition of the rubber on the corroding surface, cleaning was only necessary when heat transfer became a problem. Some older plants still use some black iron piping, which might create formulation problems at some time in the future because of dissolved iron ions. As stated earlier, there should be no copper, brass, or bronze involved in any processing equipment. The preferred materials are still glass or stainless steel.

Heat transfer is a major problem with large reaction vessels, as a runaway reaction blown through the roof from the rupture disk vent can cover a large area. The mass effect requires the reduction of initiator level (sometimes by a factor of ten) to maintain the reaction rate of a 1-L reaction at 1,000 gal. Hence, scaling a laboratory reaction to production can be a time-consuming sequence of increasing from 1 gal to 5, to 20, to 100, to 500, and finally to 3,000 gal. I have seen a bottle-reaction series of statistically designed experiments increased to 200 gal, resulting in recommended changes for the first run to make a factory-producible second run. The initiator level governed not only the rate of reaction, but also the prefloc level. A slight reduction in the initiator level converted an overly vigorous 200-gal reaction into a well-controlled polymerization.

Production systems frequently include heating or cooling water available for the jacketed reactor, and the exotherm may be controlled by judicious cooling at the appropriate moment. Runaway vinyl acetate polymerizations may be controlled with a small amount of styrene. As noted earlier, the basic structure of the vessels for production is about the same as that for high pressure laboratory vessels. However, vessels used for ambient pressure polymerizations may have thinner wells, and there may be other differences as well. The production vessels are usually housed in a shell building for weather protection.

Most production vessels are too big to allow the whole top to be opened as in lab pressure vessels. The top is usually bolted or welded in place. However, special openings, called manholes, in the tops allow entrance for cleaning, recoating, or other maintenance. Special precautions for entry are in order, as there is poor air exchange to the outside, and a threat of suffocation exists, although poisonous gases are absent. No one should enter a vessel without outside attention, and breathing apparatus is recommended for the worker inside the vessel. Information on the detail of safe entry to such vessels may be obtained from insurance companies and from the Occupational Safety and Health Administration.

The top of the production polymerization vessel is equipped with sight glasses, injection ports, and the stirrer bushing. The ambient pressure production vessel may also have an opening for a reflux condenser or a distillation condenser for concentration and stripping. Modest pressure vessels may be fitted with distillation condensers for condensation and stripping, and may be rigged for vacuum to aid those operations. Ganapathy and Elango [3] designed a scheme for such pressure/vacuum vessels.

Some manufacturers employ their own in-house constructions to modify the stripping and concentration process. Wiped-film vacuum evaporators are energy efficient for stripping and concentration. General Tire has a patent for an effective heat exchanger for heat transfer in a wiped-film system. It gave them a commercial advantage in processing efficiency over competitors for many years.

The stripping operation is the source of recovered monomer, a major cost savings during monomer shortages. The recovery of monomer requires vacuum distillation from an inhibited mixture of the organic materials gathered from the stripping process. In addition, all pressure polymerizations that stop short of complete conversion will be made more economical by venting the unused monomer gases (butadiene, ethylene) to a cryogenic recovery unit. Some older plants flared their vented butadiene into a flare stack that burned any organics that passed through. This may no longer be economical.

The bottom of the production vessel is normally fitted with a drain connection of large size. Occasionally, a smaller sampling port is also installed in the bottom, to remove a quart or pint of latex when a check on composition or solids content is needed.

Filtration systems in production facilities may involve such simple cheese cloth bags at the end of a pipe in a tank-to-tank transfer to something more sophisticated. In in-line filtration systems, such as Ronigen–Pettinger, a filter bag or mesh cylinder is installed within the pipeline. A simple design for continuous filtration of difficult materials is shown in Figure 4.7. In another design, offered by at least two manufacturers (Sweeco and Vorti-Sieve), the latex falls

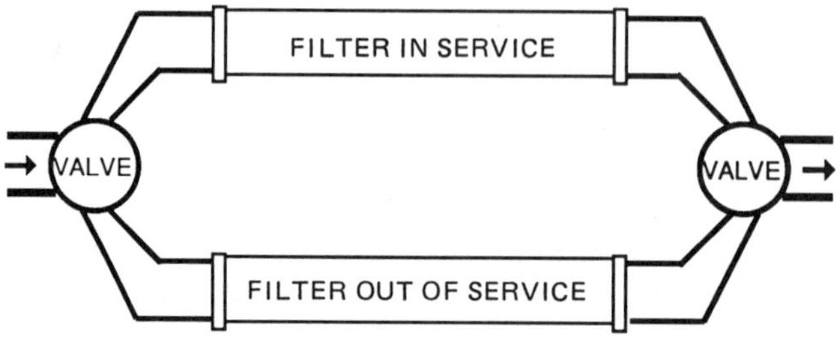

Figure 4.7 Diagram of dual filter in pipeline. Strainers or cartridge filters in a pipeline need not disrupt continuous flow when cleaning or repair needs arise. This array is recommended where continuous flow is mandatory in a process, be it latex or a formulated coating/adhesive product.

onto an open moving circular filter pan that is slightly off the horizontal plane and slowly moving in a circular pattern. This scheme is effective for removing microflocs which may clog a simpler filter design.

Compounding of latexes is done in large storage tanks. Blending several different production batches equalizes small variations between batches. Several 3000–5000-gal batches are stored together in one 20,000 gal tank, and compounding will take place when all the scheduled production are added to the tank. Alternatively, compounding may take place in the reactor as soon as the reaction is complete, followed by transfer of the latex into the larger storage tank.

Storage tanks are designed with a large very slow stirrer (anchor or paddle) that may run only for 10 min every hour or two. The tank has a "water seal" to prevent skin formation and surface evaporation. A typical design for such a water seal is shown in Figure 4.8. The air pulled into the tank is humidified, and still has a seal maintained so the latex does not form a skin.

These storage tanks also have piping through the roof or walls for filling or emptying, and manholes for inspection or entry for repairs or cleaning. The safety aspects of entry are important here, as they are for reactors. It is not uncommon to assemble large storage tanks out-of-doors in a tank farm. However,

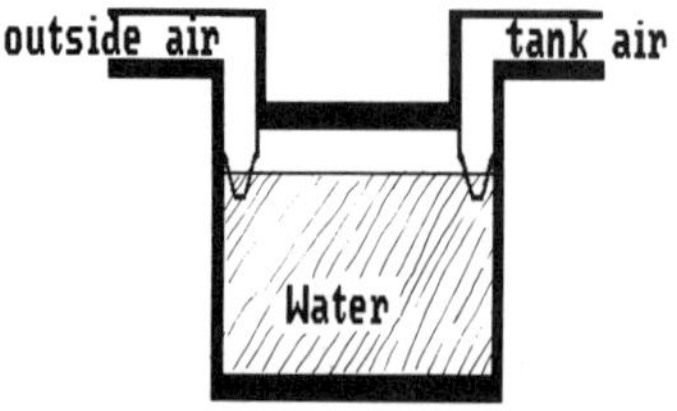

Figure 4.8 Diagram of a latex storage tank water seal.

weather problems must be dealt with in terms of keeping the latex from freezing, lightening strikes, drainage of rainwater, and so on.

IV. PROCESS EXPLANATION

Polymerization is a *dynamic* system. Many events take place, such as adsorptions and desorbtions, absorptions and desorbtions, initiator decomposition and recombination, and initiation, polymerization, termination, and transfer reactions. The monomer droplets are in motion, being first broken up by the agitation and then recombining. Ions are in motion, and organic molecules are desorbing from one site and migrating through the water to another site. If surfactants are present, they are aligning on any available surface, adsorbing and desorbing to pass through the water to another site for adsorption or forming new micelles.

To make a dispersion of polymer during a polymerization requires a dispersion medium in which the polymer is insoluble (water in most cases), the monomer, and an initiator. Stirring at high speed assures small polymer particles. The dispersion is only metastable for two reasons: Stokes' Law is the first, unless the density of the polymer and dispersion medium are well matched. The second is that the particles agglomerate with time and by settling.

The stabilization of the polymer dispersion takes place with the incorporation of special surface-active additives. Protective colloids, surfactants,, and dispersants aid in the stabilization of the final product, but also in the polymerization. So-called soap surfactants form micelles, that is, agglomerates that can imbibe monomer and thus be a site for polymerization. The polymerization within the micelle creates a particle, which grows from monomer migration from much larger monomer droplets that are suspended by stirring or other agitation.

Initiation usually occurs at the micelle or particle interface with the water-soluble initiators, in spite of their oil insolubility. This should be no surprise, as the salts are adsorbed at the surface of the micelle or particle, as are all salts, to some degree. The driving force for that salt migration to adsorb on that surface is the ionic concentration gradient. The quantization of that concentration gradient is the Nernst equation for a concentration cell, where the concentration of ions within the micelle or particle is zero. The more ionic materials (surfactants, salts, dispersants) are present, the higher the ionic strength, and the stronger the driving force for adsorbtion.

Now we have a formulation (recipe in the trade) with four components: water, monomer, initiator, and colloidal stabilizer. There may be more than one monomer and more than one stabilizer, even more than one type of stabilizer. But there are yet other additives. We discuss most of those in Section IV of the book.

A useful additive for the polymer chemist is the chain-transfer agent, commonly called the regulator or modifier in the trade. Its function is to reduce the molecular weight of the polymer. It is commonly used in bulk, solution, suspension, or emulsion polymerizations. However, in emulsion polymerizations, it is usually an oil-soluble, water-insoluble material. The Sulfole materials (mixed aliphatic C12+ mercaptans from Phillips Petroleum) are a typical example, though carbon tetrachloride is also commonly used. There are several arguments against use of water-soluble chain-transfer agents, especially their bad odor. They also are less likely to migrate into the particle where they are needed. They may also consume initiator, forming a much less active disulfide. Indeed, even the mixed tertiary tridecyl mercaptans give indication of direct reaction with initiator, as I have observed a reaction rate enhancement comparing with a mercaptan-free run at same monomer-water-initiator levels.

REFERENCES

1. L. T. Fan and J. S. Shastry, *Macromol. Rev., 7,* 155–187 (1973).
2. A. Penlidis et al., *J. Ctgs. Technol., 58(737),* 49 (June 1986).
3. V. Ganapathy and R. Elango, *Chem. Engr.,* 143 (May 19, 1980).

BIBLIOGRAPHY

D. C. Blackley, *High Polymer Latices* Palmerton, New York (1966).
C. R. Noller, *Chemistry of Organic Compounds*
Morris et al., *J. Poly Sci. A, 4,* 985 (1966).
T. R. Paxton, *J. Coll. Interface Sci, 31,* 19 (1969).
F. Tokiwa and N. Moriyama, *J. Coll Interface Sci., 30,* 338 (1969).
A. *General Introduction To Emulsion Polymerization* A. Raave, *Organic Chemistry of Macromolecules,* Chap. VII, 1977.
B. *Effects of Colloidal Stabilization Systems on Polymerization and Film Properties* Vinyl Acetate Butyl Acrylate Copolymer, E. Leving, W. Lindlow and J. A. Vona, *J. Paint Technol., 41,* 531 (1969).
"Emulsion Polymerization with ULTRA Surfactants," Witco Chemical Co., Bergen Mall, Paramus, N.J.
C. Inverse Emulsion Polymerizations
J. W. Vanderhoff, et al., *Polymerization and Polycondensation Processes,* p. 32. No. 34 in ACS Advances in Chemistry Series. Br. Pat. 1168551
D. Emulsion Polymerization Kinetics Studies
J. W. Vanderhoff, *Polymerization and Polycondensation Processes* 34 in ACS Advances in Chemistry Series.
J. L. Gardon, *J. Polym. Sci, A1, 6,* 2453, 2859, and articles referenced therein.
W. C. Griffin, *Encyclopedia of Chemical Technology,* Vol. VIII, Wiley, New York
J. Ugelstad and F. K. Hansen, *Rubber Chem. Tech., 49,* 536 (1976).
E. Model Describing Observed Stages
W. D. Harkins, *JACS, 69,* 1428 (1947).
W. V. Smith and R. H. Ewart, *J. Chem. Phys., 16,* 592 (1948).

F. *Determinations of Critical Micelle Concentration and Micelle Size* above C.M.C., S. Glasstone, *Textbook of Phys. Chem.*, 2nd, ed., 1946.

F. "Typical" Polymerization recipes

W. R. Sorenson and T. W. Campbell, *Preparative Methods of Polymer Chemistry*, pp. 238, 249, 313, 315, Interscience, New York.

J. Megee and G. Totty, *TAPPI, 56(11),* 118 (1973).

G. Surfactant Information

Detergents and Emulsifiers Annual, J. W. McCutcheon, Inc. (Yearly)

HLB system proposed by Atlas Chemical workers, W. C. Griffith, *Encyclopedia of Chemical Technology,* Vol. VII, p. 131, Wiley, New York.

Section II

The Monomers

The monomers are the building blocks of the dispersed polymer particles called latexes. Since each monomer has a unique chemical structure, the implications of that structure in the final polymer properties are important. An analogy would be the bricks on a facade of many different colors, sizes, and strength properties. These bricks can only fit together in certain geometrical patterns to make a solid wall having real strength. With the building of the polymer, the strength, color, or transparency, and chemical or water resistance all depend on the chemical structure of the monomers used.

There are four main properties the polymer builder considers in the choice of the monomers:

1. solvent resistance and hydrophobicity;
2. hardness and flexibility;
3. strength; and,
3. optical properties.

Each of these final polymer properties will be governed by the choice of monomer(s) used to make the (co)polymer. There are additional controlling factors, such as manufacturing technique, but the, monomer contributions are the usual starting point.

Hydrophobicity is one of the major considerations, as many of the latex polymer systems are required to resist water in laundering, scrubbing, or even corrosion protection. Strictly speaking, hydrophobicity is also related to *solvent resistance,* a solubility-parameter related phenomenon, since water is a simple solvent, albeit very polar. Several factors are involved in the resistance to water or solvent, and there are several levels of interaction with the polymer. No polymer is impermeable to a gas or a fluid, in that the free volume allows migration of even highly polar water through the nonpolar polyethylene or other hydrocarbon polymer. The choice of monomers to minimize free volume to enhance permeation resistance is limited, but possible. The next higher order of interaction is the association of solvent with polymer, via a Solubility Parameter match. Thus resistance to a particular solvent (e.g., dry cleaning fluid or gasoline) can be accomplished by building the polymer to be a mismatch to that particular solvent. Cross-linking will also increase resistance to a solvent with a close Solubility Parameter match, and therefore cross-linkable monomers must be chosen in the design of the polymer. The next higher level of permeation and penetration is mechanical, with droplets migrating through pores. Herein, the monomers must be chosen to enhance film formation without porosity.

Hardness and *flexibility* are another related pair of considerations that depend on the glass transition temperature (T_g), a unique property for each monomer. Indeed, these considerations are more likely to be thought of in terms of T_g by those who understand the rubbery-to-plastic transition. The changes in flexibility and hardness as a function of degradative attacks is dependent upon monomer structure. However the T_g of the hydrophilic monomer is reduced as water is absorbed, the water acting as a plasticizer.

A third consideration is *strength,* which is related to cross-linking ability, T_g, associative phenomena, and the like, as well as molecular weight. The structure of the chosen monomer(s) in the polymer form will contribute to degradation mechanisms that reduce or increase strength as well.

Optical properties are a further consideration, in that the contributions of the monomer refractive index to the polymer may yield a transparent opaque film, or film that absorbs UV light, fluoresces, or even degrades and discolors in time due to heat, UV, or hydrolytic attack.

We shall primarily describe the large-volume commercial monomers or smaller-volume monomers in common usage. Many monomers have been made as laboratory experimental materials, but they are rarely used for commercial products.

The data shown for monomers will be an aid in the design of latex systems. We show refractive indices for those aiming at polymer transparency or opacity. The reactivity ratios shown may help in designing the copolymers by monomer addition staging and sequences. The T_g is shown to aid design of hard or soft copolymers.

5

Vinyls

I. INTRODUCTION

There are essentially three major monomers that are commonly used in commercial emulsion polymer production. Other commercial monomers exist (the vinyl ethers, for instance), but they are generally polymerized in solution by a cationic mechanism for use in adhesives or other products. The generally recognized list of emulsion polymerized vinyls includes:

1. vinyl acetate;
2. vinyl chloride; and,
3. vinylidene chloride.

The properties of these monomers are shown in Table 5.1. The discussion of their polymerization and use is in the sections that follow. However, we add one more monomer (for convenience), ethylene.

A. Vinyl Acetate

This monomer is a structural isomer of methyl acrylate, having the same number of hydrogen, carbon, and oxygen atoms. Some properties of the homopolymers are similar (moisture absorption, for example), others differ (T_g). Both monomers are readily attacked by acid and alkali for hydrolytic degradation, the acid catalysis of vinyl acetate hydrolysis is autocatalytic because of the acetic acid liberated.

Table 5.1 Properties of Vinyl Monomers

Monomer	Refractive Index	T_g, C°	Alfrey-Price Values Q	Alfrey-Price Values e
Vinyl acetate	1.467	29	0.026	−0.22
Vinyl chloride		81	0.20	0.044
Vinylidene chloride		−19	0.36	0.22
Ethylene	1.490	−26	0.17	−0.020

Vinyl acetate is relatively inexpensive and easily homopolymerized or copolymerized. Copolymers with vinyl chloride form the basis for many vinyl emulsions used by the paint industry. They are less hydrophilic than a vinyl acetate homopolymer would be. The copolymers with acrylates or methacrylates are commonly used in coating and adhesives applications. Patella [21] also encourages use of the polyfunctional acrylic monomers for cross-linking during polymerization to further the development of strength. The development of ethylene–vinyl acetate (EVA) copolymers about 25 years ago was an important advance, as these copolymers weather (resisting thermal and UV discolorations) as well as the acrylates. Copolymers with the curing monomers work well in commercial applications. Such a curing monomer might be an unsaturated carboxylic acid [1] such as crotonic.

Polymerizations of vinyl acetate as homopolymer or with acrylic or curing comonomers may be done at ambient pressures with a simple reflux condenser system for removing the heat liberated [20]. A "run-away" reaction is easily controlled by adding a few drops of styrene. The reactivity ratios of the styrene–vinyl acetate pair favor the formation of styrene homopolymer, and therefore styrene is an effective short-stop for the reaction that is out of control. Fujii [8] detailed the temperature rise expected in vinyl acetate homopolymerizations, when to expect the monomer reflux to start and stop, and rationalized the dropwise addition technique as a means of temperature control. Friis and Nyhagen [9] described the reaction and particle size dependence on initiator and surfactant concentrations, while Vijatendran and associates [29] characterized the absorption of surfactants onto poly(vinyl acetate) or poly(vinyl acetate-*co*-butyl acrylate) as functions of surfactant molecular weight and functionality.

The formulations for vinyl acetate polymerizations are simple, based on the same surfactants and dispersants discussed earlier. Initiators may be of the organic peroxide–hydroperoxide varieties or perxydisulfate salts. A typical recipe is shown by Sorenson and Campbell [14], whereas the work of Bataille and coworkers [32] shows that metal salts accelerate the polymerization when nonionic emulsifier was used and persulfate (peroxidisulfate?) initiator. They also found that silver ion accelerated the polymerization in the absence of nonionic emulsifier. Work at the Lehigh Emulsion Polymer Institute showed that (methacrylic acid) was an excellent protective colloid for vinyl acetate polymerizations [17], whereas a patent [19] disclosed that poly(*N*-vinylpyrrolidone) is useful as an emulsion stabilizer, too. Litt and co-workers [18] hypothesize migration of stable radicals from the growing emulsion particles as part of the polymerization mechanism for vinyl acetate, whereas Zollars' model [31] best fits a mechanism wherein radical terminations occurred within the polymer particles.

The ethylene–vinyl acetate copolymers are made in pressure reactors usually at high pressue [25], but Russ Meincke of General Tire developed a process for "cold" polymerization of EVAs. The added ethylene reduces the hydrophilicity, and the film does not absorb as much water as the vinyl acetate homopolymer would. The reactivity ratios of ethylene and vinyl acetate favor vinyl acetate homopolymerization. The high pressure of the ethylene forces a high initial ethylene monomer concentration, and the ethylene is incorporated at about 10–20% in the final copolymer. After most of the vinyl acetate is consumed, the excess ethylene is recovered for the next reaction. The Meincke cold process takes advantage of the lower pressure needed to emulsify enough ethylene at 5°C centigrade to incorporate 15–18% into the final copolymer. Other hydrocarbon monomers have been copolymerized with vinyl acetate in emulsion as well [30].

Processing normally pushes the polymerization of vinyl acetate homo- or co-polymers with a small excess peroxide and some extra heat to consume all the residual vinyl acetate. Concentration or stripping was not commmon in my experience, but residual monomer can cause odors (acetaldehyde and acetic acid) due to hydrolysis. Fossick and Tompsett [2] used gas chromatography to monitor residual vinyl acetate in latexes. Since the residual vinyl acetate has a characteristic odor, deodorants are commonly used.

There are two major differences that may appear in vinyl acetate polymerization formulation, though. The first is that they are generally buffered to near neutrality to reduce the rate of hydrolysis during and after the polymerization. Simple alkali carbonates are commmon buffers, and I also like diammonium phosphate for some systems. It acts as buffer and methylolamide cure catalyst. One patent claims that polycarboxylic acids (tartaric, citric, maleic) stabilize vinyl acetate copolymers against weakening in a dry wall adhesive [22].

The other major difference lies in the use of the protective colloids, ranging from poly(vinyl alcohol) and hydroxyethyl cellulose synthetics to simple dextrins and starches. These protective colloids aid in making larger-particle-size latexes, and may be grafted to the growing polymer in some instances. The grafting of the protective colloid makes titanates effective thickeners by an associative mechanism hypothetically based on hydrogen bonding at the interface [3]. The incorporated protective colloid also makes the final film more hydrophilic, which however can be a detriment in some coating applications. This can be confirmed by casting a film of the latex and allowing it to dry to good clarity. If a water drop makes the film blush (i.e., hazy or opaque), water absorption is rapid and substantial.

Since the vinyl acetate polymer has a T_g near room temperature, plasticizers are often added to help film formation. Indeed, the famous Elmer's white glue for school children is a simple vinyl acetate homopolymer with dibutyl phthalate incorporated. A client of mine, making a binder for a temporary parking-lot striping paint (rain was supposed to wash it away), added the plasticizer to the emulsion polymerization with no ill effects. The formulation contained a large amount of protective colloid to ensure breakup upon wetting.

There are other sets of commercial vinyl esters, mostly used in Europe. They are based on synthetic short-chain branched fatty acids, termed Versatic acids by Shell Chemical. Their copolymers with vinyl acetate and unsaturated acids are given as examples in the patent by Grommers and Vegter [15]. They increase hydrophobicity (lower oxygen content) and are hard (high T_g) monomers.

Vinyl acetate-based emulsions are used in a wide variety of applications. Interior latex paints are frequently based on this emulsion, as are may adhesives. The adhesive sealing edge envelopes may be based on a vinyl acetate emulsion. One expert I saw making such a sealer tasted all the ingredients to make sure that no bad taste would prevent acceptance by the customer. All the components were approved by the Federal Drug Administration (FDA), but still I was surprised to see this procedure. The paper industry uses large amounts of vinyl acetate emulsions in coatings, mostly formulated with starch and kaolin (clay) to make publication grades of paper and some label stock. The textile industry uses curable vinyl acetate copolymers in nonwovens and some specialty coatings. The ethylene–vinyl acetate copolymers find their way into nonwovens, paints, and paper coatings as well.

Review books or symposia papers on vinyl acetate polymerizations are available [33,34].

B. Vinyl Chloride

Another inexpensive monomer, vinyl chloride (VC), is commonly homopolymerized in bulk [37], emulsion, or suspension [38]. These processes yield different products in terms of incorporated double bonds, carbonyl, or hydroperoxide groups [42]. The products of bulk or suspension processes are used as dry PVC powder for extrusion or molding processes, or in plastisol–organosol formulations. Although suspension processes are aqueous, the use of surfactants, dispersants, and protective colloids is minimized to facilitate isolation. Size, shape, and performance of the suspended particles depends on the surfactant used [39]. Indeed, one process uses what we would normally think of as coagulant salts in the suspension polymerization of VC with methyl cellulose, alkyl sulfate, and barium or magnesium salts [36]. One process adds plasticizer and water-soluble initiator to the VC dispersion formulation in the absence of surfactant or dispersant [45]. Meeks [44] described computer modeling of the VC suspension polymerization process.

Emulsion polymerizations are more to our interest. The polymers may be used in coatings for cans, textiles, paper, and other materials. Emulsion polymerizations appear to follow the standard models used for hydrophobic monomers [40]. Berens [47] derived a continuous-stirred-tank reactor model from batch model equations.

Shilov and co-workers used a nonmicelle-forming surfactant (a wetting agent or hydrotrope in a later chapter) to make poly(vinyl chloride) (PVC) emulsion polymers [46]. Seeded polymerization techniques may be used [41]. Copolymers (random [48] or graft [49]) are claimed to be useful in patents. One emulsion polymerization patent claims a stereoregular PVC. Calgari and associates [16] claimed tetrachloroethylene is required for low-temperature, (−30°C) emulsion (dispersion?) polymerization.

Processing the vinyl chloride from monomer to polymer is a low pressure, but pressurized, polymerization. The boiling point of the VC monomer is lower than room temperature but not by much. That means the loading of this monomer is much like that discussed for the butadiene systems. However, the toxicity of this monomer imposes several personnel protection constraints on the processing. The latest information on required handling techniques must be obtained from the monomer supplier. The toxicity concerns also impose a special low residual monomer requirement on the product latex. Hence, stripping and procedures for monomer degradation are needed. Head space analysis may be used to determine the toxicological heat from residual monomer (see Chap. 13 on Class-II tests).

Since PVC is hard, it must be plasticized if flexibility is desired. The common plasticizers (alkyl phthalates, for instance) are useful if volatility or leaching is of no concern. Alkyl or aryl phosphate esters confer flame retardance, but are leachable and volatile to some degree. Two approaches are used to avoid the problem of leachable, volatile plasticizers: copolymerizing to "internally" plasticize the PVC or incorporating a polymeric plasticizer.

Copolymerization with VC is a common practice, and copolymers with ethyl or butyl acrylates as well as ethylene or vinyl acetate are commercial products. Oxygenated comonomers reduce the hydrophobicity, but copolymerization offers an even greater advantage. Dehydrohalogenation degrades PVC creating double bonds. If the dehydrohalogenation produces several double bonds in a row, the normally clear colorless polymer turns yellow, and orange or red upon further degradation (remember the structure of carotene!). However, if the comonomer is statistically distributed well enough to keep the VC sequences at four or less, much less color will develop upon dehydrohalogenation. Hence, one must design the copolymer accordingly.

The use of polymeric plasticizers in PVC homopolymer latexes is common, too. Indeed, preplasticized commercial products are available from latex suppliers. In some cases, butadiene-acrylonitrile latex is used as plasticizer. If film formation is critical, IR spectroscopy or other analytical techniques can be used to govern the ratio of plasticizer.

The PVC and copolymer latexes contain curing functionality, if curing is desired in the formulation. Indeed, there are self-cross-linking PVC-based systems for use in textiles, nonwovens, and other materials.

C. Vinylidene Chloride

I disagree that vinylidene chloride (VDC) should be included in this group, but industry organizations choose to call vinylidene chloride a vinyl in their groupings. Rigorously, the vinyls have an unsymmetrical structure, $X-CH=CH_2$, while the structure of vinylidene chloride is 1,1-dichloroethylene. This symmetry of vinylidene chloride is responsible for many of its unique properties. For instance, it is an excellent barrier resin, and is used as an oxygen barrier in potato chip bags. The symmetry allows the monomer segments to align closely, yielding little free volume for the oxygen molecules to migrate through to the potato chips. This same symmetry is responsible for the success of butyl rubber (based on isobutylene) as inner tube stock for so many years.

The T_g for vinylidene chloride is $-19°C$, but the homopolymer is a hard plastic due to crystallization (another consequence of the symmetry). Hence, most uses of this monomer require it to be a copolymer with a monomer that will suppress the crystallization tendency. Supplier literature [4] shows that vinylidene chloride copolymerizes will with most monomers, and especially well with ethyl acrylate in most ratios. Krishan and Margaritova [6] described the emulsion homopolymerization characteristics of vinylidene chloride in their work with methyl methacrylate and chloroprene. Seeded polymerizations to make poly(vinylidene chloride) (PVDC) were patented by Smith and Peterson [7]. Wessling and Gibbs [10] compared PVDC homopolymerization (of a nonswelling particle) to the copolymerization of VDC with butyl acrylate (where the mixture forms a swelling particle) in continuous monomer addition polymerizations. Lyon and Coover [12] patented a two-stage copolymerization of VDC with acrylic monomers (seeded?). Covington [13] gives examples of carboxylated VDC copolymer latexes in his patent wherein terpenes are incorporated. More recently, Friedl and Keillor [35] described coating latex polmerizations using VDC and acrylates.

The barrier resin property is of particular interest to coating formulators.

Since water and oxygen are the prime sources of metal corosion, VDC barrier properties are of prime importance when designing a corrosion-protective coating. The nonpolar structure of vinylidene chloride makes it hydrophobic.

Because of its high chlorine VDC is of interest to coatings formulators and textile chemists. Since it contributes to flame retardance. Indeed, about 20 years ago, we examined such a resin as binder for a disposable nonwoven, as it would eliminate the need for flame retardant salt saturation of the cellulosic fiber segments. However, the dehydrohalogenation upon drying and curing made the binder locations immediately visible as a brick red-orange. In systems with good temperature control or that dry and cure at ambient temperature the vinylidene chloride-based polymer systems can work very well. Indeed, the State of California specifies water-borne vinylidene chloride-based binders in some corrosion-inhibitive coatings [5]. But here is a caveat here, in that dehydrohalogenation, if and when it occurs, produces corrosive HCl.

D. Ethylene

Although ethylene is an inexpensive monomer, it requires higher capital expenditure for production facilities because of the elevated pressure required in commercial practice. It adds hydrophobicity to oxygenated copolymers, and flexibility (as we have already seen) to vinyl acetate or vinyl chloride copolymers. The ethylene–vinyl acetate or ethylene–vinyl chloride copolymer latexes are the only systems produced in large volumes, though other copolymers are made for bulk or solution applications. Monsanto has a useful series of 1:1 copolymers with maleic anhydride (the EMA resins), which are good water-soluble polymers in dispersant and thickener applications. Ethylene–acrylic copolymers are also marketed by Union Carbide. Dupont was assigned a patent on E (ethylene)/DM (diene monomer) or E/P (propylene)/DM latexes [11].

Another approach to making water-borne poly (ethylene) systems is to disperse the microfine powders, such as the Microthene F from USI. McSherry and co-workers [23] described such a process for coatings applications, as did the USI brochures. But Dees and Gabbert [24] patented an emulsion polymerization formulation using salt and mixed emulsifiers. Helin and Mantell [26] patented the emulsion polymerization in water–*t*-butanol mixtures, claiming better stability during the polymerization. Burkhart and Zutty [27] claim that especially designed suractants support the emulsion polymerization of ethylene. Stryker and associates [28] published a series of articles on the emulsion polymerization of ethylene.

REFERENCES

1. S. Masuda et al., Jap. Pat. 74 41412 (Cf: *CA*, **74,** 92520t).
2. G. N. Fossick and A. J. Tompsett, *JOCCA, 49,* 477 (1966).
3. Jon Hall, 1986 FSCT Meeting presentation, Atlanta, GA.

4. Vinylidene Chloride Monomer, Dow Chemical Co. Technical Bulletin No. 102–243–84.
5. PWB 145 and PWB 146, CA Dept. of Transportation Paint Specifications.
6. T. Krishan and M. Margaritova, *J. Polym. Sci., 52,* 139 (1961).
7. D. R. Smith and H. Peterson, US 3424706 (28 Jan. 1969).
8. M. Fujii, Y. Ohtsuka and Y. Kohko, *Chem. High Polym., 25(283),* 742; and *25(284),* 817 (1968) (in Japanese).
9. N. Friis and L. Nyhagen, *J. Appl. Polym. Sci., 17,* 2311 (1973).
10. R. A. Wessling and D. S. Gibbs, *J. Macromol. Sci.– Chem., A7,* 647 (1973).
11. R. O. Becker, US 3379666.
12. H. R., Lyon and H. W. Coover Jr., US 3379665.
13. E. R. Covington, US 3236818 (22 Feb. 1966).
14. W. R. Sorenson and T. W. Campbell, *Preparative Methods of Polymer Chemistry,* p. 238, Wiley-Interscience, New York, 1968.
15. E. P. Grommers and G. C. Vegter, US 3294727 (27 Dec. 1966).
16. S. Calgari, A. Matera and P. P. Rossi, Ger. Offen. 1940095 (12 Feb. 1970) (Cf: *CA* 82 90996t, 1970).
17. W. E. Daniels, J. G. Iacoviello and C. T. Enos, *ACS Colloid Surf. Sci.* Prepr., No. 110 (Sep. 11–14, 1978).
18. M. H. Litt et al., *ACS Colloid Surf. Sci.* Prepr. No. 129 (Sept. 11–14, 1978).
19. H. Grubert, W. Druschke and W. Sliwka, Can 1018690 (4 Oct. 1977) (Cf: *CA* **88** 51515w).
20. P. J. McDonald, *J. Water Borne Coatings,* 11 (May 1982).
21. R. F. Patella, *Mod. Paint Ctgs.,* 47 (July 1978).
22. Anon. (assigned to Chuo Rika Kogyo), Jap. Pat. 76 028300 (18 Aug. 1976) (Cf: Derwent 41550w/25).
23. J.J. McSherry et al., *Kunstoff, 59,* 545 (Sep. 1969).
24. O. S. Dees and J. D. Gabbert, US 3380945 (30 Apr. 1968).
25. H. Hopf, S. Goebel and C. W. Rautenstrauch, US 2342400.
26. A. F. Helin and G. F. Mantell, US 3226352 (28 Dec. 1965).
27. R. D. Burkhart and N. L. Zutty, US 3296710 (3 Jan. 1967).
28. H. K. Stryker, G. J. Mantell and A. F. Helin, *J. Appl. Polym. Sci., 11,* 1 (1967), for instance.
29. B. R. Vijayendran, T. Bone and C. Gajria, *J. Appl. Polym. Sci., 26,* 1351 (1981).
30. F. P. Glintz, US 3738954.
31. R. L. Zollars, *J. Appl. Polym. Sci., 24,* 1353 (1979).
32. P. Bataille, H. Bourassa and A. Payette, *J. Ctgs. Technol., 59* (753), 71 (Oct. 1987).
33. T. Tsuruta and K. F. O'Driscoll, *Structure and Mechanism in Vinyl Polymerizations,* M. Dekker Inc., New York (1969).
34. M. S. El-Aaser and J. W. Vanderhoff, *Emulsion Polymerization of Vinyl Acetate,* Appl. Sci. Publ. Ltd., Englewood, NJ (1982).
35. H. R. Friedl and C. M Keillor, *J. Ctgs. Technol., 59(748),* 65 (May 1987).
36. N. I. Sokoleva and A. A. Panfilov, *Plast. Massy,* No. 6, p. 7 (1973); Cf: RAPRA Abstr. 26298L, (30 Sept. 1974).

37. J. Chatelaine, *Br. Polym. J.*, *5(6)*, 457 (Nov. 1973).
38. Ye. P. Sharev, Z. S. Zakharova and S. A. Nikitina, *Polym Sci. USSR*, *17(1)*, 82 (1975).
39. ICI America SPAN Bulletin No. 761–33 (1975).
40. V. I. Lukhovistkii, *Polym. Sci. USSR*, *17(3)*, 598 (1975).
41. Belg. Pat 833251 (13 Sep. 1974), assigned to Ugine Kuhlmann; Cf: Derwent 22653X (1976).
42. E. N. Zil'berman, E. M. Perepletchikova, E. N. Getmaneko, V. I. Zegelman, T. S. Molova and Yu. A. Z.-vereva, *Plast. Massy*, No. 3, p. 9 (1975).
43. Jap. Pat. 76021674 (3 July 1976), assigned to Sumitomo Chemical KK; Cf: Derwent 8363W/04 (1976).
44. M. R. Meeks, *Polym. Sci. Engr.*, *9(2)*, 141 (1969).
45. Belg. Pat. 803681 (18 Aug. 1972), assigned to BP Chemicals Ltd; Cf: Derwent 15517V (1972).
46. G. I. Shilov, A. I. Kirillov and A. I. Krymova, *Vyosokomol. Soyed. A15(6)*, 1378 (1973).
47. A. R. Berens, *J. Appl. Polym. Sci.*, *18*, 2379 (1974).
48. Brit. Pat. 1450463 (22 Sep. 1976), assigned to Wacker Chemie; Cf: Derwent 57110V/32 (1976).
49. Brit Pat. 1450511 (22 Sep. 1976), assigned to Eurane Europ Polyur; Cf: Derwent 48102W/29 (1976).

6

Styrenes

The styrenes are all based on the vinyl-substituted benzene ring. The common major commercial monomers are shown in Table 6.1, but few latexes are made from anything but the parent unsubstituted styrene. Opportunity? The other substituted styrene monomers are most commonly grafted onto drying oils or drying-oil alkyds for solvent-borne coatings, or solution polymerized as homo- or copolymers for lacquer-type coatings. Their advantage is that they have better solubility in aliphatic solvent systems, as the aromatic solvents are not desirable in many applications because of their odor. However, there are research reports of vinyltoluene emulsion polymers, for instance [1].

There are other styrenes of interest as specialty monomers. Chlorostyrene [45] and vinylbenzyl chloride [46] are (or were) available from Dow Chemical. The latter is interesting as an intermediate to otherwise functional styrenes (e.g., methylol) [46]. The tetra- and pentachlorostyrenes do not copolymerize.

The latex systems based on styrene benefit from the huge amount of research on polymerizing this monomer as a model hydrophobic material. As styrene is among the least expensive of the commercial monomers, it is used widely in many industries besides our own latex applications. Seymour [2] reviewed styrene as a component in plastics and coatings from its original synthesis in 1839.

Styrene homopolymers are transparent, hard plastics. The T_g is high, and the material can be brittle unless the molecular weight is high. These materials are seldom applied as homopolymer latexes, though I made a solvent-based overprint varnish sometime ago that met alkali-resistance requirements. The material can be softened by external plasticizers (an inexpensive way to formulate), but the plasticizers can be volatile or bloom to the surface, so one can lose the softening effect in time. Chlorinated waxes or phthalate esters have been used to plasticize poly(styrene)s.

Since styrene is a hydrocarbon, it is not subject to hydrolytic attack by alkali or acid, and is an excellent hydrophobe to keep moisture from penetrating. However, the emulsion polymerization process (using the standard salts and surfactants we have discussed in Chapter 13) will give the desired latex stability, however, with the penalty of a possible increase in hydrophilicity. If the salt and surfactant additives are chosen to be incompatible with the bulk polymer on drying, the additives will migrate to the surface to be leached off upon weathering and the bulk film will be as hydrophobic as possible.

Poly(styrene) latexes (or copolymer latexes) are often used to assess the effectiveness of protective colloids and surfactants. Volkov and Danyushin [55] used polystyrene latex to test the effectiveness of polyether nonionic surfactants. In the opposite vein, Ahmed et al. [56] at Lehigh University "cleaned" the surfactants from such latexes by serum replacement. Tadros and Vincent [57] described the influence of salt on poly(vinyl alcohol) adsorption to such latex. Vincent et al. [58] used such latexes to define the salt concentration and temperature interdependence for coagulation sensitivity, finding temperature a minor factor at low salt content but of increasing importance with increased salt concentration. Ono and Jidai [59] used styrene–acrylonitrile (SAN) latex to show nonionic surfactant stabilization against magnesium salts. They found that acrylonitrile content increased stability, and that electrophoretic mobility decreased with increased nonionic surfactant additions. They and others [60] also used the SAN latexes to test electrophoretic mobilities against calcium ions. Saunders [61] used monodisperse poly(styrene) latex to assess adsorption of methylcelluloses via serum refractive index changes. Force and Wilson [62] used styrene–butadiene copolymers to test the degree of fatty acid adsorptions. Gorenkova and Nieman [63] reported that mixed emulsifiers, one having high CMC value and one having low CMC value, yield a nonlinear "soap titration" that can be rationalized. Aleksandrova et al. [64] varied latex particle size and surfactant content in electrolyte coagulation studies, and found that small particles irreversibly agglomerated, whereas the larger particles (more than 105 nm) formed a reversible longer-range aggregation. Rabold [65] used styrene–butadiene latexes and sodium dodecyl sulfate to demonstrate the advantages of nitroxide spin-probe analysis to characterize CMC and the types of micelles formed, and correlate it T_m and T_g of the polymers [66]. Everett and Gultepe [67] used ultrasound to characterize salt and surface sulfate (initiator derived) on poly(styrene) latexes, and finding salts affected ultrasonic vibrations absorbed.

The kinetics of styrene polymerizations have been often studied, as noted earlier. Piirma et al. [6] used peroxydisulfate initiation and a variety of emul-

Table 6.1 Properties of Styrene Monomers

			Alfrey-Price Values	
Monomer	Refractive Index	T_g, C°	*Q*	*e*
Styrene	1.05 to 1.065	105	1.0*	−0.8[a]
t-Butylstyrene	1.53	132	1.30	−1.00
p-Methylstyrene		106	1.08	−0.91
α-Methylstyrene	1.61	155	0.72	−0.010

[a]By definition, the standards from which all others are calculated.

sifiers, finding agreement with Smith-Ewart theoretical description of the polymerization. Galbraikh [22] used gas chromatography to determine the rate of disappearance of styrene during emulsion polymerization. Nomura et al. [7] used carbon tetrachloride or tetrabromide, or the ethyl, butyl, or dodecyl mercaptans to determine the effects of chain transfer on radical polymerization kinetics; they found that initiator relation to number of particles disappears and emulsifier dependence increases.

A Russian patent [15] shows that injection of chain-transfer agent, after the polymerization of butadiene and styrene has started, improves the properties. C. A. Uraneck, of Phillips Petroleum, has published material on this injection technique, as well [69]. A wax incorporated into the styrene emulsion polymerization was the subject of a patent by Wiley [23]. Ogo and Sano [35] described the effect of high pressure on the emulsion polymerization of styrene.

The effects of surfactants in emulsion polymerization and afterward are discussed in many contributions to the literature. For instance, Piirma and Chang [36] found a nonstandard polymerization rate and a bimodal latex particle size distribution in a styrene latex synthesis based on a monoionic emulsifier. Wolf [47] patented a starch ether as a colloidal stabilizer for styrene or copolymers, while Medvedeva et al. [48] used poly(vinylpyrrolidone) and emulsifiers to increase the polymerization rate; the emulsifier is necessary to support polymerization. Alcohol–surfactant mixtures were investigated by Chou et al. [49] and Ugelstad et al. [50]. Turck [51] reports styrene–butadiene copolymerizations (in combined nonionic–anionic emulsifier systems) show particle size also depends in comonomers (e.g., acrylamide) and peroxydisulfate concentration. Emulsifier-free polymerizations of styrene have been reported by English universities [52,53].

Poehlein and many co-workers [8–10] modeled continuous-stirred-tank polymerization reactions for styrene emulsions, and Panina et al. [14] developed a mathematical model describing a peroxydisulfate–triethanolamine initiated laurate-emulsified styrene polymerization. More recently, Bataille et al. [33] developed a digital simulation of isothermal styrene emulsion polymerizations to recommend some changes in assumed theoretical details such as termination constant and number of particles. They continued this work showing that silver(I) ions accelerate and ferrous ions decelerate the polymerizations initiated with peroxydisulfate, while reducing the polymer molecular weight by acting as chain-transfer agents (ions in the hydrophobic polymer particle?) [34]. Hawkett et al. [37] have suggested a significant amount of aqueous termination in styrene emulsion polymerizations.

The particles developing in the polymerization of styrene have been the object of considerable research. The concern is whether or not such phenomena as autoacceleration (runaway polymerization rates at high levels of conversion) or core-shell grafting of one monomer onto a "seed" latex can be rationalized by the model most used in describing the mechanism of the polymerization. Marten and Hamielec [39] based a kinetic model for bulk high conversion polymerization on free-volume arguments. Keusch et al. [11] concluded that encapsulation does occur, even where the monomer may be compatible with the polymer (that is, the monomer dissolves in the polymer). John Vanderhoff at Lehigh University has reported samples of ethyl acrylate shell–styrene core latexes, in his research on the theory of Mie scattering of light [13]. Matsushita et al. [24] described conditions for making styrene–ethyl acrylate core-shell latexes. Friis and Hamielec [12] argue that models using core-shell modification of the Smith-Ewart theory do not fit the observed autoaccelerations that occur occasionally, and Piirma et al. [6] did not see evidence for core-shell phenomena in their work. Williams [18] and Gardon [19] published an exchange of opinions on particle morphology when saturated with monomer or during polymerization. Min et al. [40] used a butyl acrylate seed emulsion for "core" with varying amounts of styrene added in the second "shell" under a variety of reaction conditions to test the final products for particle morphology and grafting, as did Okubo et al. [54], who published pictures verifying blob grafts rather than core-shell. The core-shell or not-from-seeded formulations may depend entirely upon initiation details, surfactants, and processing conditions. Lichti et al. [42] examined particle-size distributions of seeded or standard emulsion polymerizations of styrene, and found kinetic agreement with dilatometric data for radical capture or desorption and polymerization rate. Monodisperse particle-size distribution polystyrene latexes have been have been made in space [44].

The core-shell concept underlies the "seeded" polymerization practiced so often in industry. A Sinclair-Koppers patent is one example [20], a Russian patent is another example [21]. The former shows why seeded systems can have low surfactant levels and significantly higher surface tension than the soap-saturated latex would have. Some markets are convinced that high surface tension is desirable where penetration into porous substrates is not desired, though viscosity may be a better controller of that penetration.

McCracken and Datyner [16] describe a uniform-particle-size latex made by what they call "true" emulsion polymerization, where they used a methanol–

water mixture to preform an emulsion of styrene with no surfactant and no monomer droplets. Ugelstad et al. [17] showed small-droplet initiation in lauryl sulfate–cetyl alcohol microemulsion polymerization of styrene. Hansen and Ugelstad [38] used their own approach to microemulsion technology, in a series of publications, seeding with and without surfactants present and examination of partially water-soluble oligomer (or oligomer radical) migrations.

Styrene copolymers are made for a variety of reasons, such as flexibilization by the same internal plasticization by "soft monomers" we noted for the PVC systems or the construction of a curable system. The flexibilization monomers include hydrocarbons (butadiene and isoprene are most common) and the acrylates. Obviously, the hydrocabon monomers maintain the hydrophobicity desired for many water–moisture barrier applications, while the more polar acrylate comonomers increase the water uptake by the film and permeation substantially. The hydrocarbon monomers are less expensive than the acrylate flexibilizers, too. Several sources show typical polymerization recipes [3,4]. Seymour et al. [5] used acrylonitrile as a comonomer to change the solubility characteristics of styrene copolymers, improving solubility in dimethylformanide (DMF). Lin et al. [43] tried modeling "azeotrope" copolymerization (read this as "constant copolymer composition," a more descriptive terminology) of styrene and acrylonitrile in emulsion, finding the desorption of acrylonitrile radicals a significant factor. Probably the most unusual of the copolymers of styrene–butadiene emulsions was the peroxide-functional latex made by Puchin et al. [31].

However, the copolymers have other properties of importance as well. The styrene polymers tend to yellow upon exposure to weathering, UV, and heat. Incorporation of diene monomers increases coloration and loss of strength with time. The butadiene comonomer has an increased tendency to resinification and embrittlement, while isoprene tends to degrade and weaken the polymer film. On the other hand, acrylates reduce the propensity to discolor on aging, and their resistance to UV, heat, and weathering is well known, provided they are formulated properly in the emulsion polymerization and in the coating. The copolymers made by grafting styrene and acrylonitrile on poly(butadiene) result in a high impact plastic, commonly known as ABS resin. The process is normally similar to seed polymerization, though a solution process with subsequent aqueous dispersion of the graft copolymer has been patented [25].

Copolymerization giving cured products usually include softening monomers. The curing systems allow development of additional strength, toughness, and solvent resistance by the coating film, as compared to the uncured materials. Again, the monomers may be carboxylated (acrylic, methacrylic, fumaric, itaconic) to be cured with metal ions (zinc, zirconium) or by melamine–formaldehyde or urea–formaldehyde resins (see Chapter 9). Other latexes are made "self-crosslinking" by use of methylolated amide monomers. My first commercial latex product was a self-crosslinking styrene–butadiene copolymer, which required only little heat to cure after drying. I later found that it attained about

80% of full cure on air drying with no heat. In other work, we found that the discoloration problems in styrene–butadiene-based nonwoven binders could be solved with the help of appropriate additives [32].

Carboxylated copolymers (for cure with polyvalent cations, or amino/formaldehyde resins) are very common. Ceska [26,27] reports that carboxylic comonomers aid stabilization during the polymerization and even in seeded systems. In Chapter 13 on electrophoretic mobility determinations, Green et al. discuss other types of stabilization. Egusa and Makuchi [41] used gamma radiation to initiate styrene–carboxylic acid monomer latex systems, and concluded that the kinetics showed initiation of polymerizations at the emulsion particle surfaces. Stone-Masui [68] used x-ray photoelectron spectroscopy to characterize freeze-dried poly(styrene) latexes, and found all the sulfur in the form of sulfate as one might expect with sulfate ester initiators.

The functional-group substituted copolymers find use in a variety of industrial applications. For instance, a German patent discloses carboxylic acids or amides, or both as comonomers in a styrene–nitrile–diolefin copolymer for textile or paper coatings [28]. Dow Chemical claims hydroxyethylacrylate (or a similar compound) to be useful in styrene–butadiene–carboxylate latexes for paper coatings [29]. Walker has described laboratory and scale-up equipment and processes for styrene copolymer latexes [30].

REFERENCES

1. P. Bagchi et al., Paper No. 74, ACS Colloid Surf. Chem. Div. (Sep. 1978).
2. R. B. Seymour, *J. Ctgs. Technol., 58(741),* 71 (Oct . 1968).
3. W. R. Sorenson and J. W. Campbell, *Preparative Methods of Polymer Chemistry,* p. 313, John Wiley, New York (1968).
4. D. C. Blackley, *High Polymer Lattices,* Vol. I, p. 294, Palmerton Publ., New York, (1966).
5. R. B. Seymour et al., *Ind. Eng. Chem., Prod. R/D, 12(3),* 198 (1973).
6. I. Piirma et al., *J. Polym. Sci., Polym. Chem. Ed., 13,* 2087 (1975).
7. M. Nomura et al., *J. Polym. Sci., Polym. Chem. Ed., 20(5),* 1261 (1982).
8. G. Poehlein et al., *Adv. Chem. Ser., 109,* 75 (1972).
9. G. Poehlein and J. Vanderhoff, *J. Polym. Sci., Polym. Chem. Ed., 11(2),* 447 (1973).
10. A. W. DeGraff and G. Poehlein, *J. Polym. Sci., Polym. Chem. Ed. A2, 9(11),* 1955 (1971).
11. P. Keusch et al., *J. Macromol. Sci., Chem., A7(3),* 623 (1973); *Macromolecules, 2(4),* 318 (1974).
12. N. Friis and A. E. Hamielec, *J. Polym. Sci., Polym. Chem. Ed., 11,* 3321 (1974).
13. J. Vanderhoff, 1975 Emulsion Polymer Symposium, Lehigh Univ., Bethlehem, Pa.
14. I. S. Panina et al., *Zh. Prikl. Khim., 48(9)* 2009 (1975); Cf. *RAPRA Abstr.* 42C21-72233.

15. L. V. Kosmodemyanskii, Russ. Pat. 486026; Cf: Derwent 59280X/31 (1976).
16. J. R. McCracken and A. Datyner, *J. Appl. Polym. Sci.*, **18,** 3365 (1974).
17. J. Ugelstad, F. K. Hansen and S. Lange. At least four related papers follow as in ref. 38, *J. Polym. Sci., Polym. Lett. Ed., 11(8),* 503 (1973).
18. D. J. Williams, *J. Polym. Sci., Polym. Chem. Ed., 12(9),* 2123 (1974).
19. J. L. Gardon, *J. Polym. Sci., Polym. Chem. Ed. 12(9),* 2133 (1974).
20. Dutch Pat. 7304764 (assigned to Sinclair-Koppers); Cf: Derwent 68124U (1974).
21. T. S. Kazakevich, Russ. Pat. 504794 (1974); Cf: Derwent 90203x (1976).
22. N. Galbraikh, *Khim. Prom-st. Ser., 1982(2),* 18; Cf: *CA 96(9),* 143357q (1982).
23. R. M. Wiley, US 3379664 (23 Apr. 1968).
24. T. Matsushita et al., *Kobunshi Robunshu, Engl. Ed., 3(9),* 1814 (1974).
25. K. W. Doak and F. E. Carrock, US 3309422 (14 Mar. 1967).
26. G. W. Ceska, *J. Appl. Polym. Sci., 18(2),* 427 (1974).
27. Ibid., p. 2493.
28. US 3867626 (24 Aug. 1976); Cf. Derwent 69111W/42 (1976).
29. Dutch Pat. 149822 (15 July 1976); Cf: Derwent 63577P/00 (1976).
30. J. L. Walker, *Chem. Indust.*, p. 69 (April 1974).
31. V. A. Puchin et al., *Polymer Science USSR, 11(4),* 889 (1969).
32. D. W. Anderson and R. D. Athey Jr., TAPPI Paper Synthetics Conference, p. 31 (1977).
33. P. Bataille et al., *J. Polym. Sci., Polym. Chem. Ed., 20,* 795 (1982).
34. P. Bataille et al., *J. Polym. Sci., Polym. Chem. Ed., 20,* 811 (1982).
35. Y. Ogo and T. Sano, *Colloid Polym. Sci., 254,* 470 (1976).
36. I. Piirma and M. Chang, *J. Polym. Sci., Polym. Chem. Ed., 20,* 489 (1982).
37. B. S. Hawkett et al., *J. Polym. Sci., Polym. Chem. Ed., 19,* 3173 (1981).
38. F. K. Hansen and J. Ugelstad, *J. Polymer Sci., Polym. Chem Ed., 17,* 3033, 3047, and 3069 (1979).
39. F. L. Marten and A. E. Hamielec, *J. Appl. Polym. Sci., 27,* 489 (1982).
40. T. I. Min et al., *ACS Org. Ctgs. Appl. Polym. Sci. Proceedings, 46,* 314 (1982).
41. S. Egusa and K. Makuchi, *J. Polym. Sci., Polym. Chem. Ed., 20,* 863 (1982).
42. G. Lichti et al., *J. Polym. Sci., Polym. Chem. Ed. 19,* 925 (1981).
43. C.-C. Lin et al., *J. Appl. Polym. Sci., 26,* 1327 (1981).
44. *Chem. Engr. News,* p. 16 (9 Aug. 1982).
45. J. Dolinski and R. M. Nowak, *ACS Org. Ctgs. Plast. Chem. Div. Prepr., 29(2),* 349 (1969).
46. "Vinyl Benzyl Chloride," Dow Chemical USA, Forms No. 171-0008-84 and 504-002-88.
47. F. Wolf, Ger. Pat. 116462 (20 Nov. 1975); Cf: Derwent 07623X/05 (1976).
48. E. S. Medvedeva et al., *Tr. Mosk. Inst. Tonkoi Khim. Tekhnol., 5(1),* 74 (1975); Cf: *CA 86,* 30153v (1976).
49. Y. J. Chou et al., *ACS Coll. Surf. Sci. Div. Meeting Notes,* Abstr. No. 130 (Sep. 1978).
50. J. Ugelstad et al., *Makromol. Chem., 175,* 507 (1974).
51. U. Turck, *Angew. Makromol. Chem., 46(692),* 109 (1975).
52. A. R. Goodall et al., *J. Polym. Sci., Polym. Chem. Ed., 15,* 2193 (1977).
53. J. W. Goodwin et al., *Br. Polym. J., 5,* 347 (1973).

54. M. Okubo et al., *J. Polym. Sci., Polym. Lett. Ed., 20,* 45 (1982).
55. V. A. Volkov and G. V. Danyushin, *Colloids USSR,* p. 215 (1975), Plenum Publishing New York (abbreviated translation from the Russian *Kolloidni Zhurnal, 37(2),* 240 (1975).
56. S. M. Ahmed et al., *52nd ACS Coll. Surf. Sci. Symp. Abstr.,* p. 179 (1978).
57. Th. F. Tadros and B. Vincent, *52nd ACS Coll. Surf. Sci. Symp. Abstr.,* p. 205 (1978).
58. B. Vincent et al., *ACS/CSJ Congress Abstr. 119* (1979); *see also Faraday Disc., 65,*
59. H. Ono and E. Jidai, *Br. Polym. J., 7(6),* 363 (1975).
60. H. Ono et al., *Coll. Polym. Sci., 253,* 538 (1975).
61. F. L. Saunders, *J. Coll. Interface Sci., 28,* 475 (1968).
62. C. G. Force and A. S. Wilson, *J. Coll. Interface Sci. 32(3),* 477 (1970).
63. G. A. Gorenkova and R. E. Nieman, *Koll. Zh., 37(3),* 450 (1975); Cf: *RAPRA Abstr*. 37157L.
64. E. M. Aleksandrova et al., *Koll. Zh. (37)3,* 537 (1975); Cf. *RAPRA Abstr*. 37158L.
65. G. P. Rabold, *J. Polym. Sci., 7A-1,* 1187 (1969).
66. G. P. Rabold, *J. Polym. Sci., 7A-1,* 1203 (1969).
67. D. H. Everett and M. E. Gultepe, *ACS Coll. Surf. Sci. Div.,* Abstr. No. 75 (Sept. 1978).
68. J. H. Stone-Masui, *ACS Coll. Surf. Sci. Div*. Abstr. No. 76 (Sept. 1978).
69. One of his several publications on this is C. A. Uraneck and J. E. Burleigh, *J. Appl. Polym. Sci., 9,* 1273 (1965).

7

Acrylates and Methacrylates

For ease of use of terminology, we will use the term acrylate to designate this whole class of monomers. We will use the term methacrylate only as a specific designation for a particular monomer. These monomers are particularly popular for latex usages, as they produce and maintain good color in paints and textile [1] and artists' applications [21]. Acrylates do not discolor on aging and are essentially transparent in considerable thicknesses. Indeed, the large aquariums (Monterey, Baltimore) use acrylic sheets of up to seven inches in thickness because glass would be too hazy in thickness enough to stand the hydrostatic pressures. The history of the acrylics was recently reviewed [2].

Emulsion polymerization of acrylates is easy. Some of the best information on lab- and plant-operating procedures and typical formulations comes from monomer suppliers, with the information from Rohm and Haas being far superior to that from the others [3–9].

I. NITRILES

The nitriles, mostly acrylonitrile, are the bases for the nitrile rubbers and the ABS resins. The nitrile rubbers made as copolymers of acrylonitrile with butadiene at high polymerization rates. About 20% of the initiator is required one might use for a similar styrene–butadiene copolymer. Nitrile rubbers were initially of interest as they had better solvent (oil, gasoline, dry-cleaning fluids) resistance than the natural rubber or synthetic styrene–butadiene rubbers. The nitrile group increases the polarity of the rubber, resulting in a poor Solubility Parameter match with the nonpolar solvents.

However, there is substantial discoloration on aging, partly due to the diene and partly due to the nitrile group. Indeed, the acrylic fibers based on polyacrylonitrile may be thermally treated to become "black nitrile." I suspect the highly colored moieties are formed by cyclization, as has been shown with other nitrile polymers. Polyacrylonitrile latexes are hydrolyzed to poly(acrylic acid salts) to provide rheology modifers (thickeners) such as the Acrysol AN-10 from Rohm and Haas. Hydrolysis converts the latex to brick-red dispersions; the color is ascribed to the cyclized neighboring nitrile groups. The final product is a transluscent; lightly orange solution.

Nitrile polymers are useful as barrier resins in some applications. The SOHIO solid resin named BAREX is an oxygen barrier, and latexes related to this product have been patented by SOHIO for corrosion inhibition in steel coatings.

The nitrile group is less hydrophilic than ester groups, and moisture barrier properties can be built into the polymer by incorporating the nitrile monomer. The nitrile is more hydrophilic than the hydrocarbon monomers, but not as hydrophilic as the ester monomers. The polarity of the nitrile monomer does affect the polymerization in emulsion. Morris and Parts [22] described the difficulties of an aqueous emulsion homopolymerization of acrylonitrile; the problems included gel formation and irreproducible rates.

The nitrile rubbers are also compatible to some degree with poly(vinyl chloride) resins. Coatings and adhesives may be made using simple blends of these resins; they mix while coalescing on heating. A liquid nitrile rubber (HYCAR 1312) is marketed as a plasticizer for PVC. Preblended latexes containing both components are commercially available (e.g., Geon 557).

The liquid nitrile rubber HYCAR is probably made as an emulsion polymer and isolated for sale as a bulk liquid. I made a similar material about 25 years ago, discovering that one had to short-stop the polymerization to obtain a product which is not a fibrous fluff. Nitrile rubbers, if they are to be used as dissolving resins, must be short-stopped at about 60% conversion; otherwise an insoluble gel forms within the resin. The problem is that the diene is graftable, that is, chain transfer to the polymer takes place during the reaction. That is easily understood with the help of the Mayo equation, as the monomer concentration is decreasing while the polymer concentration is increasing. This graftability is useful in some products, like an ABS resin, though ABS resins are usually started from the butadiene homopolymer latex.

Nitrile rubber latexes may be made with functional groups for cure or other purposes. (The functional group monomers are treated in Chapter 9.) For instance, Stefferes and Rothenhaeuser [42] made a coating system of diene, acrylonitrile, and methacrylic acid, while Mayer-Mader [18] used the Diels-Alder

Table 7.1 Properties of Methacrylate Monomers

Monomer	Refractive Index	T_g, C°	Alfrey-Price Values	
			Q	*e*
Methyl	1.485–1.500	105	0.74	0.40
Ethyl	1.484	65	0.72	0.48
n-Propyl	1.484	35	0.65	0.44
i-Propyl	1.473	78	0.74	0.45
n-Butyl	1.483	20	0.71	0.41
i-Butyl	1.477	48	0.68	0.43
s-Butyl		45	0.72	0.24
t-Butyl	1.464	107	0.71	0.32
n-Hexyl	1.481	−5	0.66	0.35
2-Ethylhexyl		−15		
Cyclohexy	1.507	104	0.73	0.41
Lauryl	1.474	−65	0.74	0.2
Nitrile	1.52		0.75	0.91
Bornyl	1.506		0.79	0.47
Isobornyl	1.500	170	0.50	0.50
Benlzy	1.568	54	0.86	0.42

adduct of butadiene and maleic anhydride to make the monomer copolymerized with acrylonitrile in latex form for nonwoven fabric binding. Other markets for nitrile-based latexes are in masking-tape base, paper saturants, and specialty adhesives.

II. Esters

Acrylic ester monomers are the basis for most of the work cited in the introduction. Inspection of Tables 7·1 and 7·2 for methacrylates and acrylates, respectively, shows that the range of glass transition temperatures will allow making any type of copolymer with good clarity and any desired hardness or softness; For example, a pressure-sensitive adhesive using 95+% 2-ethylhexyl acrylate as the copolymer.

The copolymers used in the paint industry come from many sources, and range from the all-acrylic (which may contain some methacrylates), the vinyl-acrylics (fairly common and not always labelled as such by some paint companies!) to the styrene-acrylics (again, fairly common), to the diene-acrylics (not common, but available for textile applications. The properties are usually described in supplier literature, among the best of which is the *Resin Reviews* [10].

Table 7.2 Properties of Acrylate Monomers

Monomer	Refractive Index	T_g, C°	Alfrey-Price Values	
			Q	*e*
Methyl	1.479	8	0.42	0.65
Ethyl	1.464	−22	0.41	0.61
n-Propyl		−48		
i-Propyl	1.456	−8	0.48	0.45
n-Butyl	1.474	−54	0.59	0.23
s-Butyl		−20	0.41	0.34
i-Butyl		−17	0.42	0.78
t-Butyl		−22	0.44	0.36
n-Hexyl		−60		
2-Ethylhexyl	1.465	−85	0.40	0.53
Cyclohexyl	1.500	16	0.45	0.38
Stearyl			0.43	0.57
Benzyl	1.565	6	0.56	0.67
Nitrile		75	0.60	1.20

Within appropriate polymerization formulations, these copolymers may be made in latex form or as aqueous hydrosols, i.e., a colloidal form approaching the solution form of "smaller" molecules). Knauss [23] compares the two forms briefly; the hydrosols are frequently free of surfactant materials, an advantage in some instances, a disadvantage in others.

The polymerization of the acrylics is well described in the brochures from suppliers. However, additional aspects are worth mentioning. Mertens [11] described molecular weight control in methyl methacrylate polymerizations. Besecke et al. [12] patented a surfactant-free emulsion polymerization of ethyl acrylate based on incorporating an sulfonate monomer. Fitch [13] described an autoemulsification polymerization from methyl methacrylate dissolved in water with peroxydisulfate initiation. Bannerjee et al. [20] argue for aqueous phase initiation of methyl acrylate with water-soluble initiators, which might be reasonable because of the high monomer solubility in water. However, I suspect that the majority of the acrylate monomers are sufficiently insoluble to promote micelle surface initiation with the water-soluble initiators. The brochure mentioned earlier [7] should also be used as a guide for initiation systems for acrylate latexes.

Other initiation systems have also been reported. Sulfonated polystyrene initiated the polymerization of ethyl and methyl methacrylates, according to Olayemi and Adeoye [14]. Calcium sulfite initiated a "soapless" emulsion polymer for Arai et al. [15]. Paul et al. [16] reported a variety of monomers initiated by sodium metabisulfite. Merlin and Fouassier [17] used UV light to photoinitiate methyl methacrylate emulsion polymerization in the presence of sugars.

Graft polymerizations based on acrylate esters are possible. A Rohm & Haas patent [19] describes grafting a high T_g set of monomers on to a preformed low T_g latex polymer. Gehman and others [24] developed a technique for making "internally plasticized" polymer particles by a stepwise polymerization with butyl acrylate as the soft element added in the later stages. "Internal plasticization" is jargon in the trade, meaning no external plasticizer is needed for later additions to the formulation, because the T_g of the system was controlled within the polymerization by choice of monomers. Yeliseyva et al. [25] made three-component systems, varying the addition sequence of the monomers to demonstrate different film properties resulting from the sequencing. Although there is controversy over the scientific aspects, a "core-shell" approach to acrylic copolymers may include stepwise additions of the monomers. Indeed, John Vanderhoff (Lehigh Univ. Emulsion Polymer Institute) demonstrated light-scattering phenomena with a monodisperse poly(styrene) which he surrounded with a shell of ethyl acrylate, combining spacer and adhesive functions in the shell.

The surfactants chosen for the polymerization formulation are important too. Gabrielli and Maddii [26] demonstrated the thermodynamic incompatibility of poly(methyl methacrylate) with stearic or myristic acids. This is a good clue to production of latex films of low water sensitivity since the surfactant materials are squeezed out as they dry and coalesce. Matsumoto et al. [27] reported that the absence of emulsifiers in poly(ethyl acrylate) films increased water resistance. Yeliseyva and Zuikov [28] found that the adsorption of surfactants onto alkyl acrylates depends on both monomer structure and surfactant structure. Hence, the optimum surfactant for ethyl acrylate may not be the optimum for 2-ethylhexyl acrylate. Zuikov continued the work with Vasilenko [34] with characterization of the interfacial tension between surfactants and polymer particles.

Azad et al. [29] show that the acrylics may be made in bead, suspension, or emulsion form, depending on the types and amounts of surfactants chosen for the formulation. Ono et al. [30] used barium chloride coagulations and electrophoretic mobility to characterize the stabilitization derived from emulsifier variations in acrylic systems. Shchepetil'nikov et al. [31] studied "live" and "dead" polymer radicals in emulsion polymerizations of several acrylic homopolymers to demonstrate the "protection" given by the emulsifier. Their model involved capture of dissolved "live" oligomers by the particle, implying water-phase initiation for these polar monomers. Several patents describe emulsifiers especially helpful in emulsion polymerizations of acrylates [35-38].

Internal cross-linking within the latex particle may help to produce the strength needed for a particular application. Hashimoto et al. [33] used divinylbenzene as internal cross-linking monomer in an acrylate–vinyl acetate copolymer. One could also use the glycol diacrylate esters or even butadiene, depending on the grafting propensity.

Another copolymerization phenomenon was demonstrated by simple monomer combinations, wherein the polymerization rate depended on the monomer ratio. We found that a fast (20 oz.) bottle polymerization of an ethyl acrylate homopolymer with peroxidisulfate initiation was complete in about 1 hr. However, additions of 1-10% butadiene in place of a like portion of the ethyl acrylate slowed the polymerization to a completion in 8 hr, and remained at that rate for any higher levels of butadiene [32].

The properties of the acrylic emulsions are useful in many applications, such as paints, nonwoven binders, textiles adhesives, and industrial coatings. A survey of food cans by IR-attenuated total-reflectance spectroscopy showed that the acrylics are commonly used in can coatings. Attempts have been made to incorporate acrylic emulsions also into paper coatings. The advantages of acrylic emulsions are their solvent resistance and relative absence of discoloration upon aging. The Rohm & Haas test fences have archival samples of painted panels which have been continuously exposed to the weather for 20+ years, and I have inspected painted surfaces on houses in California that bear out that stability.

There are difficulties with the acrylics, however. They are subject to hydrolysis by alkali and acid (acid rain [39, 40] or smog). The acrylic monomers do not hydrolze as readily as vinyl acetate, and buffering is not as necessary for the protection of the emulsion during or after polymerization, buffering range can be wider, too. Monomer prices are high for some consumers. Many polymer makers prefer less expensive monomers, at least as diluents, in their acrylic formulations to increase their profits.

III. Other Acrylic Monomers

Other monomers are based on acrylic–methacrylic moieties. Those with additional functional groups (amine, carboxylate, hydroxy) are dealt with in Chapter 9 on functional monomers. However, the fully acrylate esterified polyols or glycols are not discussed in that chapter, as they are a minor component in the emulsion polymerizations and do not participate in curing reactions upon drying and film formation. As noted above, they may be added to the emulsion polymerization and will yield some additional strength. However, their main market lies in radiation-cured coating systems, a topic beyond the scope of this treatise.

REFERENCES

1. W. S. Zimmt, *ChemTech,* 681 (Nov. 1981).
2. S. Hochheiser, *Today's Chemist,* 8 (June 1988).

3. Rohm & Haas, "Preparation, Properties, and Uses of Acrylic Polymers," Brochure CM-19.
4. Rohm & Haas, "Emulsion Polymerization of Acrylic Monomers," Brochure SP-154.
5. Rohm & Haas, "Storage and Handling of Acrylic and Methacrylic Esters and Acids," Brochure CM-17 L/ca.
6. M. Salkind et al., *Ind. Eng. Chem. 51,* 1232 (Oct./Nov. 1959); reprinted as Rohm & Haas brochure.
7. Rohm & Haas, "Catalysts for the Polymerization of Acrylic Monomers," Rohm & Haas Brochure SP-159.
8. Rohm & Haas, "Acrylic and Methacrylic Monomers," Brochure CM-16.
9. E. H. Riddle, *Monomeric Acrylic Esters,* Chap. 7, Reinhold Publ., New York,; reprinted as Brochure SPR-182 by Rohm & Haas.
10. Periodicals available from Rohm & Haas, Independence Mall, Philadelphia, PA.
11. R. Mertens, *J. Ctgs, Technol., 53(677),* 45 (June 1981).
12. S. Benecke et al., Eur. Pat. Appl. EP 48320; *CA 97,* 6987b.
13. R.M. Fitch, 1975 presentation to the Akron Polymer Lecture Group.
14. J. Y. Olayemi and I. O. Adeoye, *J. Appl. Polym. Sci., 26,* 2639 (1981).
15. M. Arai et al., *J. Polym. Sci., Polym. Chem. Ed., 20,* 1021 (1982).
16. T. K. Paul et al., *J. Appl. Polym. Sci., 27,* 1501 (1982).
17. A. Merlen and J.-P. Fouassier, *J. Polym. Sci., Polym. Chem. Ed., 19,* 2357 (1981).
18. R. Mayer-Mader et al., Br. Pat. 1091480 (15 Nov. 1967); Cf: *CA 68* 13987f.
19. Br. Pat. 1450175 (assigned to Rohm & Haas, 22 Sep. 1976); Cf: Derwent 26389V/14.
20. M. Banerjee et al., *Polymer, 22(12),* 1729 (1981); Cf: *CA 97,* 6815u.
21. R. L. Feller, *Art Journal, 37(1),* 34 (Fall 1977).
22. C. E. M. Morris and A. G. Parts, *Makromol. Chem., 177,* 1433 (1976).
23. C. J. Knauss, *Indust. Fin.,* p. 55 (June 1986).
24. D. R. Gehman et al., Ger. Pat. 2811481 (21 Sep. 1978); Cf: *CA 89*81322n.
25. V. I. Yeliseyva et al., *Vysokomol. Soyed., A20(6)* 1265 (1978).
26. G. Gabrielli and A. Maddii, *J. Coll. Interface Sci., 64(1),* 19 (Mar. 1978).
27. T. Matsumoto et al., *Kobunshi Ronbun., 32(4),* 229 (Apr. 1975).
28. V. I. Yeliseyva and A. V. Zuikov, *Vysokomol. Soyed., A19(11),*2617 (1977).
29. A. R. M. Azad et al., *Colloidal Dispersions ABD Micellar Behavior,* Chap. 9, K. L. Mittal, ed., ACS Symposium Series 9, Washington, (1975).
30. H. Ono et al., *J. Coll. Interface Sci., 49(1),* 155 (Oct. 1974).
31. B. V. Shchepetil'nikov et al., *Yyoskomol. Soyed., A20(9),* 2097 (1978); Cf: *CA 89,* 180448w.
32. R. D. Athey Jr., unpublished.
33. S. Hashimoto et al., Jap. Pat. 7884092 (25 July 1978); Cf: *CA 89* 180808p.
34. A. V. Zuikov and A. I. Vasilenko, *Koll. Zh., 37(4),* 640 (July/Aug. 1975).
35. Jap. Pat. 49054486, assigned to Lion Oil and Fat Co. Ltd. (7 Sep. 1972); Cf: Derwent 78204V/45.
36. Jap. Pat. 76028120, assigned to Daiichi Kogyo KK (24 May 1976); Cf: Derwent 31560V/17.
37. Br. Pat. 1438449, assigned to Nippon *Snyth. Chem. KK (9 June 1976); Cf: Derwent 21062X/12.

38. J. E. O. Mayne and H. Warson, Br. Pat. 648001 (28 Dec. 1946).
39. R. Babian, ed., *Materials Degradation Caused by Acid Rain,* ACS Symposium Series No. 318, Washington (1986).
40. C. M. Balik, *Review and Evaluation of Scientific Approaches for Determining Paint Damage Due To Acidic Deposition,* Report submitted to EPA from ARSL, Research Triangle Park, NC (1986).
41. V. I. Eliseeva et al., *Koll. Zh., 37(1),* 152 (Jan./Feb. 1975); Cf: RAPRA Abstr. 33410L (12 May 1975).
42. H. Steffers and B. Rothenhaeusser, E. Ger . Pat 152342 (24 Nov. 1981); Cf: CA *97(1),* 6985z.

8

Diene Monomers

I. INTRODUCTION

The common commercial diene monomers are all simple derivatives of 1,3-butadiene. Most generally used in emulsion polymerization are the unsubstituted 1,3-butadiene, the 2-chloro derivative chloroprene(2-chloro-1,3-butadiene), and the 2-methyl derivative isoprene(2-methyl-1,3-butadiene). Some properties of these monomers are shown in Table 8.1, and other properties were summarized by Yaws [22]. Other dienes include 1,3-pentadiene, 2,3-dimethyl-1,3-butadiene, cyclopentadiene, hexachlorocyclopentadiene and so on; these are marketed as specialty monomers or intermediates.

The diene monomers are used to obtain the "resiliency" of natural rubber, and initially to promote sulfur curability in vulcanization. Indeed, the initial curves for styrene-butadiene (SBR) and nitrile rubber latexes were based on the addition of sulfur dispersions to the latexes. However, there were problems with that technology (staining, discolorations, odor), and alternative cures were sought (to be covered in Chapter 9).

Another advantage of hydrocarbon dienes is their relative low cost. They are fractions from petroleum distillation with such high usage, that they have to be specially made in excess of the normal from simple distillation. Even so, they are an approach to lower cost latexes. However, the propensity of dienes to react with air and especially with ozone is a distinct disadvantage and substantially shortens their useful life. Discoloration takes place especially upon overheating.

One important aspect of the diene polymerization is based on the position of the monomer unit within the polymer backbone. These dienes can be incorporated in three or four different ways, each retaining the residual double bond for the subsequent sulfur curing reaction. Figure 8.1 shows these structural variations, which also contribute to flexibility, T_g, and resilience. Although we list the T_g for the homopolymers in Table 8.1, each of the addition isomers shown in Figure 8.1 has its own unique T_g. For instance, the most resilient of rubbers, *cis*-1,4-poly(isoprene), is natural rubber; the *cis*-1,4 array is responsible for the very low T_g and excellent resilience.

In radical polymerizations, as in emulsion polymer syntheses, the dienes are incorporated as a blend of three or four isomers. The proportion depends on the polymerization temperature. Cold polymerizations (at $-5°C$) were developed to increase the cis isomer in order to improve resilience. The polymerization temperature for an unknown diene copolymer can be determined from the infrared spectrum data on cis/trans ratio [18–21].

II. BUTADIENE

The lowest-cost diene monomer is butadiene. It finds wide usage in nitrile rubbers, ABS resins, and styrene–butadiene rubbers, much of which is not used as latex. However, the syntheses are similar to those of the latexes used in paper coatings, earlier paint latex (for Super Kem-Tone by Sherwin-Williams), carpet-backing adhesives, and nonwoven fabric binders.

As noted in Chap. 5, butadiene requires modest pressure vessels. In addition, the explosive limits for air–butadiene mixtures are as wide as those for ether and air, and much care is needed to handle this monomer; toxicity regulations, should be strictly observed.

We also described earlier the handling techniques for the gaseous monomer. Another precaution for butadiene condensation is to handle it as a liquid. Simple ice water will serve as cooling agent and dry ice–acetone may be used if absolutely necessary. *Do not use liquid nitrogen!* Liquid nitrogen will condense some air oxygen along with the butadiene, forming an explosive mixture in a closed vessel. A man carrying a sealed flask of butadiene condensed with liquid nitrogen was seriously injured by an explosion. Fortunately he survived and recovered as a member of the "Wise Owl" club because he was wearing safety glasses.

Freshly prepared or freshly distilled butadiene should be used. The Diels-Alder dimer of butadiene, vinylcyclohexene, retards the polymerization of radical reactions [1]. The distinctive odor of hot butadiene polymerizations is due to the vinylcyclohexene by-product, because the Diels-Alder reaction occurs readily. Indeed, the higher the polymerization temperature and the lower the reaction rate, the more vinylcyclohexene is formed. It may, however, be easily stripped out. Because of its odor, it is commonly mistaken for styrene or acrylonitrile.

Physical properties are related to the composition of the copolymer. For instance, tire wear and braking traction indexes are determined by the ratio of

Table 8.1 Properties of Conjugated Diene Monomers

Monomer	Refractive Index	T_g, C°	Alfrey-Price Values *Q*	*e*
Butadiene	1.515	−102	2.39	−1.05
Isoprene	1.519	−67	3.33	−1.22
Chloroprene	1.558	−45	7.26	−0.020

styrene to butadiene [2]. Krotky and Tokarzewska [3] showed that the dynamic viscosity of butadiene or butadiene–styrene latex polymers varied with percent conversion, when they tried to obtain high conversion percentages. They related viscosity to the degree of cross-linking within the emulsion particle occurring at high conversions (as noted earlier) by chain transfer.

Since butadiene is a hydrocarbon, one would expect the emulsion polymerization to occur readily with the appropriate ratios of water and emulsifier. That, indeed, has been my experience. But alternative emulsion polymerization technology can also be found in the literature. For instance, Pryakhina et al. [4] used casein as emulsifier for butadiene polymerization, although they had not expected it to work. They reported that water-soluble initiators (peroxydisulfate) usually do not work with surfactant-free aqueous polymerizations. A Belgian patent [5] claims that butadiene polymers may be made with very large particle sizes (e.g., for subsequent use in ABS resin grafting) by incorporating a poly(ethylene oxide) into the colloidal system in addition to the normally used surfactants. They call these additives "agglomeration aids," but the term flocculants would have been just as accurate.

It is not unusual for the diene latex to be made, sold (sometimes), and then used as a starting material for ABS resins. Some trickery is involved, though when you are aware of the basics of emulsion polymerization and colloidal

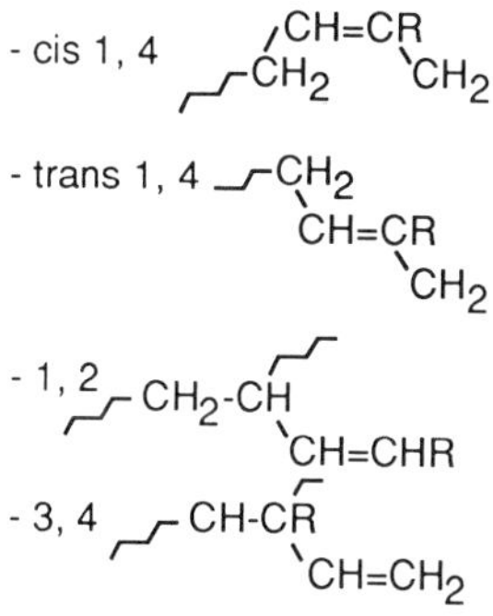

Figure 8.1. Polymerization Additions of Dienes [22].

stabilization dynamics, all seems rational. The poly(butadiene) latex is placed into the reactor with excess water, styrene, acrylonitrile, and initiator. Surfactant is only added to stabilize the poly(butadiene), but not enough to reach the critical micelle concentration for that surfactant. In this way, few (if any) new polymer particles are started, which forces the styrene and acrylonitrile monomers to swell the butadiene and polymerize.

III. ISOPRENE

Isoprene boils at slightly above room temperature, and a lab polymerization may be formulated in such a way to make it polymerize at a slight reflux in an open ambient-pressure reaction vessel. However, for polymerization at higher temperatures, it is convenient to use the "coke" bottle reactors found in butadiene labs. Since the monomer boils above room temperature, one may want to pressurize the polymerization bottle to avoid cap failure due to vacuum at the late stages of polymerization.

Again, the monomer should be freshly distilled to remove the inhibitors used for shipment. An alkaline wash will remove some of the inhibitors such as methylhydroquinone, but distillation is easy enough.

Isoprene forms emulsions with most surfactants, and therefore the "typical" emulsion polymerization formulation will work well. Ugelstad et al. [6] mentioned work by Stearns that showed isoprene polymerization occurring in the monomer droplet as well as in the growing polymer particle. However, in commercial systems most of the polymerization occurs within the growing polymer particle.

An alternative preparation of a carboxylated poly(isoprene) was described in a Kuraray patent [7]. A hydrocarbon solution of the polymer was dispersed in water with a surfactant, and the hydrocarbon was stripped off. This is not a common technique, but it has been used long enough to have been termed the "artificial" latex process. Blackley [8] described the technique of making an "artificial" latex of natural rubber, an all-*cis* poly(isoprene) by milling the dry rubber on a two-roll mill, adding oleic or stearic acid (about 5 phr) to the mixing band, followed by aqueous ammonia. The aqueous phase mixes in as a colloid dispersed in the rubber and neutralizes the carboxylate of the emulsifier until the aqueous phase becomes predominant. At that point, the system undergoes a phase inversion, and the latex runs off the mill into a container. One might consider this as an exercise in futility since natural rubber is a latex, but there are good reasons to utilize this reemulsification process. The molecular weight is reduced, resulting in a product suitable as an adhesive component. In addition, all the natural protective colloids (sugars, starches, proteins are no longer there, and the formulation may now be made compatible with materials that the natural latex would not accept. One could envision the same mill process using cationic or nonionic surfactants as well.

IV. CHLOROPRENE

Chloroprene is the precursor monomer for the Dupont NEOPRENE rubbers. Their latexes are mostly nonfunctional and are based on fatty acid or rosin acid emulsifiers. However, some grades are based on synthetic emulsifiers, and some even have incorporated carboxylic monomer for curing. Snow [9] described contact adhesives based on carboxylated polychloroprene latex.

Chloroprene is a liquid obtained by a relatively straightforward synthesis process. Acetylene is dimerized to vinylacetylene and hydrogen chloride is added to form chloroprene. The commercial product contains a considerable amount of inhibitor. Lab quantities need to be washed with aqueous alkali to decolorize and remove the inhibitor. Lab personnel should wear face masks and rubber aprons and gloves for the washing operation in a separatory funnel because the monomer is toxic. The toxicology has been reviewed [10], and current OSHA and NIOSH regulations provide a guide to safety practices.

The hydrocarbon monomer will emulsify well in aqueous surfactant systems, and emulsion polymerization proceeds well. Since the monomer boils above room temperature, one may want to pressurize the polymerization bottle to avoid cap failure due to vacuum at the late stages of polymerization. Krishan and Margaritova [11] used chloroprene for their investigation of the dependency of emulsion polymerization rates on initiator and surfactant concentrations. They showed that there are limits upon the standard equations describing those kinetics.

As noted earlier, the common emulsifiers for this monomer are the anionics, but cationic emulsifiers are also used [12]. A DuPont patent describes mixed anionic emulsifiers [13], carboxylate and sulfonate. The emulsifier can have interesting effects on end use properties. We found that a rosin acid emulsified roof coating was tacky for three days after drying, until the rosin totally dissolved in the formulation. That shouldn't be a surprise, as rosin materials are often sold as tackifiers for adhesive systems.

Another patent [14] describes "peptization" of poly(chloroprene) latexes during polymerization with sulfur and iodoform or diisopropylxanthogendisulfide. Peptization is a term formerly used for reducing the molecular weight of a polymer by mechanical action (natural rubber, for instance) or by a chemical reaction. This patent, though, discloses the emulsion polymerization synthesis of end-group functional liquid poly(chloroprene) by a chain-transfer process. Such polymers are called telechelic polymers. They have been reviewed [15], and a critique of this process based on chain-transfer kinetics was also published [16].

Poly(chloroprene) latexes are used in many applications. Unpublished investigations showed that asphalt–poly(chloroprene) emulsion blends had among the lowest of water-permeation rates of roof-coating compositions. Protective gloves are dip formed form the latex, and we have already noted the adhesive usage. The formulation of glass tire cord adhesives may contain poly(chloroprene) latex [17]. Possible dehydrohalogenation may be the only drawback of poly(chloroprene) as formulation component. But many latex formulations contain magnesium, calcium, or zinc carbonates or oxides to inhibit the autocatalysis by any hydrogen chloride generated by degradation.

V. REPRISE

The hydrocarbon diene monomers add several desirable properties to the emulsion copolymers. Resilience will improve impact strength and crack resistance and water resistance is excellent. But there are disadvantages, as well. The residual unsaturation can result in discoloration upon weathering and aging and ozone resistance can be poor. The end use will dictate the usage.

REFERENCES

1. E. R. Meinke (an old *Rubber Age* or *Rubber World* article; not found).
2. Figures 6 and 7, Elastomerics, p. 26 (Mar. 1985).
3. E. Krotki and M. Tokarzewska, *Polimery Tworzya Weil., 21(10),* 20 (Jan. 1976); Cf: *Internat. Polym. Sci. Technol., 3(6),* 1269 (1976).
4. E. A. Pryakhina et al., *Vysokomol. Soyed.*, 16A(2), 299 (Feb. 1974); Cf: *RAPRA Abstr.* 22393L (3 June 1974).
5. Belg. 817505 (10 Jan. 1975), assigned to Rhone Progil.
6. J. Ugelstad et al., *J. Polym. Sci., Polym. Lett. Ed.*, 11, 503 (1973).
7. US 3971746, assigned to Kuraray KK (27 July 1976); Cf: Derwent 69153W/42.
8. D. C. Blackley, High Polymer Latices, Palmerton Publ., New York (1966).
9. A. M. Snow, *Adhesives Age,* p. 35 (July 1980).
10. J. W. Lloyd et al., *J. Occup. Med.*, 17(4), 263 (Apr. 1975).
11. T. Krishan and M. Margaritova, *J. Polym. Sci., 52,* 139 (1961).
12. Jap. Pat. 76020558, assigned to Electrochem. Ind. (29 June 1976); Cf: Derwent 74843U/49.
13. US 3981854, assigned to DuPont (21 Sep. 1976); Cf: Derwent *75524X/40.
14. US 3984609, assigned to Distugil (5 Oct. 1976); Cf: Derwent =04696T/03.
15. R. D. Athey Jr., *Prog. Org. Ctgs., 7(3),* 289 (Aug. 1979).
16. R. D. Athey Jr. et al., *J. Polym. Sci., Polym. Chem. Ed.*, 15, 1523 (1977).
17. M. M. Girgis, US 4164485 (14 Aug. 1979).
18. W. B. Trewmann and F. T. Wall, *Anal. Chem., 21,* 1161 (1949).
19. E. J. Hart and A. W. Meyer, *JACS, 71* 1980 (1949).
20. J. E. Field, D. E. Woodford and S. D. Gehman, *J. Appl. Phys., 17,* 386 (1946) and *Rubber Chem. and Tech., 19,* 1113 (1946).
21. R. R. Hampton, *Anal. Chem., 21,* 924 (1949).
22. C. L. Yaws, Chem. Engr., p. 107 (Mar. 1, 1976).

9

Curing Monomers

I. INTRODUCTION

The monomers built into copolymers that allow them to cure upon some later treatment are an important aspect of polymerization technology for the specialty markets. The discussion of the chemicals used as additives to cure these copolymers will appear in Chapter 18. However, incorporating these functional groups (capable of cure) within the copolymer is a major task of the polymerization chemist.

There are many functional groups that may lead to cure. The carboxylate and hydroxy monomers are the main ones commercially used, followed by the methylol amides. Epoxy, amine, and isocyanate monomers are commercially available but less often used. We will deal with each of these to aid your understanding and usage. Table 9.1 includes Alfrey-Price *Q* and *e* values [13] and glass transition temperatures. The T_g data are probably calculated from copolymer T_g values, using the Flory-Fox equation [10], since the T_g for these (mostly hydrophilic) monomers is seldom determined on the homopolymer. We will not discuss the diene monomers (see Chapter 8) or the sulfate–sulfonate monomers (they are used only in trace amounts to increase colloidal stability).

Curing functional groups do not have to be derived from monomers in a copolymerization. One may use chain-transfer agents or initiators to make end-group functional polymers called telechelic [14]. H. J. Harwood of the University of Akron is developing such technology.

II. MAJOR COMMERCIAL MONOMERS

A. Carboxylates

The main commercial carboxylated monomers are acrylic acid, maleic acid, and methacrylic acid. Others finding some usage are itaconic acid, fumaric acid, and crotonic acid.

Dicarboxylic acid half-esters find some usage (half maleate or half itaconate), according to patent literature [55], but their usage is not common or in large volume. The *Polymer Handbook* [15] shows ethyl acid maleate (a half-ester) among the monomers listed for Alfrey-Price values. Zimmerman [8] made latex thickeners using nonionic emulsifier half-esters, so they would be associative poly(acrylic acid) copolymers. One would have to make the monomer from the acid or anhydride, isolate them immediately (from an equilibrium of acid and diester), and use them; the by-products may be a problem. Upon prolonged storage, the liquid-half ester disproportionates to the equilibrium mixture of acid, alcohol, and two esters, but solid half-esters may be more stable if isolated free of alcohol, acid, and water.

Analysis and testing of the carboxylate monomers is important, as impurities could have bad effects on the copolymer syntheses. ASTM methods [1] are available for monomer analyses (acidity, color, inhibitor content). The inhibitor analyses is important for polymerization rate control. Acrylic acid will, upon standing, form a Michael addition dimer (3-acryloxypropionic acid), and there is an ASTM method for determining the dimer content of the monomer [2]. The dimer will reduce the polymerization rate of acrylic acid homo- or copolymerizations [3].

The addition of the carboxylate monomer to the standard latex polymerization (acrylic esters or other hydrophobic monomers) will serve two functions. It can act as the cure site upon addition of a variety of curatives [6]. It also will aid the colloidal stabilization at pH 6 or above, provided carboxylate is exposed at the polymer particle surface. Many examples of stabilization are found in the literature [7].

There is unpublished work on the kinetics of the conversion of carboxylate monomer to the latex copolymer by dialysis and titrimetry as a function of conversion percentage in diene–styrene or diene–acrylonitrile systems. In the main, these monomers are incorporated late in the polymerization, because of their water solubility and the insolubility of the other monomers. They are incorporated well enough that 20% or less of the charged carboxylate monomer is detectable as free monomer in polymerizations that succeed to 95+% conversion. The remaining monomer is stripped out or is innocuous if retained. The more hydrophobic the monomer, the earlier it is incorporated. Expect incorporation of methacrylic acid to be faster than that of acrylic acid. Adjustment of the pH to 5 or less aids incorporation. Since the maleate is highly soluble, expect

Table 9.1 Properties of Curing Monomers

Monomer	T_g, C°	Alfrey-Price Q	Alfrey-Price e
Carboxylates			
Acrylic	103	1.15	0.77
Crotonic	?	0.013	0.45
Itaconic	?	0.76	0.5
Maleic (anhydride)	?	0.23	2.25
Methacrylic	230	2.34	0.65
Amines			
Dimethylaminoethyl Methacrylate	18	0.49	0.86
Dimethylaminopropyl Methacrylate	?	0.289	−0.052
2-Vinylpyridine	?	1.30	−0.5
4-Vinylpyridine	?	0.82	−0.2
Amides			
Acrylamide	165	1.18	1.30
Methacrylamide	?	1.46	1.24
N-methylol Acrylamide	?	0.31	0.36
Methyl ether of N-methylol Acrylamide	?	0.31	0.36
Hydroxylates			
Hydroxyethyl acrylate	?	4.08	1.51
Hydroxypropyl acrylate	?	0.87	0.78
Hydroxyethyl methacrylate	55	0.80	0.40
Hydroxypropyl methacrylate	73	0.79	0.20
Epoxies			
Glycidyl acrylate	?	0.55	0.96
Glycidyl methacrylate	46	0.85	0.10

it to perform poorly, if it does not coagulate the latex during polymerization. Maleic acid is mostly used in organic solvent-based solution copolymerizations, as the anhydride.

The surfactants chosen for the polymerization will affect the incorporation of the carboxylate monomer. One must choose a surfactant that will support the polymerization at the low pH needed. Witt [55] patented a mixed emulsifier system (mixed micelle) using a fatty acid surfactant, but stability during polymerization was poor. Our subsequent work used a carboxylate emulsifier with lower pKa to make a more producible latex [16].

In a thought analysis of the types of carboxylation that can appear in the copolymerization, we hypothesized five carboxylated species, and devised an analytical technique for their determination [4]. These species include:

1. unreacted monomer,
2. water-soluble oligomer copolymer,
3. water-insoluble surface-active oligomer copolymer,
4. latex particle surface exposed copolymer, and
5. latex particle buried copolymer.

Since the first three may be removed by an ion exchange chromatography column (which also separates them for analyses) one may obtain the "clean" latex having only surface carboxyl groups exposed. One may titrate those groups directly with alkali (add a nonionic surfactant to assure mechanical and chemical stability in the processing). Adding excess alkali and boiling will raise the temperature above the polymer, and the buried carboxyl groups will surface at random, be neutralized, and stay at the surface. Ionic groups would be more comfortable in water than buried within the hydrophobic interior. Additional detail may be found in Chapter 13.

One may get around this late incorporation of the carboxylate by several techniques. For example, a small amount of hydrophilic monomer (nitrile or ester) gathers the carboxylate earlier in the polymerization. In the seed latex technique, the carboxylate monomer is present only in the seed latex. For instance, if the objective is a 2% methacrylic acid latex in the final product, prepare a styrene–methacrylic acid 90:10 seed latex at ca. pH 4 to complete conversion, neutralize with ammonia, and add enough styrene–butadiene and more initiator to obtain 2% acid in the final product. Residual acid monomer after the polymerization may be less with these techniques, compared to a simple batch process. Greene discussed processing control of acrylic acid location in the latex particle [57]. We discussed pH effect on acid incorporation, as did Bando and Minoura [58].

An unexpected result of this surface carboxylation is a perceived particle size larger in the water than one would find from electron microscopy of the dried particles. The water swelling (only at the surface by a high degree of hydration of the ionic carboxylates) increased the observed particle size. We found this by accident when comparing latex particle size measurement techniques in a collaborative study [5]. Basset and Hoy [56] described this swelling.

B. Nitrogen-containing Monomers

1. Amines

The amino monomers and all other functional monomers are used much less than the carboxylates in emulsion polymerization. However, some amine monomers are used in large amounts. Tire cord adhesives, for example, require large quantities of vinylpyridine. The amine monomers and some T_g values are given are summarized in Table 9.1. Nyquist and Yocum [37] found that 2-dimethylaminoethyl methacrylamide and 2-(aziridinyl)ethyl acrylate promoted coating adhesion.

Secrets to incorporating these monomers into a polymer include keeping the pH high (to reduce water solubility). Doronin et al. described the hydrolysis of dimethylaminoethyl methacrylate during copolymerization with butadiene [35] and found [36] a useful surfactant. Avoid peroxy initiation wherever possible; it causes discoloration and amine oxide formation from the monomer. Azo initiation is preferred, if the color and the free amine is important.

2. Amides

The commercial amide monomers are usually acrylamide and methacrylamide. However, the reaction products of methylolation are very important for the emulsion polymer world. These *N*-methylol derivatives are the basis for the self-cross-linking systems used in coatings and textiles applications. The unsubstituted amides are mostly used in aqueous polymer syntheses for flocculants and oil-well drilling-mud additives. Other substituted amide monomers, such as *N,N*-dimethylacrylamide and *N-t*-octylacrylamide do not take part in curing reactions. Their utility is small, and they are not part of the major emulsion polymer world. Table 9.1 shows some of the amide monomers "Q" and "e" values and Tg (or at least the contribution made to Tg by these monomers in copolymers).

The methylolation of the amide is a reversible reaction, catalyzed by weak bases. It is essentially an addition of the active hydrogen attached to the nitrogen atom across the double bound of formaldehyde. However, addition does not have to stop there. Condensation with another amide molecule, with a loss of water, leaves a methylene bridge between two amides, yielding methylenebisacrylamide (a desirable cross-linker in some formulations).

The cure of the latex based on *N*-methylolacrylamide is a condensation reaction. The methylol group looses water with any active hydrogen functional group in the polymer or on the substrate. It is essentially the same reaction that occurs with the methylolmelamine curative. Hydroxy, carboxy, amine, and amide active hydrogens cure to the *N*-methylolacrylamide copolymer. Indeed, some formulations use *N*-methylolacrylamide cure to themselves with a loss of water to make the bismethylene ether with acid catalysis, or the methylene bridge with a loss of formaldehyde. Some Rohm & Haas work showed the involvement of an important water-soluble oligomer in textiles cured in *N*-methylolamide copolymer latexes. Yeliseeva described surface orientation effects to improve methylolamide latex cures [24], and commented upon the interaction with soluble species. Bufkin and Grawe [38] reviewed these cures, among others.

The latex syntheses with *N*-methylolamides is very sensitive. The pH is critical. If you want to avoid gelation during polymerization, buffer at about pH 7–7.5. Steam stripping is prohibited, as the concentrated heat will cause gelation which is difficult to clean up. Rohm GMBH patented a staged polymerization procedure where the methylolamide monomer was mostly in the early stage [51].

Stability can be improved by using the methyl or butyl ethers of the *N*-methylolamide monomer. Acrylic–styrene copolymers with such an ether have been patented [52]. These ether monomers are made in alcohol solvent, and the methylolation reverse reaction (demethylolation?) is blocked. However, these monomers require a higher temperature to effect the cure. The methylolamide attains about 80% cure upon air drying the free film on glass, as shown by a tensile strength study as an example of low temperature curability. Acid catalysis is a help to speed up these cures. With increased catalyst acid strength cure rates increase (e.g., ammonium bromide increases cure rate more than ammonium chloride as a latent acid catalyst [9].

One may simply use the amide form of the monomer and add formaldehyde and catalyst for in situ methylolation and cure. However, current regulations will not allow the needed excess of formaldehyde to assure complete methylolation. Indeed, the methylolamide-derived latexes are no longer used in many textile applications because of the presence of the free formaldehyde upon cure.

The *N*-methylolacrylamide copolymers have been used for some time; originally they were developed for textile applications. Copolymers have been made with vinylidene chloride [39], ethylene–vinyl chloride [40], vinyl acetate [41], acrylate–methacrylate–styrene [42], ethylene–vinyl acetate [43], acrylic or methacrylic acids and nitriles [44], butadiene–acrylate [45], and other combinations. Such copolymers have been useful as mineral-wool fiber-board binders [44], steel coatings [46], titania dispersants [47], hot-melt label adhesives [48], nonwoven fabric binders [45,49,52], pressure-sensitive tape adhesives [50], and more.

An alternative to these methylolamides is diacetone acrylamide, another "self-cross-linking" monomer. It has been patented for use in solid curing compounds where it was mixed with a saturated polymer [20]. A copolymer was used to treat the surface of calcium carbonate for incorporation into polymers [21]. Curable pressure-sensitive adhesives were based on this monomer [22], and it was used in water-soluble wire enamel formulation [23].

3. Isocyanates

There is only one commercial isocyanato monomer available for free-radical polymerizations, the 2-isocyanatoethyl methacrylate (Dow Chemical). One cannot use it in emulsion polymerizations as the isocyanate would disappear. We are looking forward to someone using methyl ethyl ketone (MEK) oxime or phenol to block it for use in emulsion polymerizations.

4. Aminimides

Some years ago, a series of aminimide monomers became available from the research of Culbertson et al. [17]. McKillip et al. [18] showed that they could be used for thermoset coatings, as the aminimide decomposed thermally to an isocyanate. Field et al. [19] used the same technology for tire-cord adhesive formulation.

C. Hydroxy Monomers

The main commercial hydroxy monomers are the ethylene and propylene glycol monoesters with acrylic or methacrylic acid. Table 9.1 shows T_g and Alfrey-Price values. Again, these are water-soluble monomers which may be incorporated late during the emulsion polymerization. The hydroxyethyl acrylate needs special handling because of skin permeation and toxicity.

These monomers improve adhesion to substrates and are curable (e.g., with melamine–formaldehyde resins). They may improve the colloidal stability of carboxylated latexes toward coagulation by salts and reduce the coacervation with starch or other hydroxylated polymer additives. Braidich [12] patented the formation of the hydroxylated ester in situ on the carboxylated latex by addition of an oxirane to the carboxylate anion. Some paper coating latexes are reputed to have hydroxy-based comonomers.

D. Epoxy Monomers

Glycidyl acrylate (GA) or methacrylate (GMA) are the best known epoxy-containing monomers usable in emulsion polymerizations. They are commercially available from several sources, but are specialty items not used in large volume. Table 9.1 shows T_g and Alfrey-Price values.

The emulsion polymerizations using these monomers present no particular problems, as long as the pH is kept near neutral. Oxiranes are opened by acid or base, and pH control is necessary to keep the oxirane functionality intact. Blemmer Chemical Company (in Japan) has developed formulations for grafting their GMA product to a wide variety of polymers. These monomers require special handling, as they are suspected toxic agents; one may develop sensitivity to them.

Yeliseeva found no surface effect to enhance the cure of GMA-based latexes [24]. Product patents made with GMA include stoving (baking) enamels [25]. water-soluble paints [26], powder coatings [27,30], crease-resistant textile coatings [28], adhesives [29], nonwoven fabric binders [31,32], wire coatings [33], and ion-exchange resins [34].

E. Acetylacetonates

Recent introduction of the acetoacetylethyl methacrylate (ACACEMA) has opened up a new field of curing options for use with polymers in general and latexes in particular. For instance, a coordination cure with copper or cobalt may be possible; the ion in the water phase and the ACACEMA in the latex particle meet only to develop cure upon drying and coalescence. The literature form the supplier [11] shows cures with formaldehyde, diamines, polyvalent cations and polyenes. The last group was the basis for the 1988 prize-winning presentation by Clemons on triacrylates as curing agents for the ACACEMA copolymers [53].

III. CONCLUSION

Although we have covered a dozen or so monomers herein, there are certainly other candidates that may be useful. They are specialty items, at premium prices, as are many of the above monomers. One can even envision making two different latexes, having two different functionalities, to be mixed and used as a single self-curing binder system, such as a carboxylated latex mixed with an epoxy-functionalized latex where the groups would not reach each other to cure until coalescence. Challenges still exist.

REFERENCES

1. ASTM D-4416-84 for acrylic acid; ASTM D-3845-84 for glacial methacrylic acid.
2. ASTM D-4415-84
3. BFG Chemical, unpublished research result.
4. R. D. Athey Jr., *Mod. Paint Ctgs.*, pp. 31-35 (Feb. 1984).
5. R. D. Athey Jr. et al., *Coll. Polym. Sci., 255,* 1001 (1977).
6. R. D. Athey Jr. et al., Golden Gate Society for Coatings Technology Technical Committee Voss Award Presentation at the 1988 National Paint Show and FSCT National Meeting (*European Coatings,* in press).
7. L. E. Kelley, US 3812070.
8. C. A. Zimmerman, US 3657175.
9. R. D. Athey Jr. and James Wonnell, unpublished.
10. T. G. Fox, *Bull. Amer. Phys. Soc., 1,* 123 (1956).
11. Eastman Chemical Products, Kingsport TN, Product brochures No. X-262, X-263A, X-280 and X-282.
12. E. V. Braidich et al., US 3873480.
13. T. Alfrey and C. C. Price, *J. Polym. Sci., 2,* 101 (1947).
14. R. D. Athey Jr., *Prog. Org. Ctgs., 7(3),* 289 (1979).
15. T. Brandrup and H. Immergut, eds. *Polymer Handbook,* 1st ed., Wiley-Interscience, New York, 1975.
16. R. D. Athey Jr., E. G. Sammak, and E. Witt, US 3591541(6 July 1971) (assigned to Standard Brands Chemical).
17. B. M. Culbertson et al., *ACS Org. Ctgs. Plas. Chem. Div. Prepr., 21(1),* 603 (Apr. 1968).
18. W. J. McKillip et al., *Ind. Eng. Chem. Prod R&D, 13(3),* 197 (1974).
19. R. E. Field et al., *J. Elast. Plast., 7(1),* 22 (Jan. 1975).
20. US 3816559 (11 June 1976).
21. G. M. Grudus et al., US 3763084(2 Oct. 1973).
22. US 3900610(19 Aug. 1975).

23. D. Laganis, US 3974115.
24. V. I. Yeliseeva, *Br. Polym. J., 7,* 33 (1975).
25. J. E. O. Mayne et al., Br. Pat. 988171(7 Apr. 1965).
26. Yu. P. Kvasnikov et al., *Lakrokas. Mat., (5),* 13 (1974).
27. Br. Pat. 1441379(30 Jun. 1976).
28. Jap. Pat. 500063094(29 May 1975); Cf: Derwent 74458X/40.
29. Jap. Pat. 760370939(13 Oct. 1976); Cf: Derwent 81429V/47.
30. M. H. Sakai et al., US 3989767.
31. R. L. Adelman, US 3081197(12 Mar. 1963).
32. N. A. Matlin and B. B. Kline, US 3074834(22 Jan. 1963)
33. Jap. Pat. 76018256(8 Jun. 1972) - Cf: Derwent 09698W/06
34. A. Wyroba, PT 75/08/385; Cf: *Internat. Polym. Sci. Technol., 2(11),* A/13 (1975).
35. A. S. Doronin et al., *Ibid.,* p. T/72
36. *Ibid., 2(9),* SK 75/09/14
37. E. B. Nyquist and R. H. Yocum, *J. Paint Technol., 42(544)),* 308 (May 1970).
38. B. G. Bufkin and J. R. Grave, *J. Ctgs. Technol., 50(641),* 41 (June 1978).
39. B. K. Mikofalvy and D. P. Knechtges, US 3748295(24 July 1974)
40. a. US 3755233; Cf: Derwent 51311T A87-F4(A18) week U37
    b. W. F. Fallwell Jr., US 38466354
41. a. P. F. Stehle et al., US 3806402(23 Apr. 1974)
    b. US 3770680; Cf: Derwent 72354T A81-G3-(A14)
42. Ger. Pat. 1644993; Cf: Derwent 70748U A82-G2-(A14)
43. M. K. Lindemann and R. P. Volpe, US 3380851 (30 Apr. 1968)
44. US 3756971 (4 Sep. 1973)
45. Neth. Appl. Pat. 6509310(31 Jan. 1966)
46. V. A. Spasov et al., *Lakrokras. Mat., 5,* 27 (1974).
47. T. S. Krasotina and P. I. Ermilov, *Kill. Zh. 36(6),* 1171 (Nov./Dec. 1974); Cf: *RAPRA Abstr.* 31591L (3 Mar. 1975).
48. Br. Pat. 1448813 (8 Sep. 1976); Cf: Derwent 69211X/37
49. Neth. Appl. Pat. 299134(11 May 1964)
50. A. M. Coffman, US 3738971
51. Br. Pat. 1447203(25 Aug. 1976) to Rohm GMBH; Cf: Derwent 52645W/32.
52. Br. Pat. 1346325(6 Feb. 1974); Cf: Derwent 61462S A87-F1.
53. R. J. Clemons and F. D. Rector, *J. Ctgs. Technol., 61(770),* 83 (Mar. 1989).
54. H. J. Harwood, ACS TV Satellite Lecture, "Making Tomorrow's Polymers," (Sep. 1988).
55. E. Witt, US 3480578 (25 Nov. 1969) (assigned to Standard Brands Chemical).
56. D. R. Bassett and K. L. Hoy, 176th ACS Meeting, Div. of Colloid and Surf. Chem, Abstr. #27 (Sep. 1978).
57. B. W. Greene, *J. Coll. and Interface Sci., 43(2),* 1949 (May 1973).
58. Y. Bando and Y. Minoura, *J. Polym. Sci., Polym. Chem. Ed., 14,* 1195 (1976).

10

Waterborne Condensation Polymers

I. ALKYDS

Though the past discussions have dealt with the standard latex materials based on addition polymers, there are classes of water-borne polymers derived from so-called condensation or step-growth polymerizations. One of the more important of these systems is the waterborne oil-based alkyd system, a modification of an old chemical technology wherein natural oils are "cooked" (reacted) with a polyalcohol and a polyacid, typically a triol and phthalic acid.

For solvent-borne coating applications, the acid and alcohol equivalents titratable from the alkyds should be low, unless the oligomers were to be cured with a melamine resin. For waterborne applications, the reaction uses an excess of carboxylic acid monomer to ensure unreacted carboxyls at the end of the reaction. This allows the final product to be water dispersible (if not water soluble) after amine neutralization to a pH of 8.5 or higher. For a variety of reasons (not the least of which is toxicological), the preferred amines for neutralization are ethanolamine derivatives. These amines can be added to the solvent cook after polymerization is complete. This mixture is then emptied into water to emulsify everything. The polymerization solvent can be stripped off at low temperature for recovery without significant amine loss.

One aspect of the reaction involving natural oils is that the ester interchange liberates some fatty acid. This is the consequence of the polymerization equilibrium reaction, but it has some implications for the user. Upon neutralization the free fatty acid will act, as soap. The surface tension of the dispersion will be lower than if the alkyd were heated to no carboxylation and then dispersed in water. However, this added surfactant helps stabilize the emulsion or dispersion without having to add external emulsifier, and may even aid the incorporation of pigments and fillers in the grinding steps of paint manufacture. In addition, the fatty acid usually has unsaturation in the carbon backbone to allow it to react with, and be incorporated into, the main curing alkyd polymer. The free fatty acid can result in polyvalent cation precipitation (soap scum, foam) and one should be careful to guard against those possibilities.

Since these resins are based on natural oils, they cure by uptake of oxygen from the air to eventually form free radicals that cross-link the many oil-based polymer side chains. This will occur (albeit slowly) without catalysis, but additions of small or even trace amounts of transition metal oxidation catalysts aid the cure. The transition metals are true catalysts, participating in the reactions but remaining free of any covalent ties to the products. Publications describe mixed metal catalysts as providing fastest cures with minimum discoloration.

The benefits to be derived from the water-dispersible alkyds lie in their chemical structures. Being based on natural products, they are independent of petroleum prices. The air oxygen-induced cure reduces the curative cost to a minimum. The high carbon and hydrogen content, with respect to more polar nitrogen and oxygen atom in the structure, result in coatings more hydrophobic than those derived from more polar acrylics, for instance. This in turn results in a lower vapor transmission and better barrier properties to moisture. Some excellent flexibility and impact-strength formulations are marketed. These systems can be sprayed or roller coated at relatively high solid content, compared to the general latex-based systems. The alkyds do not need coalescents, but may need cosolvents to increase water miscibility.

The drawbacks of these materials lie in their structures. Additives to prevent bacterial attack on the liquid coating will be needed, as well as mildewcides for the final coating film. Discoloration upon aging may be a problem, depending upon the mixture of natural oils and on the transition metal catalysts. Although some formulations are baked in industrial usage, the baking may not be a speedy cure or dry other chemical systems. Scratch resistance and adhesion can present problems. The amines used to help dissolve (or colloidally disperse) the alkyd may inhibit (or even block) the curative effects of the transition metal catalysts. In addition, the carboxylate anion of the resin can tie up and even precipitate the transition metal catalyst.

Since these resins are condensation polymers (that is, they split water off during the polymerization), the alkalinity needed to dissolve or disperse them can reverse the polymerization by a hydrolysis process (the ester bond is broken by water). The hydrolysis reaction should be slow, unless both the amine and the water can reach the inner parts of the dispersed molecule. Clues to the hydrolysis reaction taking place may be development of turbidity, loss of vis-

Table 10.1 Ingredients for Waterborne Polyurethane Synthesis[a] [2]

Material	Weight Percent
TONE 0305, caprolactone-based polyol	15.17
TONE 0200, Caprolactone based polyol	16.85
2,2-Dimethylolpropionic acid[c]	1.43
4,4′-Diphenylmethane diisocyanate[d]	15.62
Methyl ethyl ketone solvent	49.97
N,N-Dimethylethanolamine (DMEA)[b]	0.96

[a]Heat the first five components at 80°C until no further isocyanate can be detected. Cool to about 80°C, and add the DMEA and stir for 30 min.
[b]Union Carbide.
[c]IMC.
[d]Mobay Chemical Co.

cosity (the opposite of acrylic or vinyl hydrolyses) and last of all (at later stages), a change of pH.

The solvent-borne oil-based alkyds hold ample precedent for the waterborne oil-based alkyds to live up to as time progresses. The problems seen (such as the skinning over or dry-to-touch without through drying) occur in both systems, unless the catalysis is correctly formulated. Although I have not seen one, I would suspect a water-borne wrinkle finish could be formulated similar to solvent-borne wrinkle finishes. This might be useful for the machine coatings now using texture coatings.

Waterborne alkyd polymer dispersions perform well in metal-finishing coatings. They should, however, be handled carefully in terms of formulation. The initial formulation must be made especially careful to avoid all the potential problems. The changes in formulation after being accepted for metal finishing coatings will require much care in evaluation, as drastic failure could occur if not all the needed requirements are met in production. Encouragement comes from the Eckler description of the comparison of water-soluble air-drying alkyd formulations, as many would do well in coating applications [1].

II. POLYURETHANES

The well-known urethane chemistry to prepare polymers is based on the reactions of polyisocyanates with diols. The diols are based on polyethers (like ethylene oxide) or polyesters having a slight excess of glycol to result in all hydroxyl end groups. The "moisture cure" systems contain polyisocyanates that react with the moisture of the air to cure. These systems are all solvent-borne because of the reactivity of the isocyanate group with water.

Many years ago, some manufacturers offered urethane polymers as latexes.

These had been polymerized in solvent and then emulsified, with a solvent-stripping process leaving the polymer behind as a latex. They were thermoplastics, somewhat rubbery, and of modest strength, as compared to the latexes made from acrylic or styrene–diene sources. Other manufacturers offered di- or polyisocyanates based on fatty acids (very hydrophobic) that could be emulsified without the water attacking the isocyanate for six months or more. None of these proved very popular with the users. These products were made by converting the dimer or trimer acids to amides and the amides to isocyanates via Curtius, Lossen, or some such reaction. Emulsification was easy, mixing the oily product into a surfactant solution. However, the water permeating into the emulsion globule converted the isocyanate to amines or urea as time passed, and the isocyanate content decreased.

More recently, polyurethane hydrocolloids (not true latexes) have been available based on two advances. The first was the incorporation of pendent carboxylic acid groups on the backbone of the polymer to aid in the colloidal stabilization when neutralized in water with ammonia or amines. A typical formulation is given in Table 10.1 [2]. The second advance was the development of the blocking technology which prevents the attack of the water on the reactive cross-linking groups. This means that the polymer cures during (or just after) the drying process or upon the application of heat. Some will even cure at ambient temperatures upon drying. There are patents describing polyurethane based latexes [3,4].

The materials are supplied in aqueous dispersion, at near neutral pH. The incorporation of carboxylate amine salts gives stable noncoagulating colloids. However, these materials will be sensitive to all the same influences the other carboxylated polymers are sensitive to. Beware of drastic pH changes and of the effect of dissolved polyvalent cations (calcium from limestone, zinc from zinc phosphate or oxide, and others).

The advantage of the polyurethanes lie in their toughness and flexibility, and adhesion can be very good. They can be useful in clear or pastel colors, dried in ambient air. They are relatively foam free in formulation or application. We have been investigating them as replacements for solvent-borne urethanes over vacuum-metallized substrates. Free films cast on siliconized paper range from leathery to stiff, and from clear to hazy.

Disadvantages may lie in their wetting some substrates, but standard compounding (dispersant or surfactant additions) can solve this problem. We have seen crawling on some aluminum substrates, and even on Leneta charts. They may dry a more slowly than latex-based systems, but that is a consequence of their semi-soluble nature. They may develop some color on heated drying, though this varies from brand name to brand name, and from item to item within the brand name. Polyurethanes have always been suspected of hydrolytic instability, based on the urethane linkage and (mainly) the ester linkages from the diol. The polyether-based urethanes have better hydrolytic stability, but the oxide link in the polyether is an aid to moisture vapor transmission. Choose the most hydrophobic polyether you can for moisture vapor protection.

All in all, I have been impressed with these materials, and even look forward to seeing how useful they are in markets other than metal coating. They might supply excellent adhesives or paper coatings.

REFERENCES

1. P. E. Eckler, *Amer. Paint Ctgs. J. 37* (Feb. 1988).
2. Union Carbide, TONE Marketing Department, 1989 Presentation, San Francisco Airport, Westin Hotel.
3. S. L. Axelrood, US 3294724 (27 Dec. 1966).
4. D. C. Bartizal, US 3988278 (23 Feb. 1980).

11

By-Products in the Latex

I. INTRODUCTION

Although we have discussed the materials used for emulsion polymerization and the expected latex product, little mention has been made of by-products to be expected. The by-products are important from many standpoints. They can import odor to the latex or the coating or adhesive film laid down by the latex, and taste to paper or can coatings. They may act as plasticizers (even if only temporarily) that slow the drying of the film.

In recent years, the food packaging industry and the FDA have become aware of the potential for long-term toxicity, and therefore require characterizations of packaging materials. An assumption may be made that one part per million may not be harmful in the extracts of food packaging materials.

Analyses are seldom made specifically for by-products, except where a particular by-product is known to be harmful. It would be an excellent exercise for the collaborative analysts (such as the TAPPI or FSCT technical committees) and for academic researchers. In the main, the by-products are organic materials, easily isolated from the stripping or concentration distillate residue mixtures. Preparative chromatography and characterization by NMR and IR would be enough to qualitatively confirm the presence of by-products.

The treatment herein could be called speculative. We have isolated some polymerization by-products, and have identified others imply by odor from the latex. However, these materials should be identified to demonstrate their significance (or lack thereof) in the polymerization process and product. The treatment herein is organized by materials charged into the reaction vessel which may be responsible for the formation of the by-products.

II. INITIATOR REACTION BY-PRODUCTS

The initiators are reactive. The species that cleave homolytically are made to split to generate the free radicals needed. The simplest of them, the peroxides, form the peroxy radicals. Ideally, these radicals add to the monomer, and the growing polymer chain is on its way to polymerization. Some of these radicals are unstable, however. Carboxy-derived radicals are particularly well known to be unstable, losing carbon dioxide, as in reaction (11.1). The remaining radical

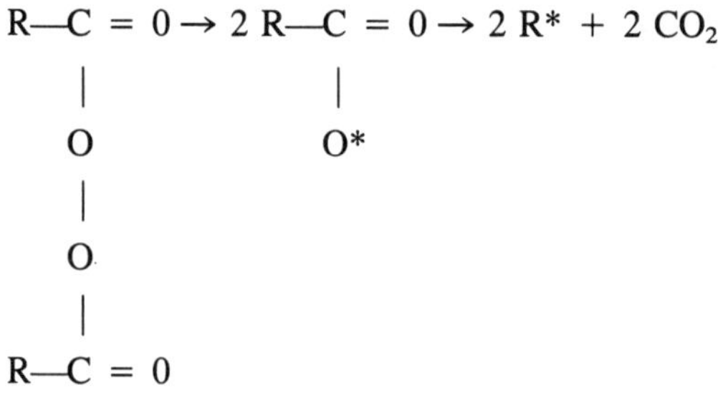

(11.1)

may yet add to monomer to form the growing polymer chain, but the carbon-based radicals are hot. From benzoyl peroxide, we expect benzene, biphenyl, and phenyl benzoate as the by-products, and I know of analyses showing benzene and phenyl benzoate resultant in commercial polymerizations. Confidentiality agreements prohibit further discussion. Similarly, from the alcohol-based peroxides, we expect ethers.

The azo initiators also generate by-products. In the telechelic polystyrene syntheses done at the University of Delaware, we found tetramethylsuccinonitrile (TMSN), isobutyronitrile (IBN), and possibly an unstable IBN–thioglycolate adduct in an azobisisobutyronitrile (AIBN)-initiated polymerization [1]. The IBN is a simple product of the IBN radical reacting in chain transfer, removing a hydrogen atom to become an inactive molecule. The TMSN is a "cage recombination" reaction product. The solvent cage around the IBN radicals (formed by AIBN decomposition and loss of a nitrogen molecule) inhibits their movement away from each other, and the two IBN radicals react to form the TMSN. Our purification of AIBN allowed us to observe that methacrylonitrile (MAN) was formed (and polymerized) in the AIBN recrystallization in methanol [2]. We argued for a chain-transfer cage reaction, combining the two IBN radicals into MAN and IBN, but alternative mechanisms may be proposed (e.g., chain transfer to initiator). The observation of MAN from AIBN has been confirmed by its incorporation into copolymers by subsequent investigators [3]. A ketimine product was identified with TMSN and other by-products from an AIBN decomposition study [4].

The peroxydisulfates yield bisulfate as their main by-product. The initiation reaction is about 50% efficient (if that), so that every 2 peroxydisulfate charged in the reaction, yields at least 1 g bisulfate. This promotes colloidal stabilization, but the bisulfate can also act as a corrosion catalyst in coatings over steel.

Since 50% (or less) efficiency applies to all initiators, one may look for benzoic acid–benzene–phenyl benzoate from the benzoyl peroxide initiations, and t-butanol from t-butylperoxy initiations. The redox initiators have two constituents that react to form one radical; the other reactant must remain in the reaction vessel in some modified form. Look for amine oxide as the by-product from the tertiary amine-induced decomposition of the peroxide. This is a side reaction of serious consequence for some formulations using a peroxide initiator and cationic monomer or emulsifier. To avoid the discoloration and high rate of polymerization with amine base monomers or emulsifiers, I prefer to use azo initiation.

Alternative decomposition routes should be a concern, too. The peroxidisulfate is a stable anion at neutral or alkaline pH, but in acid it can become the persulfate anion and an epoxidation reaction pathway can result. Equation 11.2 shows the conversion of the sulfur species to Caro's acid. The in situ formation of an epoxy species (from Caro's acid and monomer) might not have only

$$S_2O_8^{=} + H_3O^{+} \rightarrow HSO_4^{-} + HSO_5^{-} + H^{+} \tag{11.2}$$

drawbacks. The epoxy by-product could be useful if the copolymer has some chlorinated monomer incorporated, as low molecular weight epoxy species are commonly used to prevent attack on polymer that can dehydrohalogenate.

III. MONOMER-DERIVED BY-PRODUCTS

We have already noted the possibility of Diels-Alder self-additions of monomers. The main case in point is the 4-vinylcyclohexene so prevalent in hot butadiene emulsion polymerizations. The butadiene adducts of acrylonitrile, styrene, or acrylate esters [6] are also likely to occur. Hence, one should look for a 4-phenylcyclohexene, or a cyclohexene-4-nitrile or carboxylate ester in the organic strippings. The acrylics can form oxygen- or nitrogen-containing six-membered ring species.

Chain transfer to monomer can also generate low molecular weight by-product. Prior and Coco [5] found a styrene dimer that not only was a product of chain transfer, but had a high chain-transfer constant of its own.

The acid monomers also self-react, though not in a Diels-Alder reaction. The addition is more like a Michael addition, as the acrylic acid dimer is an acrylic ester of 3-hydroxypropionic acid [7]. This dimer, formed slowly in ambient storage of the acrylic acid, substantially reduces the polymerization rate.

IV. BY-PRODUCTS FROM CHAIN TRANSFER AGENTS

In our attempt to make a telechelic polystyrene, we found oils remaining in the liquid after isolation of the solid polymer. Analysis of these oils allowed us to isolate several species of mono- and distyrene adducts of methyl thioglycolate which we had seen on the gas chromatography traces in our attempts at kinetics studies [1]. These products, upon hydrolysis to the acid, showed some merit as corrosion inhibitors [6].

The mercaptan materials often used as chain-transfer agents are very reactive. They may be participants in the initiation reaction, as I have seen rate increases in SBR latexes made with peroxidisulfate when a long chain mercaptan is added. In any case, one can look for the disulfide from whatever mercaptan is used, and for the adduct to one or two monomers. In alkaline systems, the mercaptide anion can add across the double bond of a monomer as in the Michael addition, giving the monosulfide product.

The mercaptans are always suspect when odor problems arise. However, other problems can be just as likely. In a case of acrylic acid polymerization with thioglycolic acid as chain-transfer agent, color on storage in tanks was a problem. I suspect the dithiodiglycolic acid by-product of mercaptide linkings was chelating dissolved iron from the storage tanks; the problem was solved by using another chain-transfer agent.

Other materials include carbon tetrachloride, an excellent and inexpensive chain-transfer agent. If its radical can make the mono- and diadducts similar to those in the first paragraph of this section, one should be able to find them in the stripping organics. Carbon tetrachloride is essentially a tetrafunctional transfer agent. One could also find bridged dichlorides and many other species. Aqueous alkaline environment may hydrolyze the tri- or dichlorides to carboxylic acids, ketones, and aldehydes.

V. COLLOIDAL STABILIZERS

The preceding series of materials (initiators, monomers, chain-transfer agents) are relatively pure. All the additives for colloidal stabilization are industrial-grade products, and carry their own by-products including salts which normally do not react further in the polymerization; alcohols in the emulsifiers and cellulosic protective colloids, which may be subject to oxidation by peroxides; and, oils from emulsifiers, which may be unsaturated and could slow polymerization.

These products may be important to the end product, and you may never identify the responsible species. In a latex for textile printing inks, we found that the shipments from six suppliers of triethanolamine lauryl sulfate were not equivalent. All but one resulted in discoloration of the ink upon accelerated aging.

The product you are using for colloidal stabilization may be affected by an inherent equilibrium. We have noted the difficulty of the cationic surfactants with peroxides; it is not the quaternary amine that causes the problem. That quaternary amine is in equilibrium with a tertiary amine and the counterion acid; the tertiary amine is the species converted to amine oxide.

Other mixtures can cause difficulties. Mixed fatty acid emulsifiers, derived from natural products, have to be screened for their composition. The lower the iodine number, the better. Unusual unsaturations can slow or terminate the polymerization.

VI. CONCLUSION

The emulsion polymer may contain low molecular weight materials that can be a problem to the latex user. Certainly, one can look to processing to eliminate odorous and toxic substances; alternative formulations may solve the problem.

REFERENCES

1. R. D. Athey Jr., Ph. D. Dissertation, University of Delaware (1974).
2. R. D. Athey, Jr., *J. Poly. Sci., Polym. Chem. Ed., 15,* 1517 (1977).
3. W. H. Starnes, I. M. Plitz, F. C. Schilling, G. M. Villacorta, G. S. Park, and A. H. Saremi, *ACS Polym. Div. Preprints, 25(2),* 75 (1984).
4. M. Talat-Erben and A. N. Isferdiyaroglu, *Can. J. Chem., 37,* 1165 (1959).
5. W. A. Pryor and J. H. Coco, *Macromolecules, 3,* 500 (1970).
6. M. J. B. Carroll, K. Travis, and J. Noggle, *Corrosion, 31(4),* 123 (1975).
7. E. H. Riddle, *Monomeric Acrylic Esters,* Chap. VI, Reinhold Publ. Corp., New York (1954).

III

Analysis and Testing

I. THE NEED

A. Quality Assurance

An emulsion polymer supplier manufactures emulsion polymers by a polymerization process, which we have discussed briefly. The amount of each component of his formulation varies from batch to batch, which can make a difference in the final product properties. The manufacturer will commonly blend several batches to smooth out the variations. He will often blend in 5-10% of some off-specification material to unload it at a selling price, rather than at some lower "distressed-product" price, or even pay to have it hauled away. Hence, your tests as the product is delivered are your assurance that the material meets your specifications.

The emulsion polymer supplier employs processability and common property test specifications. He ordinarily will give the test results to the customer, but does not commonly use customer-based tests for his qualification of an emulsion polymer for shipment. Hence, your own tests for your application, which might relate to fiber–binder polymer adhesion, pigment–binder polymer adhesion, substrate–binder polymer adhesion, web tensile, or other properties could be deficient. Sometimes the emulsion polymer meets the producer specifications *but not yours*. The producer will often take the product back and sell to a less-demanding customer. Alternatively, the producer may offer to let you keep the material at a reduced price to make up for your need to reformulate.

Thus we see a variety of reasons for you to "qualify" an incoming emulsion polymer shipment, as self-protection against the production of an off-specification end product that may not bring the premium price desired. The price to be paid depends on the cost versus benefit. Obviously, you do not test the 30 gal of emulsion polymer received to make the 100 yd of web or 25 gal of paint to fill a once-a-year order. However, the delivery of the weekly tank truck might be very profitably tested frequently to ensure there are no mistakes in the shipment scheduled, or that the viscosity is not too high to pass a spray nozzle.

B. Product Development Information

Development of a new product is a challenge. You have a variety of candidate fibers and emulsion polymers to choose from to make a nonwoven web (or pigments and emulsion polymers to make a paper coating), and you build an internal library of results you have obtained in the past to guide you to your new products. A new polymer combination is usually needed to meet new specifications.

You can survey the literature to learn the experiences of emulsion manufacturers polymer and web and paint makers. Your own experience, combined with the experience of others, may be your guide. Commonly you find that no information source gives enough detail to make the task easy for you. I recommend making a tabulation of web or coating information, relating emulsion polymer and fiber–pigment properties. One can then better plan experimentation to fill in gaps and approach the desired new product property.

One must have adequate emulsion polymer information to ensure that the history referred to above is complete. One can retain web or emulsion polymer samples to obtain some of the information, but occasionally the web or coated substrate has discolored and stiffened. Old latex or coating samples may be moldy. Getting the information the first time available is somewhat easier, and eliminates the need for storage.

C. Process Development Information

The development of a new process is a pioneering effort, and consumes a large amount of time. The material used must be well characterized to ensure smooth operation. Viscosity, pH, glass-transition temperatures, and cure conditions must be considered.

The statistical design of experiments separates the important controlled variables from the insignificant, and measures precisely the progress of operation.

The mathematical model can be used to vary machinery, personnel, and materials. We have shown, for instance, that self-cross-linking latex systems are catalyzed equally well by a variety of latent acid catalysts and that the pKa of the acid is a factor [1]. Another study showed that press-cured samples have tensile strength as a function of press time and temperature, whereas applied pressure is not a significant factor [2]. In the same publication, we showed significant differences in web cures in press, forced air oven, and a waffle iron [3]. We have used the mathematical modeling technique to screen coating formulation components for effectiveness of corrosion protection and moisture resistance [4]. However, one must have a good data base characterizing emulsion polymer, fiber, and pigment to use these variables in process development modeling.

II. REPRISE

The testing of emulsion polymers has several driving forces. The need may be the simple assurance that the emulsion polymer will work in certain process. The need may be more complex, as in a product or process development problem. In any case, the testing of the emulsion polymer should be done only as needed on a cost-effective basis. The testing should balance the reasonable predictions of final web or coating properties with the emulsion polymer polymer properties as derived from both formulation and structure. Our survey of test methods should guide your own search through the appropriate literature for the methods you can use profitably. We suggest that the statements on accuracy and precision be used as guides until your own experience gives you a better picture.

REFERENCES

1. R. Athey and J. Wonnell, unpublished work, 1970.
2. R. D. Athey Jr., "Saturated Paper Characterization of Nonwoven Fabric emulsion polymers: Method Description and Characterization", *TAPPI, 60* 118 (1977).
3. R. D. Athey and C. C. DePugh, *TAPPI 60* Table III) (1977).
4. R. D. Athey Jr., "Latex Coating Formulation Evaluation of Organosilane Treated Talcs: A Statistically Designed Study", *J. Water Borne Coatings, 8(1)*, 7 (Feb. 1985) and *8(2)* 10 (May 1985).

12

Class-I Tests for Emulsion Polymer Systems

I. DEFINITION

There are many tests one may apply to latex systems, but only a few that should be done on all latex products received, whether they are raw uncompounded latex or a compounded formulation. These are the tests that quickly tell whether the supplier is providing what he says he is. They are simple and inexpensive and provide critical information about the utility of the latex before formulation or processing. There are standard methods recommended for these tests [1,2].

II. TESTS

A. Hydrogen Ion Concentration, pH

As you remember from your freshmen chemistry, the pH scale is a logarithmic representation of the hydrogen ion concentration. the pH affects a number of aqueous-formulation properties, such as the drying rate for carboxylated systems, corrosion of metal process equipment, cure in curable systems, and even mechanical stability or storage stability for the raw latex or formulations. The smell of vinegar from poly(vinyl acetate) suggests that someone did not store the latex at the correct pH. Jaquith [3] told me long ago " . . . that you have to play ball with the hydrogen ion if you are to succeed with any chemistry." That message has carried me through many problems over the years. The effectiveness of a water-repellent formulation containing chromium stearate and a wax emulsion was related to application pH on a paper laminate nonwoven, for instance [4].

The measurement of pH has been simplified in the past 30 years with the help of the glass electrode. It is a simple matter of dipping the electrode into the latex and taking the reading. However, there are caveats that are often ignored in the rush of laboratory operations. These pitfalls are:

1. The glass electrode *must* be kept wet when not in use, as the thin glass membrane at the bottom can be permanently damaged upon drying. This will lead to erroneous readings or no response at all, depending on the extent of damage. A source of deionized or distilled water, and a container in which the electrode may be stored, is the common recommendation.
2. The electrode must be thoroughly cleaned between readings, as one wants no contaminating ions to migrate thorough the wetted glass membrane and destroy the delicate electochemical measuring system. Again, a dedicated wash bottle at the instrument is recommended, and washing the electrode (even under any rubber sleeve that may be present) until no further opacity shows is the best technique.
3. The solution *inside* the electrode must be maintained, as dilution or drying out can cause error or malfunction.
4. The instrument should be standardized against a calibrated buffer no further than two full pH units from the samples before and after a series of measurments (against the buffer afterward guards against possible drifting in a series of measurements).

Dicker [5] presented a good detailed argument for the correct methods of calibrating, storing, and cleaning the pH meter and electrodes. Okamoto [10] describes rejuvenation of an electrode that becomes erratic when someone walks past, as the AgCl in it has plugged the reference junction and it acts like an antenna. Chematrix publishes a number of useful booklets, such as *pHault Finder*, *pH in Plain Language*, and *pH Control is Easy* [6]. There are standard test specifications for determining pH on styrene-butadiene (SB) latexes [7], on natural rubber latex [8], and general aqueous systems [9]. The instruments chosen from any number of suppliers may be used without much thought about their precision or accuracy. However, ruggedness of instruments and electrodes, and repair ratings should be considered when pricing these devices.

B. Viscometry

Viscosity is the measure of flow for a liquid system. The viscosity of a latex can be very complex, as the latex is a hetrophase system. One hopes that the viscosity measurement will give an indication of flow through piping, tanks, spray nozzles, and into the substrate upon coating or spraying, although the shear at these various processing points may differ. Unfortunately, the heterophase systems, especially when compounded, often give different viscosity measurements with different measuring devices, which often use different shear rates. This can create problems for the process engineer.

It would be useful to define several terms common in rheology that describe variation of viscosity as a function of shear rates, so that the user may understand why these measurement are needed to relate to different shear rates found in processing. The terms are:

Newtonian flow: Flow that maintains the same viscosity no matter what shear rate is applied.
Dilatant flow: The perceived viscosity increases with increased shear rate or shear thickening.
Pseudoplastic flow: The viscosity measured decreases with increased shear rate or shear thinning.
Rheopectic flow: The viscosity increases with time at constant shear.
Thixotropic flow: The viscosity detected decreases with time at constant shear.
Bingham plastic flow: The viscosity is too high to allow any flow until a critical point in increasing shear rate is reached and flow will occur at any higher shear rate.

The engineer or chemist operating in a lab or production facility probably knows instances of most of these viscosity phenomena from personal experience. However, detecting these phenomena and modeling them entails rheology, the study of flow. Our concern here is not the study of flow for characterization but viscometry (in the case of single measurements) or rheometry, and making those measurments useful.

It is beyond the scope of this review to discuss the details of viscometry, the basics can be found in a standard reference [10]. Several reviews touch on the variety of measuring instruments available to determine viscosity of fluids [11,12]. Collaborative testing has shown that the results obtained with just about any viscometer are highly operator dependent, and care should be exercised [13].

The most common instrument in use for latex viscosity measurement is the Brookfield device, specified in rubber industry standards [14,15]. It can be used with the expectation of about 10% precision [16]. It will indicate any variation of viscosity with shear rate (spindle speed) [17] as a warning sign to those who would model a high shear-rate process. Again, there are pitfalls in the use of this instrument. They are:

1. The spindles must be kept rigorously clean, as any deposit or corrosion of the surface will yield an erroneous reading.
2. The spindles must be handled very carefully, as surface damage or bending of the shank will cause error.
3. The instrument must be handled carefully while taking apart or putting together, as the spring indicator attached to the shank can be damaged by rough handling.

4. The instrument must be checked for leveling by looking at the bubble leveler prior to any measurement.
5. *Always* use the clutch while changing speeds, as the rotator gears may be damaged.

These devices are recommended for quality control laboratories and development work.

Another pair of instruments may be useful in quality assurance testing or in line-process monitoring. They are the Nametre and Dynatrol instruments, which apparently operate by measuring the damping of vibration generated by the instrument. They essentially measure at a single shear rate, and do not determine non-Newtonian behavior. However, some paper coaters use these to monitor the viscosity of a compounded latex fluid in process.

High shear devices are not commonly necessary (except for modeling high shear processes), and could be considered Class II tests (see Chapter 13). For convenience, they are described and recommended here. The ICI Cone and Plate is a simple instrument to use; it is relatively inexpensive and has different shear rate measurements to indicate non-Newtonian viscosity, but does not give indications of the Bingham plastic yield point. The Hercules High Shear Viscometer and the Haake Rotovisco continuous recording models will give demonstrations of all non-Newtonian flow behavior. These instruments have caveats associated with their usage, however. The cups and bobs and other liquid-contact surfaces must be kept scrupulously clean and undamaged by scratches or dents.

It would be wise, no matter what instrument is used, to maintain a viscosity standard for calibration of the instruments at reasonable time intervals (e.g., a Brookfield silicone oil). Remember that it is convenient to use the Brookfield instrument at two different speeds, and that will identify the Newtonian materials. The more sophisticated instruments may be fruitfully used on the more complex rheometry problems.

C. Total Solids Content

One may determine the water content in a system or the dry solids (nonvolatiles to the paint chemists) content. The total solids content (TSC) may be calculated by Eq. (12.1).

$$\begin{aligned} \%\text{TSC} &= 100 \, \frac{\text{dry sample weight}}{\text{wet sample weight}} \\ &= 100 - \text{percentage of water in sample} \end{aligned} \tag{12.1}$$

The simplest determination in concept is the gravimetric determination, that is, weighing a small amount of the latex (1–2g) into a disposable aluminum or plastic weighing pan and drying to constant weight. The plastic weighing pan is preferred for highly alkaline materials (some latex thickeners at high pH) or for microwave oven drying.

The sample may be dried by any one of several techniques. Infra-red lamps, 105°C forced-air ovens, 50°C vacuum ovens, and even hot plates are used. The standard for paints is 1 hr at 110 ± 5°C [18], whereas the rubber industry standard is 125 ± 5°C at 6in. Hg vacuum for 45–60 min [19]. Rather than a constant time, constant weight is recommended as the measure of dryness. There are pitfalls even in this. Some samples will attain a minimum weight, and start oxidizing and gain weight. Other samples will lose weight continuously. Indeed, a sample of a poly(propylene oxide) disappeared entirely during an overnight solids determination, as it is rapidly oxidized and degraded. Therefore, one must watch the sample carefully to ensure that the film does not char, degrade, or even spatter from boiling. The latter may be avoided by coagulating the sample with a volatile organic coagulant, such as methanol, isopropanol, acetone, or acetic acid. The coagulating solvent should be a nonsolvent for the polymer, so one obtains crumbs which dry without skinning over, bubbling, and spattering.

Microwave ovens are very good for this determination, as the microwaves are preferentially absorbed by the water in normal cases. However, if some coalescent cosolvent has been added to the system as a temporary plasticizer for the polymer, the results may differ between microwave oven and thermal oven. The coagulation solvents recommended above may also evaporate from the sample very slowly in a microwave system. Manufacturers of especially designed microwave drying ovens, maintain that you can ruin a standard kitchen microwave oven by heating a sample beyond the point where the water is gone. Apparently, the irradiation components burn up when the absorptions of the microwave are very small. Some companies have combined the microwave with some other type of heating (bottom surface?) to make sure that the organic volatiles will evaporate during the non-volatile determination.

There are a variety of instruments available to facilitate this determination by automation. An automatic taring system gives a microelectronic readout of percent nonvolatiles from a small sample in an infrared heated chamber with an air flow over the sample. In one instance, the chamber is microwave irradiated to drive off the water.

Other devices measure the water in a sample. The Brabender C-Aqua Tester converts the water to acetylene by reaction with calcium carbide and determines the acetylene. A Dupont instrument uses a phosphorus pentoxide absorption to measure the water content in a sample. These devices are expensive, however.

Another technique that involves reaction of the water to form a measurable species is the Karl Fisher titration. Completely automated instruments are available from Photovolt, Sybron-Brinkmann, Beckman, Cosa, and Fischer Scientific. The reagents are available from J. T. Baker, VWR Scientific, and Riedel-de

Haen (through Crescent Chemical Co.). The Karl Fisher determination is based on the conversion of an iodine–pyridine complex to the pyridinium–hydroiodide salt with the concomitant conversion of a sulfur dioxide-pyridine complex to a sulfur trioxide-pyridine complex. This is a redox reaction, and a number of chemical interferences can be due to the latex constituents. The actual determination is coulometric, as the instrument oxidizes the hydrogen iodide species back to iodine, and the number of amps required is converted to percent water in the sample. Iron salts (common redox initiation catalysis for low temperature latex syntheses) interfere, according to Italian investigators [20]. Consumption of iodine by any unsaturated materials by simple addition to the double bond, followed by facile dehydrohalogenation in a polar basic medium, could be an expected false high water value indication from a latex system containing residual monomer. The highly polar media (the cellulosic, protein, or PVOH protective colloids or the swollen hydrophilic latex particle surfaces themselves), which contain water may be expected to yield the water slowly, as they did with sucrose and other species for Yap et al. [21]. Since methanol is a common carrier for samples in the Karl Fisher reaction system, any aldehydes or ketones (from self-cross-linking comonomers in the latex formulation) in the sample may be converted to hemiacetals or acetals, with water elimination to give another sort of false high reading, albeit slowly. Given these possible or even likely interferences, and the necessity of handling the odoriferous reagent, I do not choose to use a Karl Fisher system unless no reasonable alternative exists.

The Praxis Corporation markets a nuclear magnetic resonance relaxation device that nondestructively determines the water-to-polymer ratio after calibration. Epsilon Meters determines the dielectric constant of a fluid, and claims the dielectric constant is sensitive enough to determine polymer to water ratio. Mapco Inc. suggest that their Nusonics Concentration analyzer could work, as sonic velocity in a fluid would depend on the concentration of dissolved or dispersed materials. Refractive index is offered by many supporters as a useful technique, and devices are available for lab and in-line plant usage. All these methods require a calibration curve for comparison, and each type or brand of latex will require a separate calibration curve. These techniques are reliable and fast for quality control.

Collaborative studies of three methods for waterborne paints could be instructive [22]. There comparison was made of a Dean-Stark trap distillation of water from a foaming sample to volumetrically estimate the water, a gas chromatography technique, and the Karl Fisher reagent titrations. The gas chromatography technique became an ASTM standard [23]. The Golden Gate Society preferred the Karl Fisher technique, as it is lower in equipment cost and easier to set up.

Precision in testing is difficult to evaluate. We recommend that gravimetric determinations be done in triplicate, and that the two or more that agree within 0.5% be averaged. For instance, determinations of 49.4, 49.8, and 48.8%, would have the two higher results averaged and the lower eliminated. The standard paint practice is 1.5% relative repeatability and 4.5% lab-to-lab variation expectation [24]. In some data developed in the Los Angeles Society for Coating Technology viscometry study [13], we found that the precision results depended upon the actual total solids content, with 67.3 ± 1.7% at one end of the scale and 1.8 ± 0.5% at the lower end of the range of fluids we tested. A number of B. F. Goodrich studies [25] have shown reproducibility around 0.5%, and that microwave ovens are just as good as (if not better than) hot plates and ovens.

D. Surface Tension

The surface tension of polymer fluids or dispersions is one of the most often neglected properties for quality or development testing programs. This property is a directly measurable thermodynamic quantity that governs wetting and adhesion. It is also a quick measure of the degree of saturation of the particulate dispersion by whatever surfactants are compounded therein. It is therefore important to product development and quality assurance on the latex coming into the plant, or on the formulation used in the nonwoven production process.

The importance of surface tension to development work or troubleshooting lies in the relation of surface tension to adhesion. Mark [26] relates surface tension and cohesive energy to polymer structure. Owens [27] relates the work of adhesion (the sum of all the surface free energies of the interfaces, air, polymer film, solution) to whether a coating on polypropylene would adhere. A chapter in a recent book on web processing dealing with corona treatment of webs [28] mentions a variety of methods for characterizing the surface tension of films, with an eye to improving ink adhesion to hydrophobic surfaces and demonstrating the effectiveness of the surface treatments. The nonwovens developer could profitably borrow or modify those techniques to ensure that the emulsion polymer will seize the fibers. Heitcamp [29] relates the ease of spraying a formulation not to viscosity (as many would think), but to the surface tension.

Measurement of surface tension is relatively easy. The standard undergraduate laboratory tests of surface tension include capillary rise, drop size or weight, and, the DeNouy Tensiometer; college texts discuss the principles. The quickest techniques to set up in a laboratory are the capillary rise and drop weight measurement. However, these depend largely on the cleanliness of the glassware and the precision of the height measurement (for the former). The easiest instrument to use for purposes of repetition is the DeNouy Tensiometer, a device that measures the force needed to pull a wire of known length out of a fluid.

The force, in grams required, is converted to dynes/cm as the standard unit. Pure water has a surface tension of about 72 dynes/cm, while a surfactant solution may have a tension of 25–40 dynes/cm.

There are commercial devices to measure surface tension. The DeNouy Tensiometer is available from your lab supply house (e.g., Fisher Scientific), at $800–$1000, it is covered by a rubber industry standard [30]. Alternatives include:

1. the SENSADYNE bubble pressure measurement device;
2. the Rosano device (like the DeNouy, based on disk to be pulled up from the fluid surface), available from VWR;
3. Tekmar's droplet diameter measurement device (*Expensive!*); and,
4. contact-angle measurement devices from Kayness, Rame'-hart, Kernco, and Imass.

Contact angle is a frequently attempted measure of surface tension. Byrne, Ross, and Roberts [31] go into details of contact angle relation to surface tension. Wetting angle determinations for webs [32] and for paper single fibers [33] or for single textile fibers [34,35] have been recorded. Huntsberger [36] offers a simplified interpretation of the diverse opinions relating contact angles, surface energetics, and adhesion. Bragole [37] points out the difficulties in adhering to nonpolar surfaces (e.g., poly(propylene) like HERCULON) and how contact angle indicates the effectiveness of surface treatments by improvement in adhesion. The dynamics of the process has received some emphasis from Schwartz [38], and Cherry and Holmes [39]. Single textile fiber measurements of contact angle offer a technique problem to the investigators, in that the fiber and wetting liquid surface orientations have been found important by Dupont workers [40]. The advice I offer from my own work is that the droplet system surface drying rate and the viscosity (and thus crawl rate due to gravimetric compression) may be factors that the experimentalist should be concerned about.

The preferred surface tension measurement technique is the DeNouy ring method, whether using the commercially available tensiometer, or a simple adaptation of the Instron tensile tester developed by Jackman at Mellon Institute [41]. Cautions on use of these methods should emphasize the handling of the ring. It must be scrupulously cleaned (burning with a Meker burner removes the organics, but the salts of a latex will deposit in time, so washing with water is also desirable). Gentle handling guards the geometry of the ring, which must remain planar and not lose circumference due to bends. The viscosity of the fluid is important in a listing of caveats, as it controls the approach to equilibrium contact angle or capillary rise. Viscosity also controls the drainage from DeNouy Tensiometer ring. In cases where the viscosity was very high (several thousand centipoise, as in a foam-applied latex emulsion polymer), it is a good idea to dilute 10:1 to obtain a relative measurement between variations in formulation.

E. Coagulum

Ideally the latex has all its solids in the form of particles in the range of 1000 to 3000 Å (100-300 nm) in size. These are small enough to cause no real problems in settling, and creaming, or clogging pipes or pumps. However, there are larger particles present to some measurable degree. These may be pieces of skins or edge-dried particles that have been broken up, or they may be evidence of some high-shear pumping action in the process or shipment, or they may just be evidence of the metastability of the particular latex in question. Hence, a major recommendation is that all latexes be filtered upon receipt and when used, if for no other reason than to remove debris and impurities.

The coagulum one tests for is not one of those unforseen phenomena referred to above, but the micro-percentage of particles one expects to find in a latex larger than some reasonable sieve size. As a manufacturer, one could expect a good polymerization formulation to yield 0.1–0.5% coagulum (or prefloc) in processing. The user should expect something similar, and can derive some quality assurance information to aid his supplier in describing the shipping process through this test. The 0.5% looks small, but amounts to significant cost in a 20,000-gal shipment.

In addition, the process engineer will want to know that all the solids are passing through the tip of the spray nozzle. He will not want any particles larger than the orifice, and would like assurance that the particles are smaller than the hole by 50% or so, to obtain a regular pattern.

Coagulum determinations are generally done on a 100-g lot, and are a simple gravimetric determination of that percent retained on a 100-mesh (or your defined mesh size) screen. I prefer a conical screen of stainless steel made from a circle about 6-8 in. in diameter. This can be put in a ring stand and the latex sample poured through quickly. Dilution with water will speed up the process for viscous latexes, provided the dilution does not cause coagulation (which can happen with low surfactant latexes). The screen is preweighed (dry), and the coagulum collected. One should rinse the coagulum and screen with water to reduce adhesion during drying. If there is a significant quantity of coagulum, it should be isolated and washed to remove occluded latex. The coagulum is dried overnight (on the screen, if there is not too much) at 105–110°C, and weighed to calculate the percentage on the original latex. The percentage reported should be on a solids/solids basis. The screen is readily cleaned by burning the residues off with a Meker burner (in a hood, for smoky, offensive nitrile- and styrene-based polymers). Occasional skin formation is of no concern but should be reported if common. One can prefilter with a 35 or 8 mesh to remove the skins, if they are rare. Your supplier will probably provide the recommended procedure. Again, there are rubber industry standards for determining coagulum in a latex [42,43]. The ASTM D1 committee is doing experimentation to devise an easy, simple, reproducible method. Some early work with that group suggested that the latexes used for the round robin may have been metastable, with coagulum forming continuously giving widely scattered results.

The size of the screen is not of prime importance. I remember a case where a blade-coated system required quick passage through a 325-mesh screen (44 μm or 440,000Å). Some latexes passed smoothly, and others very slowly, leaving little residue. The feeling was that the slow ones had coagulum "extruding" through the mesh. We checked commercial latexes from several suppliers, to find that all had microflocs at about the same concentration by Coulter Counter, but the 325-mesh screening extruded some microflocs faster than others. We were never able to establish that the swift passage through the 325-mesh screen was connected with processing.

In spite of the above experience, the determination of coagulum content can be important. Instances in my work have shown that an occasional material generates coagulum slowly over a period of time, and that it is worth while to both filter and monitor to keep sensitive processes clean.

REFERENCES

1. *Testing of Adhesives*, TAPPI Mongraph, TAPPI Press, 1963 and 1976.
2. ASTM D1076 and D1417.
3. R. Jaiquith, University of Maryland, private communication.
4. D. York, private communication.
5. D. H. Dicker, "The Laboratory pH Meter," *Am. Laboratory*, p. 73 (Feb. 1969).
6. S. Okamoto, "Caring for Your Electrodes," *Chematrix Quarterly, 2,* 1 (May 1984). (Chematrix Inc., Hillsboro, OR).
7. ASTM D1417-75, paragraph 6.
8. ASTM D1076, paragraph 15.
9. ASTM E70.
10. R. B. Bird, W. E. Stewart, and E. N. Lightfoot, *Transport Phenomena,* Wiley, New York, 1960.
11. J. R. van Wazar, J. W. Lyons, K. Y. Kim, and R. E. Colwell, *Viscosity and Flow Measurements*, Interscience, New York, 1963.
12. R. D. Athey Jr., "Flow Properties Make All The Difference," *Chemtech,* p. 308 (May 1981). Note that there are errors in the designation of units in the equation on p. 308, and the axes are reversed on Figs. 2b and 5.
13. P. Shaw, "Viscometry, Myth and Reality,“ 1982 FSCT presentation for the Los Angeles Society for Coating Technology Technical Committee, *J. Coating Technol., 56(718),* 59 (Nov. 1984).
14. ASTM D1417-75, paragraph 8.
15. ASTM D1076-77, paragraph 11.
16. P. A. Shaw, *op cit.,* noted this was the ASTM statement in 1979, withdrawn in 1982!

17. ASTM D2196-81.
18. ASTM D2369-81.
19. ASTM D1417-75, paragraph 4.
20. G. Serrini and G. Serrini-Lanza, "Water Determination in the Presence of Iron Chlorides: Critical Examination of Various Methods," in Italian, Commission of the European Communities, Luxembourg, 1981; Cf:NTIS PB82-147745 (1982).
21. W. T. Yap et al., "Estimation of Water Content by Kinetic Method in the Karl Fisher Titration," *Anal. Chem., 51,* 1595 (1979).
22. Golden Gate Society for Coatings Technology, "Comparison of Methods to Determine Water content of Emulsion Paints," *J. Ctgs. Technol., 51(659),* 45 (1979).
23. ASTM D3792-79.
24. ASTM D2369.
25. Unpublished work by D. J. Keller and others, B. F. Goodrich Chemical, Avon Lake, OH.
26. H. F. Mark, "Cohesive and Adhesive Strength of Polymers," *Adhesives Age,* p. 35 (July 1979).
27. D. K. Owens, "Some Thermodynamic Aspects of Polymer Adhesion," *J. Appl. Poly. Sci., 14,* 1725 (1970).
28. W. M. Collins, "Surface Treatment," Chap. *14, Web Processing and Converting Technology and Equipment,* D. Satas, ed., Van Nostrand Reinhold, New York (1984).
29. "Surface Tension 'tells WB And HS Sprayability'," *Industrial Finishing,* p. 53 (May 1979), reporting a talk given by A. Heitcamp at the 1979 Water Borne and Higher Solids Coating Symposium.
30. ASTM D1417-75, Paragraph 7.
31. K. M. Byrne, J. R. H. Ross, and M. W. Roberts, "Contact Angle Studies of Polymer Surfaces" Chap. 2, *Adhesion* 1, K. W. Allen, ed., Applied Science Publishers Ltd., London, 1980.
32. L. Agbezuge and F. Wieloch, "Estimation of Interfacial Tension Components for Liquid-solid Systems from Contact Angle Measurements," *J. Appl. Polymer Sci., 27,* 271 (1982).
33. J. H. Klungness, "Measuring the Wetting Angle and Perimeter of Single Wood Pulp Fibers: A Modified Method," *TAPPI, 64(12),* 131 (1981).
34. T. H. Grindstaff, "A Simple Apparatus and Technique for Contact-Angle Measurements on Smaller Denier Single Fibers," *Text. Res. J.,* p. 958 (Apr. 1967).
35. J.-I. Yamaki and Y. Katayama, "New Method of Determining Contact Angle Between Monofilament and Liquid.," *J. Appl. Polym. Sci., 19,* 2897 (1978).
36. J. R. Huntsberger, "Interafacial Energies, Contact Angles, and Adhesion," *Adhesives Age,* p. 23 (Dec. 1978).
37. R. A. Bragole, "Factors Affecting the Adhesion of Paints to Non-Polar Plastics and Elastomers," *J. Elastomers Plastics, 6,* 213 (July 1974).
38. A. M. Schwartz, "The Dynamics of Contact Angle Phenomena," *Adv. Colloid Polym. Sci., 4,* 349 (1975).
39. B. W. Cherry and C. M. Holmes, "Kinetics of Wetting Surfaces By Polymers," *J. Colloid Interface Sci., 29,* 174 (1969).

40. W. C. Manning, "A Common Error in Contact-Angle Measurements on Fibers Using the Level-Surface Method," *Textile Research J., 19,* 760 (1975).
41. V. Jackman, "Special Techniques Used in the Study of Material Properties," Instron Application Series X-3, Instron Corp., Canton, MA.
42. ASTM D1417-75, paragraph 9.
43. ASTM D1076, paragraph 13.

13

Class-II Tests for Special Problems

I. DEFINITION

Class-II tests include special tests and analyses that are time-consuming, use expensive equipment or personnel, and are seldom needed for quality assurance. These specialty tests are used primarily for product or process development in cases where the properties measured may be related to the process or a product problem.

II. TESTS

A. Particle Size

The common range of particle size of latex systems is 1000–4000 Å, but one can make 10,000-- or 500-Å latexes if needed. Carmichael [1] showed that latex particle size made a difference in paper-coating properties, and I know of similar earlier unpublished work by E. Witt and D. S. Heller. We suspect that particle size may be a possible factor in observed differences in saturation pickups at identical total solids content (TSC) in identical web [2]. Certainly the Einstein equation relating viscosity to hydrodynamic volume fraction of dispersed particles suggests that one should be concerned with particle sizes of all constituents in a formulation where the viscosity may a controlling factor in processing or product properties. Paint companies often blend pigments and latexes of different particle sizes to minimize the viscosity, a practical necessity at high solids content based on geometrical considerations developed by Farris [3].

Part of the problem of particle-size determinations is that there is normally a broad distribution of particle sizes in a latex, unless it is specifically formulated (during the polymerization) to produce a narrow range. Description of these particle sizes can be a simple average (one number), or a description of the distribution of sizes (requiring several numbers). Smith and Jackson [4] discuss how to gather and display particle-size data normally found in latex systems. Just as in molecular weight measurements, the particle size distributions may be number averages or weight averages, but there are specific other averages, such as the surface average. When the particle size distribution is narrow, these averages are just small numbers apart, and it does not matter which is reported. However, when the distributions are broader, the specific properties of the latex may be affected (e.g., the turbidity is most affected by the larger particles).

The measurements of particle size is an art requiring considerable technique and dedication by the practitioners, who tend to be specialists. The work demands time and concentration on detail, as well as facility with numbers, and more recently, a computer. There are many ways one may choose to measure particle size of a latex, and the way chosen may bias the result. For instance, in a collaborative study to find a suitable technique for quality control in a latex manufacturing facility, we found that carboxylated latexes (containing less than 5% of the charged acid comonomer, very much like the commercial latexes in use today) give a result about 20-30% lower or by electron microscopy than with turbidimetry, disk centrifuge photosedimentometry, or hydrodynamic chromatography. Our explanation for the finding was that the three latter methods measure the true hydrodynamic volume of the wet particles, while the dried particles shadowed by some technique were measured by the electron microscope [5].

Table 13.1 summarizes the methods along with some of their advantages and drawbacks. Several good reviews of particle size measurement have been published [6–9]. We will only briefly discuss some of the techniques herein, in order of their relative expense for equipment and personnel.

The easiest method is called soap titration. It is described in detail by Maron et al. [10–14] and extended by others [15,16]. The method assumes that no surface-active materials are present on the latex particles, and that you know the area covered by the surfactant molecule. You simply measure the surface tension of the weighed latex sample (of which you know the total solids), and titrate the sample with surfactant until you reach a minimum surface tension (four or five readings at the same low surface tension). You then determine the critical micelle concentration (CMC) for your surfactant by titrating the surfactant into pure water until you get the same low surface tension that you had in the latex sample. The difference in the surfactant concentration used to saturate the latex and the CMC is the measure of the surface area, as the number of molecules used to cover that area times the area per surfactant molecule is the total area.

Table 13.1 Latex Particle Size Measurement Techniques[a]

Measurement Method	Capital Expense	Technologist Expense	Time to Run	Results, Dist'n or Avg.
Light scattering	Medium to high	High	High	Distribution, if particle shape is known
Turbidity	Low	Low to medium	Low	Special average
Hydrodynamic chromatography	Medium to low	Medium to low	Medium to low	Distribution, poor resolution
Disc centrifuge	Medium	Medium	Low to medium	Distribution
Electron microscopy	High	High	High	Distribution, if enough counted
"Soap" titration	Lowest	Lowest	Lowest	Surface average

[a]*Source:* R.D. Athey, Jr., *Met. Fin.*, p. 46 (March, 1986).

The conversion of area to radius is a simple geometric derivation, and the particle size obtained is the surface average. Caveats do exist. The choice of surfactant is critical (a co-worker showed me experiments where there were two and three minima in steps on some soap titrations [17], and the cleaning of a latex of surface active species prior to the determination can destabilize and even coagulate the latex. Collins et al. [18] warn that you must be sure that there are not two surface-active species present to compete, and the recent discussion of competition by adsorbed species by Stenius et al. [19] documents the displacements of surfactants and polyphosphate dispersants from latex and clay surfaces. The technique is also a simple qualitative tool to determine whether the latex particles are saturated with surfactant, which we used to solve a mechanical stability problem experienced with an acrylic latex in processing [20].

Turbidimetry is an inexpensive technique, requiring little time for sample preparation and determinations, and can be done with the simplest of instruments (e.g., the Spectronic 20). Indeed, for quality assurance, turbidimetry is a very good fast tool, when the care needed for good accuracy and precision is exercised. For this method, one must know the total solids content very accurately, and a calibration curve is needed for the calculation of particle size (a quantity close to, but not exactly equal to the weight average particle size). The refractive index of the particle is needed for the initial setup of the particle size calculations, and one should be aware that pH adjustments and changes in the polymerization formulation or latex blends can affect the refractive index. Collins et al. [21] discuss other possible problems, and give a simplified operating procedure for the determination. A basic reference on the technique was written by Kerker [22] and the publications of Josip Krathovil at Clarkson should be reviewed.

Another relatively easy technique developed within the past few years is hydrodynamic chromatography (HDC). The technique was first announced by Small [23], and pursued by Vanderhoff's group at the Lehigh Emulsion Polymer Institute [24].

Hydrodynamic chromatography is a size-exclusion technique, and is fast enough to analyze a sample in an hour of elution. Ten to twelve samples per day can be analyzed, as one can stack the samples on the column without the risk of mixing. A sample of known latex particle sizes is analyzed for calibration purposes. The detectors are either UV absorbtion flow cells or refractive index flow cells. The elution time is estimated as the time span between the absorbtion of a marker salt (which elutes first, a chromate salt, perhaps) and the later latex absorptions.

Although there was no equipment specifically designed for this technique, a standard liquid or gel permeation chromatograph could have been easily adapted by appropriate choice of columns and packings. The cost of the initial instrument used by the Lehigh Emulsion Polymer Institute group was between $5000 and 7000; it was assembled from bits and pieces. More recently, MicroMeritics licensed the technology from Dow Chemical and has been producing the devices for two or three years.

Critiques of the technique concern the relative insensitivity to blends of particle sizes very similar in size, the lack of basic principles, and the lack of agreement of results with other measurement or mathematical systems describing particle size distributions. The method is practical and easy to use. It can be relatively inexpensive to set up, compared to others, and is simple to maintain and operate. It gives measures of the distribution of particle sizes. The theoreticians will adequately explain it in time to our more rigorous colleagues, as was the case for gel permeation chromatography for molecular weights in the early 1960s.

A relatively inexpensive system for a variety of materials is a disk centrifuge with a photometer attached. The photometer is the detector for monitoring particle movement in the field by turbidity measurement with time at a particular point on the transparent disk. The results obtained are a measure of the distribution of particle sizes. Although it may be desirable to use standards to obtain a calibration curve, one may use Stokes[1] law to derive the particle sizes from first principles. Joyce-Loebl offers this device commercially; it requires that the particles' refractive index and density be different from those of the centrifuging dispersion medium. Some adjustments can be made on the latex for measurement, e.g., cesium salts can be added to the water to increase density, or an organic solvent miscible with water to reduce the density, preferably an organic solvent not imbibed by the polymer particles. However, such additions may destabilize the latex, and a nonionic surfactant may be needed. The technique has advantages and disadvantages listed by Collins et al. [25]; the discussion by Provder and Holsworth [26] may also be useful. This technique may be applied to latexes containing fillers and pigments, and is often used in large paint industry laboratories.

The classical techniques for particle size analysis are electron microscopy, light scattering, and ultracentrifugation. Optical microscopy will not resolve the individual latex particles to be seen, and the electron microscope must be chosen if that is desired. The relative expense of this instrument and the operating personnel required can be a serious hindrance to usage by the industrial purchaser of emulsion polymers. Table 13.1 and the review by Collins et al. [27] are sufficient coverage for our concerns. Nicomp Instruments has recently introduced an instrument that is purported to yield a measure of particle-size distribution by a computer-monitored light-scattering technique.

Another instrument measuring latex particle sizes is the classic Coulter Counter. This device has been used for many years in medical laboratory work on blood cells, and less frequently in pigment labs for particle sizing. It has the capability of determining distributions and is relatively fast. The underlying principle is that a dilute suspension of particles, in a fluid of known electrical characteristics, is passed through a capillary that monitors the change in these electrical characteristics. The volume of fluid displaced by the particle reduces the measurement indication, so one reads the particle size as an equivalent sphere volume. That is not a bad assumption for latex particles, but agglomerates should give skewed readings. The problem with aggregated latex particles mentioned in the section on coagulum determination in Chap. 12 was investigated by Coulter Counter, which in those days did not measure any particle less than a micrometer. Their current claim is down to 3 mm (30Å). Pitfalls in the use of the Coulter Counter would include agglomeration of latex particles due to the salt needed in the water (for the electrical detection) or the dilution water; the capillary must remain free of agglomerates.

Kinsman [28] claims that the measurement of nonwoven binder latex particles can be done at the rate of 3000 per minute in the Coulter Electronics device. Another such instrument, called the Elzone, is available from Particle Data Inc. Franczek [29] used a multichannel Coulter Counter with varying orifices to characterize flocculations of paper fiber fines, Avicel (a rayon shaft fiber used in paper making), and clay by a variety of salts and organic cationic coagulants (see section on critical coagulation concentrations).

As the most cost-effective approach, we recommend that these particle size measurements be purchased by the industrial emulsion polymer purchaser from an external laboratory service. Having participated in a collaborative study of particle size measurement, I can assure all that the results are of similar precision from external labs compared to the emulsion polymer manufacturer internal labs staffed with graduate-degree professionals. The check of precision and accuracy in particle-size measurements by the laboratory doing the analysis (done externally by another laboratory or internally by your own) may be done with samples of known particle size, such as the standards available from Dow Chemical, Duke Scientific Corporation, or Polysciences, Inc. The small investment in the few particle-size determinations for the process modeling is worth while, so the recommended particle size measurements needed for process or product quality assurance or control may be specified based on an economic and technical basis.

B. Mechanical Stability

The mechanical stability of a latex is its resistance to particle agglomeration by high shear motions. Although the manufacturers of latexes recommend against gear pumps, one can imagine two gears trapping individual latex particles between the teeth and mashing them together to form agglomerates. Latex handling systems are designed to minimize the mechanical stresses placed on the particles. However, the resistance of latexes to mechanical stresses differs, according to compositions, additives and total solids contents.

Measurement of mechanical stability is difficult, as one should limit the test to the maximum stress the latex would experience in processing. However, this is generally not known and some standard tests have been developed to give some indications to the user. However, the measurements indicate when we need to add something to build up the mechanical stability; typically, a charged surfactant is added according to the guidance offered by the latex makers.

The simplest of the mechanical stability tests is the rub test. One puts a small amount (one half-inch to one quarter-inch in diameter is sufficient) of the latex or formulation in the palm of one hand, and rubs it in a circular or back-and-forth motion until it dries or coagulates. One counts the cycles of rubbing until coagulation. I stop after about ten rubs and call it rub stable at that point. The test is fairly repeatable, provided one starts with surfactant- or perspiration-free hands. This test only indicates, by detection of a low number of rubs, that a mechanical stability problem in processing may occur. Passing this test does not guarantee that the latex will process well.

Attempts to mechanize this test for standardization have led to practices such as stirring with high shear mixers for a standard time, and measuring the amount of coagulum generated as a percent of the initial sample. The Hamilton Beach milkshake mixer is specified indirectly in one rubber industry standard [30]. One can as easily use the Waring blender for comparisons within your facility alone. The time of shearing and the weight percent coagulum are normally reported. I would caution that long mixing times (30 min.) will generate a temperature rise from Joule heating, but there are water-cooled Waring blender jackets available that will reduce or eliminate this problem. We found the coagulum generated in an acrylic self-cross-linking binder did not vary, when we compared the cooled determination to one that rose to 180°F in the half hour of shearing [31].

Other attempts to find a shearing device more perfectly modeling the nonwovens industry usage have been an aim of the TAPPI Nonwoven Binders Committee. Some feel it desirable to see a pad-bath roll test in miniature. The ASTM D-1 committee on paint could parallel that work by testing the coagulum generated in paint dispersers, sand mills, and pebble mills. We look for the committee to recommend standards in the future.

C. Foaming

Foaming of latexes or their formulations can be a processing problem, or the aim of the process. Foaming in a pad bath, a size-press coater, or a spray system can generate webs or coatings with some binder-poor areas that do not meet specifications for weight, coverage, opacity, or strength. However, when a foamed coating is specified, the foam density controls the amount of binder added per square yard, and without control of the foam density, the weight variation per square yard can be lower or higher than specified. Foaming techniques are common for application of the adhesive systems to laminated carpet systems and some types of nonwoven fabrics.

There are several distinctions between frothing and foaming latex formulations for application to a web, but the best distinction to my mind lies in the fact that the froth is supposed to collapse, leaving binder-starved areas, while the foam is supposed to maintain air bubble integrity upon drying, allowing no air passage upon cure (where a dried froth would). Characterization of foamability poses a problem, and I know of no standard tests, though the TAPPI Nonwoven Binder Committee was trying to develop one a few years ago for help in antifoam–defoam applications. Nothing has been published on the product or process development methods for foamability testing for correlation to foam or froth application methods. One would hope the antifoam–defoamer suppliers might standardize a test for their customers.

Foaming can be characterized two ways in any test. The first is the amount of foam generated at a specific time in the foaming device (or at equilibrium maximum with time unlimited), and the second is the foam-decay rate. In cases where the foam is undesirable, the former is a measure of the efficiency of the antifoam agents, and the latter is a measure of the defoamers' efficiency. Plotting a physical measure of foaming versus time for a test or series of tests gives good indications for comparisons.

One test can be classified as the foam height test. A number of ways can be used to generate the foam. The sample may have air (or some other gas) blown into it through some subsurface nozzle(s) or frit(s), or it may be whipped as in a Waring or Hamilton Beach blender. The former method might be difficult to keep as scrupulously clean, as necessary from sample to sample. The latter method subjects the formulation to a higher mechanical shear which might be harmful. I prefer a method under investigation by the aforementioned TAPPI committee. It uses the Red Devil Paint Shaker to mix sample with air, and produces less shear on the latex particles, and does not heat up at long mixing times. Although there is no standard, one can operate with the same method for some time and gather sufficient data to make fair comparisons and draw conclusions about formulations. A description of my technique follows:

Use one-pint glass jars covered with poly(propylene) netting (to guard against the breakage), add 25 mL of formulation, and measure the initial height of liquid from the bottom of jar, and record that height. Place the sample on the shaker, making sure the lid is on tight and rubber matting covers the shaker jaws in contact with the jar. Shake for 5 min, stop, and record the height of liquid and foam from the bottom of the jar. Shake for 5 more min, and remeasure and record; Repeat above step four more times. Set jar on level surface, and record foam height every 5 min for one-half hour. Subtract initial formulation height from all subsequent measurements, and divide resultant differences by initial formulation height to obtain the foam fraction. Multiply the foam fraction by 100 to get percent foam.

Plots of percent foam versus time give an indication of the dynamics of foaming and foam decay. We observed the addition of a wax emulsion to a coating doubling the foam generated and extending the foam decay more than twofold because of the surfactants in the wax dispersion [32].

Another measure of foamability originated in the foamed carpet-backing industry, and may be useful to those who apply a foam to the nonwoven web for bonding. One uses a relative low shear mixer, such as the gallon-size cake mixers used in restaurants, and a sample perhaps of 1-4L in volume. A sample of the formulation is weighed in a tared 25 mL beaker or cup, and returned to the mixer pot. Mixing is started and a sample taken every 5 min to be weighed in the same size cup, and returned to the mixer. The mixer is turned off, and the foam allowed to decay with sampling every five minutes, but now the samples should be discarded. These cup weights (after subtracting the empty weight of the cup) represent formulation plus air after the mixing has started. One can calculate the air–formulation ratio with ease. Plotting this ratio or the percentage air incorporated as a function of time to foam or decay, may yield useful processing information.

D. Critical Coagulation Concentration

This determination is based on the observation by early colloid scientists that ionic materials coagulated colloidal dispersions. The observations were formalized in the Schulz-Hardy rule, which states that a coagulation of a colloid by a salt depends on both the concentration of the salt and its formal charge. Thus, 1*M* sodium chloride may coagulate a latex, whereas 0.01*M* calcium chloride may suffice, and 0.0001*M* alum will likely suffice, all for the same latex.

The practical implication of the sensitivity of latexes to polyvalent cation coagulation suggests that any dissolved iron from corroding pipes or tanks could destabilize the latex. Other sources of salts should be concerns, such as formulation with zinc oxide curatives, calcium carbonate fillers [33] and even flame-retardant salts like ammonium phosphate or sulfamate. The critical coagulation concentration (CCC) indicates whether the formulation needs additional stabilization (e.g., by nonionic surfactant).

There are reports on the stability of latexes in the presence of polyvalent cations. Matijevic and Force [34] describe the pH sensitivity of the CCC using aluminum ions. Greene [35] showed that copolymerized acid comonomers helped protect against salt coagulations, with increasing acid comonomer content increasing the salt concentrations necessary to coagulate the latex. Ono et al. [36] showed that CCC depended upon the styrene–acrylonitrile ratio, with decreasing stability toward salts at increased acrylonitrile content. Nieman et al. [37] found that the CCC dependence on pH could be correlated with the surfactant stability at that pH.

The techniques for determining CCC are relatively simple identifications of the concentration of salt that causes every interparticle collision to lead to agglomeration. One of the best descriptions of the technique was by van den Esker and Pieper [38]. The experimenter sets up a sequence of test tubes (I use about five 2-oz. jars with lids) and proceeds to make up the following formulations.

Constituent	A	B	C	D	E
Latex, mL at 0.5% TSC	20	20	20	20	20
Water	20	18	16	14	12
Salt solution	10	12	14	16	18

The procedure is as follows: Prepare latex at 0.5% TSC; place 20 mL latex in containers; add prescribed amount of water and swirl gently; simultaneously start stopwatch and add metal ion solution with swirling; observe swirling mixture until it shows chunks of coagulum or increases in opacity (whichever is easier to see), and stop watch when that event occurs; and, finally record the time on a chart that contains the formulation record for that variation.

It is always easier to start with high concentrations of the salt in question and instantaneous coagulations, and use more dilute salt solutions until the coagulation is just visible after about 300 sec, an arbitrary cutoff for the convenience of the experimenter. These data are plotted as the log (time to coagulation) on the *Y*-axis and log (salt concentration) on the *X*-axis. Use the instantaneous coagulations at 1.0 or 0.1 seconds, and the plot will be L-shaped with the vertical segment leaning toward the *Y*-axis. The data for coagulations taking from 10 to 300 sec are extrapolated to the 1.0- or 0.1-sec line intercept, and that intersection point is the CCC.

One may mechanize this; Shute [39] devised a technique using unstirred latex in a Cary 14 or 18 IR spectrophotometer by injecting the salt solution and watching the change in transmittance. We used a Brice-Phoenix visible spectrophotometer with stirring [40], and later a Spectronic 20 visible spectrophotometer with no stirring gave reasonable results [41]. The important point is that all devices had recorders attached to follow variation of transmittance with time.

Commercially instruments are available that will record coagulation rates (such as those from Seinco Inc. and Medical Laboratory Automation Inc.), which were devised for blood agglomerations. The recorder trace is composed of three segments: the initial decrease in transmittance; the minimum transmittance plateau; and the last increasing transmittance segment with much scatter and sawtoothing as large particles pass through the light beam. The first two segments are of interest. The minimum transmittance plateau is a measure of precision of formulation of each variation, assuming the agglomerates are all of the same size in all variations of the CCC experiments. The average minimum should be drawn in on all strip-chart records to act as an intercept (this is drawn parallel to the paper long direction, the time axis). The initial coagulation rate is determined by drawing a tangent at the point of salt injection to pass through the line representing the average minimum transmittance. The projection of the tangent line on the time axis line represents the time to coagulate in that particular experiment. Plot each of these times (four or five determinations should suffice) in the log–log plot scheme discussed above.

Another mechanization scheme was presented by Franczel [42]. Therein, sodium and barium chlorides were compared to organic coagulant materials based on poly(ethylene imine) and poly(dimethyldiallylammonium chloride), for flocculating paper fiber fines or a textile fiber with or without clay. The work was done using a Coulter Counter having multiple channels with differently sized orifices.

Use this information to screen surfactants for effectiveness as chemical coagulant protectors. Use it to screen pigment or filler interactions with latexes, as they can also act as coagulants due to adsorbed or dissolving salts. Attapulgus and Kaolin clays have been used to coagulate a vinyl acetate copolymer stabilized with poly(vinyl alcohol) [43]. Any additive that causes latex coagulation may be tested with this technique. That includes the hydroxylic thickener that coagulates the carboxylated waterborne polymer [44], or the vinyl acetate latex stabilized with poly(vinyl alcohol) [45]. On the other hand, some of our experiments showed the inordinately high stability of emulsion polymers made for the paint industry as they survived 100 times the calcium and zinc ion concentrations that coagulated a standard abietic acid emulsified poly(chloroprene)[46]. Earlier work showed that styrene–butadiene–carboxylic monomer (acrylic?)–hydroxylic monomer (hydroxyethyl acrylate?) copolymer for paper coatings resisted molar quantities of sodium chloride. The latex formulation should remain fluid instead of flocculating or solidifying, for most nonwoven, paint, or coating applications.

E. Functional Group Characterization

The latexes used for nonwoven binders, paints, or adhesives normally contain a small amount of comonomer(s) to modify properties, aid cure, or otherwise serve some specialty function. The literature and the brochures from the supplier may give the monomers or clues. The specification of formaldehyde in the vapor space over the latex suggests the use of *N*-methylolacrylamide or methacrylamide, while epichlorohydrin suggests a glycidyl acrylate or methacrylate. Analyses for these are particularly difficult, although the methylolated amides can be analyzed by a simple Kjeldahl or Dumas determination of nitrogen on a free film, provided there are no other sources of nitrogen in the polymer (acrylonitrile? amine monomer?).

Infrared spectra (transmission on very thin films or attenuated total reflectance on coatings) may be useful, but one may need a known standard to confirm qualitative analyses; quantitative analyses require many standards. Industrial competitive produce analyses for acid copolymers provide sophisticated differentiation on acrylic or itaconic acid content; heating the film yields the anhydride [47]. O'Neill and Christensen [48] reported the results of a seven-laboratory collaborative IUPAC study on analyses of paint resins containing carboxy and hydroxy monomers; the hydroxy comonomer could only be identified by a very careful IR or NMR technique. Combining some results (nitrogen analysis and IR) was critical in finding the amide monomers.

Conductometric titration is preferred for the determination of carboxylic or other acid species on a latex particle because the end points are easier to see. In the following discussion, we assume that amine functionality should be treated like acid functionalities, and therefore will say nothing further on amine groups. A basic example of the conductimetry approach to characterizing weak acids was given by Gaslini and Nahum [49]. Although potentiometry is the basis for many automatic titration systems, the plot generated by the instrument (a sigmoid or "S" curve) makes it difficult to see an exact end point for low concentrations of weak acids. Kangas [50] reported significant differences between latex-bound acrylic or methacrylic acids and their homopolymers, and apparently improved the comparative accuracy between conductometry and potentiometry to the point where he preferred the latter.

We devised a system of cleaning a latex and directly titrating the surface carboxyl groups. We then added an excess of base with heat for a later back titration with strong acid; this procedure identified surface and buried carboxyl groups of the latex particle [51]. It was based on work done at the Lehigh Emulsion Polymer Institute by Vanderhoff, El-Asser, and others. The conductometric titration may be interfered with by latex "plate out" on the electrode with coagulum, but this can be avoided by adding a nonionic emulsifier to the latex prior to titration.

Standard conductivity probes and instrumentation are available from Beckman, Uniloc, YSI, Wescan, Thornton, Lazar or Leeds, and Northrup. One could also try the noncontact "electrodeless" conductivity monitors (the Foxboro Model 1200 systems), although I know of no trials of this system. Scientific Glass Engineering has recently introduced a small conductivity meter and electrode setup that may reduce the tedium of conductimetry.

The potentiometric equipment for acidimetry in latexes is the standard equipment common to large laboratories. Systems are available from Radiometer, Beckman, ECO and others. We offer again the advice on electrodes given in the pH section in the preceding chapter and that offered by Dicker [52].

Titrimetry requires that the latex system be cleaned of interfering substances, such as the salts derived from initiators, surfactants and dispersants. One technique used is fiber dialysis where water is flowing through the fibers in a container, but there is considerable dilution. Dilution is not bad, as the viscosity of the latex rises as the low molecular weight constituents are removed, because the Helmholtz double layer extends further into the dispersion medium (water), a phenomenon called the "secondary electroviscous effect." Campbell and a coworker, using a low molecular weight poly(acrylic acid), showed that the fiber dialysis system swept out water-soluble oligomers, such as poly(acrylic acids) of MW = 5000, at the rate of about 10% per hour [53]. A Japanese patent claims that dialysis will clean latexes commercially, reducing residual acrylic acid monomer content to parts per million [54], whereas Hearn et al. report problems with residual styrene monomer (conversion to benzaldehyde) in poly(styrene) latexes cleaned by dialysis [55]. There is apparently a distribution-coefficient problem between hydrophilic and hydrophobic bodies.

Dialysis is a relatively older technique, and the equipment for bag or fiber dialysis is readily available. Cole-Parmer, Schleicher & Schull, and others sell collodion bags and associated apparatus. Spectrum Medical Industries markets lab-sized fiber dialysis unit, called Spectrapor; RIA Research Corp. has larger-sized units as well as electrodialysis equipment. Romicon, a division of Rohm and Haas, has production-size hollow-fiber ultrafiltration units used in the electrodeposition paint cleanup.

An alternative to the dialysis technique is simple ion exchange. Double-bed resins, containing both anionic and cationic sites for exchange, are available from Bio-Rad and Diamond Shamrock, whereas Rohm and Haas and Dow sell the resins in single-functionality form. The Bio-Rad double-base resin yields conductivity quality water, by stirring in ion exchange (watch the stirring rate, as the resin beads break up to a fine powder if you stir too fast), or column-ion chromatography to isolate fractions of latex and soluble constituents [56]. Dionex, ESA, and others offer column ion chromatography systems. The Lehigh Emulsion Polymer Institute group actively pursued this technique of cleaning for latex characterizations [57,58]. One concern in using ion exchange resin is that one could lose the acid species which tied to the base sites on the resin. Foster et al. [59] show Freundlich isotherm values and equilibria for carboxyl-functional proteins and standard basic resins. The pH has a dramatic effect, according to their work, so perhaps a single-bed approach is desirable to control necessary variables to recover all the acid species in the sample. There are at least five types of carboxylic acid species in a latex made by copolymerizing an acid comonomer [60], and their determination may be important to a product or process development scheme that involves colloidal stability, viscosity, or cure (rate or extent).

There are functionalities on latex particle surfaces other than those derived from comonomers. A number of publications deal with the sulfate groups derived from the initiator species [61–66]. Since the peroxydisulfate initiators are preferred by industry, the analyst should be aware that they can show up on the latex particle, albeit in a concentration very small compared to the carboxylates from comonomers. Sakota and Okaya [67] prepared amine-functional self-stabilizing (surfactant-free) latexes derived from an amine-functional azo initiator molecule.

F. Electrophoretic Mobility

The latex particle is charged on the surface, normally anionic. The Helmholtz double layer is comprised of those surface charges and their counterions (for electroneutrality) at some distance from the particle surface. In an electric field, the latex particle is attracted to and will migrate toward an electrode of opposite charge, at a rate that depends on the electric field strength, the particle surface charge (related to the zeta potential), and the viscosity and dielectric constant of the dispersion medium (water). With the known or controlled viscosity, electric field strength and medium dielectric constant, the migration rate of the particle is a measure of the zeta potential of the particle. This zeta potential is related to the amount of neutralized carboxylate comonomer on the surface and the anionic surfactant and salts adsorbed, as these are the anionic charge-generating species in the formulation. Anionic-charge generators are important in determining the resistance to mechanical coagulation, and give some Schultz-Hardy coagulation resistance. Zeta potential measurement can become important as a formulator's tool to assess the effectiveness of salt or surfactant additions, or of surface carboxylation. The salts in a latex may be added for good reason, as Blackley [68] showed in a relation of latex viscosity to electrophoretic mobility changes with added sodium sulfate. In a production facility where the polymerization that normally took 12 hr was speeded up to 9 hr to increase productivity, product properties were adversely affected, and zeta potential measurements on cleaned latexes showed less surface carboxylation [69].

Sennett and Oliver [70] published a review of the zeta potential concept. Sennett et al. [71] discuss clay–starch paper-coating formulations for paper coatings, whereas Stratton and Swanson [72] interpret the uses of zeta potential in papermaking (to aid formulations for drainage or retention) with cautions as to why some results are not as expected. Smith [73] presents an excellent discussion on experimental detail for electrphoresis cell usage. Temperature dependence of latex conductivity [74] and surface charge on colloids [75] have been discussed. These authors raise concerns both in the determination of zeta potentials and in relation of the information to processing details (e.g., what happens to the formulation during drying). The zeta potential governs the electrodeposition of a colloidal dispersion, such as the electrodeposited primers on car bodies or the web bonding from a Kendall patent by Carlos Samour.

The effects of surfactant addition on zeta potential are frequently discussed in the literature [76]. Studies showed that nonionic emulsifiers imparted some anionic charge on latex particles [77], but there seems some controversy on this point [78,80]. I look for the changes in zeta potential as nonionic surfactants are added to colloids with two possible mechanisms for the two effects:

1. Drops in zeta potential occur when the nonionic surfactant displaces salts or surfactants from the colloid surface.
2. Increases in zeta potential occur when nonionic surfactants are adsorbed because part of the water associated with the polyoxyethylene chain is comprised of hydroxyl anions hydrogen-bonded to the ether oxygens.

Zeta potential can be determined with equipment purchased from Komline-Sanderson, Zeta-Meter, Inc., Pen-Kem Inc., and M/K Systems Inc. These systems use the classical technique of microelectrophoresis and some optics to follow the moving particles. Since the latex particles are smaller than can be seen by standard microscopes, laser imaging has become popular. The limitations of these instruments lie in the fact that the formulation must be drastically diluted (possibly with serum equivalent in composition to the dispersed phase of the formulation).

Another option is the mass-transfer device, available from Micromeritics Instrument Corporation, which measures the movement of formulation components from one cell segment to another by weight changes in the cell segments. One must watch for heating effects in such a large system, and there are modifications of the system that you may want to make [81].

A zeta potential-related device has recently come on the market from Matec Instruments, which measures the momentary sound in the electric dipole moment that occurs when an electrical pulse is piped through the dispersion. This acoustical emission is the converse of a principle first proposed by Debye [82] in 1933.

A simpler, less precise, method for determining the average zeta potential for colloids uses the so-called moving boundry technique. A simple U tube is constructed with electrodes at each open end and a central sample introduction valve and container. The outer arms of the tube are filled with serum from the formulation (e.g., obtained from ultracentrifugation), and the formulation is introduced without disturbing the interface between latex and serum. The current is turned on and the movement of the latex–serum boundry is monitored. Caveats include the guarding against electrode-heating effects (causing thermal swirling that can disturb the boundry) and the care needed in introducing the sample without disturbing the interface. If there are distributions of particle size and surface functionality or other charge-influencing species, you will get a measure of the mobility of only the fastest of the particles.

The measurement of zeta potential is complex, depends on the technique, and requires trained professionals to conduct the experimentation. The common nonwoven or paint developer will not want to set up such experimentation, but will turn to an outside service. However, the technique can be very useful in guiding the formulator of binders to the system that has maximum stability to processing or coagulants.

G. Fluid Property Aging

Fluid property aging is not normally of concern to the nonwovens manufacturer, as the formulations are generally made up and used in the next 24 hr or so. The fluid property aging problem is more likely to be of interest to the household paint manufacturer, whose cans may rest on store shelves six months before the consumer decides to buy. Most printed instructions state: "Use within six months," without giving a reason. The Lehigh Emulsion Polymer Institute has samples of polystyrene latexes that have remained unopened and stable for 20+ years. But the latexes we use in commercial products are likely to have reactive groups for cure, and many have hydrolyzable ester and amide groups. Admittedly, the hydrolysis rates for esters and amides is a function of pH, which we always keep near neutral to protect against hydrolyze more slowly. However, we do know that many latexes contain methyl methacylate or vinyl acetate, which hydrolyze quickly under the right circumstances. Their hydrolysis is autocatalytic, since each hydrolysis step liberates another acid molecule.

Perhaps a check on storage stability might be in order for the tank car of latex received every third month. Strictly speaking, the tank car and tank truck shipments have probably been in the manufacturer's storage for a week or less, and are probably stable for another few months. However, the drum and can shipments you receive may be older. Perhaps we might be able to find out how much older, if there is some marker chemical or property that indicates deterioration.

The key signs of deterioration are most commonly increased coagulum content, increased or decreased viscosity; and change in pH.

A Hercules brochure [83] makes the important point that formulations of latex paints and thickeners can be degraded by enzymes or oxidants and reductants from the polymerization recipe. They recommend, as the tool to identify enzymatic or chemical viscosity loss, the addition of aqueous polyacrylamide to a degradable suspect sample and addition of aqueous hydrosyethylcellulose to another suspect sample with storage of both at 45°C for a week with intermittent viscometry. Enzymes will reduce the viscosity of the latter sample by 30% in a week, with no loss in viscosity by the sample compounded with polyacrylamide. Chemical attacking agents (e.g., residual peroxydisulfate initiator) will reduce the viscosity of both samples by 30% in a week. Enzymes in a waterborne formulation have a bacterial origin, and destroying bacteria present in a batch may not solve all your problems. You have to keep the bacteria out altogether, through rigorous cleaning and destruction before the problem arises.

More serious signs of polymer or colloid aging, such as loss of cross-linking efficiency or color change, can only be detected by evaluation in web. The odor of the latex is often the first sign of a change, as with the vinegar smell over vinyl acetate-containing latexes or with the formaldehyde odor over some self-cross-linking latexes. Just look for something different in the latex sample stored on a shelf for six months to a year, in monthly or bimonthly examinations.

Thermal techniques (e.g., an Arrhenius plot) to speed up the test have advantages and disadvantages. If you are sure that a simple chemical reaction is the course of the aging phenomenon, the storage of samples at some three different temperatures can give you the data quickly and efficiently. But there are just as many instances of polymer degradations that do not follow the Arrhenius plot as there are those that do [84]. In addition, other reactions may occur in the formulation that do not occur at the lower temperatures. For instance, there are formulations that are stable at room temperature but coagulate at elevated temperatures because the stabilizer has an upper consolute temperature in water. There are practically insoluble filler or pigment that are adequately soluble at the elevated temperature to cause a Schulz-Hardy coagulation on the latex.

Aging stability may be a problem that must be dealt with, and it takes some time to prove it. If it is important, it can affect the product selling lifetime and the economics of production. The early check on formulation aging stability is well worth the cost and time in the product development cycle.

H. Residual Monomer

Residual monomer is a concern for a variety of reasons. Product odor can cancel repeated purchases of nonwovens or interior house paints. In some instances (cup coatings or can coatings), the material contained can extract any residual monomer to cause an off flavor. Any number of regulatory agencies express concern over the exposure of customers or workers to volatile monomer fumes. Residual monomer can also degrade to form a more odoriferous molecule (e.g., (benzaldehyde from styrene). The monomer compatible with the polymer acts as a temporary plasticizer, and may make the adhesive or coating tacky.

Residual monomer is not a new concern, as the SBR latex rubber standards contain a determination for styrene [85]. The method assumes that all unsaturates in the analyses are styrene, although there is also appreciable vinyl-cyclohexene, the butadiene Diels–Alder dimer, present in all butadiene hot polymerized latexes. The method is a wet chemical method involving bromination of the double bond, and back titration of excess bromine with a standard iodine solution.

Current recommendations for an easier, cleaner, and faster technique are based on gas chromatography (GC). Fossick and Tompsett [86] used the method. Austin and Cunningham have used the technique for acrylonitrile [87] in paints, while Rygle [88] used a capillary GC system. Injection of latexes or polymer solutions onto the column has always worried some chromatographers, due to the suspicion that a slow degradation would give an ever increasing baseline, or that the deposited solids would plug the column. The sample solids always fell into the injection port, in my experience, and could easily be removed. Only an experienced user could give details on baseline stability. Jordan [89] has much confidence in simple on-column injection to follow poly(acrylic acid) syntheses for residual monomer analysis.

The development of headspace analyses (analyses of the vapor space over the latex in a container) avoids some objections. Steichen [90] stresses the methods of forcing polar monomers to the vapor space from solutions for determinations at ppm levels. Krockenberger and Gmerek [91] note that results of their analyses for vinylcyclohexene and styrene were reproducible within 10%, and that the water content of the latex affected the analyses. Pausch and Sockis [92] also determined residual monomer in rubbers by headspace chromatography. Lattimer and Pausch [93] describe the determination of vinyl chloride monomer over powdered PVC samples. Conditions of analysis must be carefully controlled, as the all above workers emphasize. The vapor pressure of the chemicals of concern are all affected by temperature, water concentration as compared to hydrophobes (polymer particles, solvents, and surfactants), and time required to reach equilibration.

We suspect that any normal gas chromatograph may be used for the analyses of residual monomer; Perkin-Elmer has considerable literature on use of their instruments and accessories for headspace analyses. Based on my own experience of following monomer syntheses and polymerizations with gas chromatography, I would emphasize that the peak height or area you measure best correlates to the weight percent of the constituent, and that the choice of marker material as an internal standard is critical to success (try to include retention time with the other measured parameters).

An ASTM D-1 committee is attempting to standardize a method for acrylate monomers based on capillary columns with apparent success, to judge from early reports. With appropriate care and choice of type of column, column packings (if any), temperature, and other factors, gas chromatograph can be used successfully.

I. High Shear Rheology

Fluid-flow processes in industrial applications provide many examples where high shear viscosity is a controlling factor in applying adhesives or coatings. The current commercially available viscometers are reviewed here with commentary on their ease of operation, the data treatment to obtain viscosity, their relation to theory and practical usage of fluid flow problems, and the range of their applications. Voscometers for laboratory or process in-line applications are included, as are devices specifically constructed for specialized uses for one industry. The practicing engineer in coatings, cement, paper, or other industries requiring viscous fluid characterization for process control should take note of the special features on measurement instrumentation as a means to better product characterization and process control.

One example of more complex fluid-flow problem, with viscosity as a control feature, is the application and usage of carpet-backing adhesives. The components are mixed and transfered to a storage tank to be withdrawn as needed. The fluid, just prior to application, may or may not have air whipped in during an in-line frothing process. At the point of application, the adhesive may be poured into a pan to be picked up by a roller applicator. A doctor blade may govern the thickness of adhesive layer on the applicator roll or on the carpet after it touches the applicator roll. A secondary backing may be laminated by the pressure of a "marriage roll." The adhesive may be under a capillary wicking stress at any point from the application until it becomes dry enough to gel. The temperature rise during drying may also induce increased flow, (Fig. 13.1) adhesive migration toward the heat source, and migration or flow due to gravity. Figure 13.1 depicts the process wherein viscosity may be a control parameter at a variety of points in the process. Careful inspection shows that the process contains elements that are common to roller coating coil steel, paper coatings, and ink deposition. The viscosity of concern, in any of these processes, is one

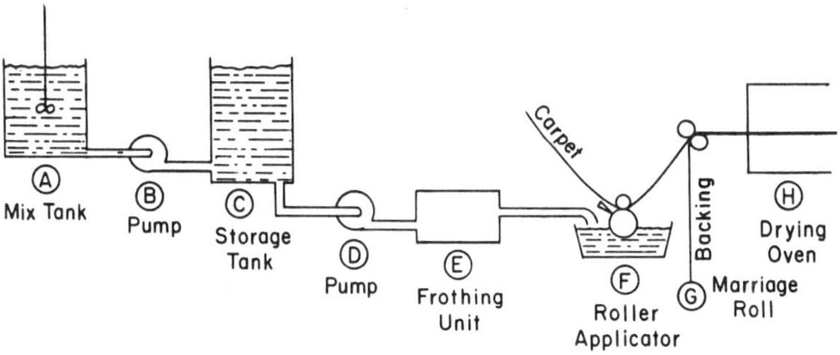

Figure 13.1 Carpet Backing Adhesive Application Process.

which is not that one normally perceives in the laboratory capillary viscometer, but a viscosity behavior that has been modified by mechanical or thermal stress. Although the emphasis is on the high rheometry measurements and devices, one cannot ignore the low shear portion of the rheogram.

The fluid-flow problems facing the engineer dealing with viscous fluids require characterization and process control viscometric instruments which yield the maximum necessary information. The value of the information required lies not in the data and reports therefrom, but in the practicable process control to maintain production rate and quality. The needs of different industries vary, and the process instruments used in viscometry will bear out this variety.

In one instance, a paper-mill engineer may only need a single low shear viscosity measurement to determine whether his kaolin-latex dispersion paper coating will flow through his gravity-feed recycle lines. In another, the engineer may need high shear viscosity characterization of non-Newtonian flow to determine a coatings behavior on a roller or blade coating machine. Time-dependent viscosity behavior at a shear may determine whether or not the latex paint is dripless, or whether the limestone-latex carpet backing adhesive may be stored overnight before use. Viscometry may indicate a process upset in a detergent manufacturing unit, or a food processing unit, and may indicate that the pigment or filler for the cement or coating needs another cycle through the grinding or dispersion unit to meet specifications.

The fluid-flow problems to be dealt with by the engineer require him to know several aspects of his process and the viscometry instrumentation available. The shear rate(s) used and how they are attained are important in both, as well as the temperature effects on flow behavior. Non-Newtonian viscosity behavior may be detrimental or helpful in the process, as may Bingham plastic yield. Berry (of Carnegie-Mellon University) characterizes fluids as having a viscosity "personality" which should be completely defined to ensure that determination of the correct aspect of viscosity is used as a control parameter in the process.

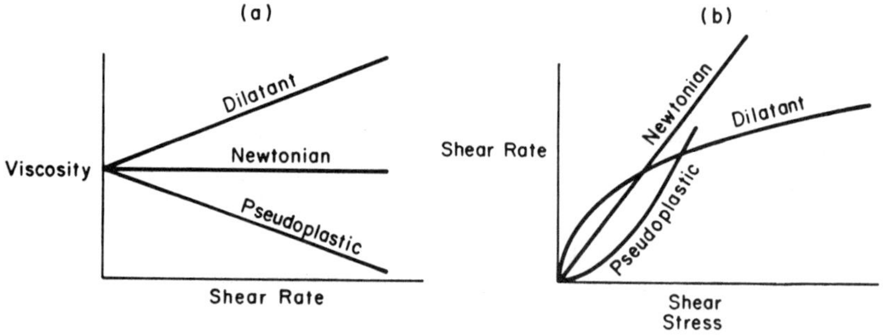

Figure 13.2 (a) Viscosity as a Function of Shear Rate. (**b**) Viscosity as a Function of Shear Stress.

Hence, the details of viscosity behavior at a variety of shear rates must be familiar to the engineer concerned with process measurement and control. A discussion of the characterization of viscosity behavior will lay the basis for a brief review of some of the current commercial instrumentation used in industrial fluid viscometry.

1. Principles

The simplest viscosity behavior (Newtonian flow) is defined by Eq. (16.1). An initial assumption necessary for its validity lies in the flow character. The flow must be lamellar, with adjacent layers sliding past each other parallel to the wall. The geometry of the measurement system is of critical concern to the instrument user, as it should

$$\eta = \frac{\tau}{\gamma} = \frac{F/a}{v/h} = \frac{Fh}{va} \tag{13.1}$$

correlate to the fluid-flow practical problem at hand. Admittedly, flow through a 4-in. pipe might be an impractical problem for laboratory study, but an engineering analysis can be made of even a relatively complex flow problem [94,95]

The variation of viscosity with shear rate is a common phenomenon to industrial practitioners. The latex adhesive or paint formulation commonly decreases in viscosity as shear rate increases, a behavior termed pseudoplasticity or shear thinning. Pigment dispersions will often increase in viscosity with increased shear rate, a phenomenon called dilatancy or shear thickening. These may most easily be remembered by the representations shown in Fig. 13.2a, the plot of viscosity vs. shear rate. They are more commonly plotted (as in Figure 13.2b) with shear stress as a function of shear rate. Recognize that these representations are idealized, and that more complex dependence of viscosity on shear rate might be determined. For purposes of definition, these representations are adequate.

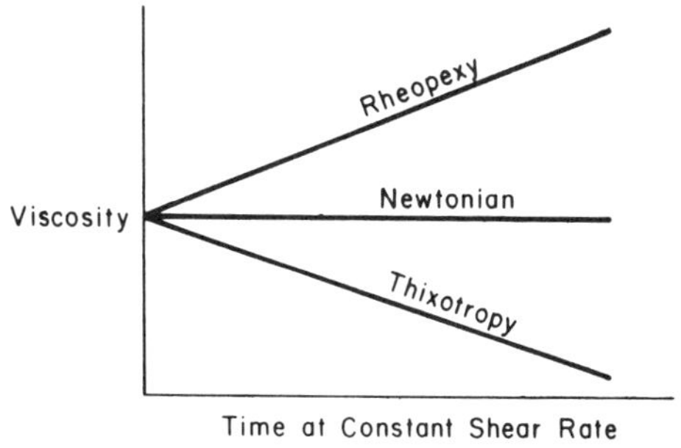

Figure 13.3 Viscosity as a Function of Time at Constant Shear Rate.

The variation of viscosity as a function of time under constant shear rate is another important non-Newtonian flow behavior. The decrease in viscosity with time at constant shear rate is defined as thixotropy, and can be commonly found in "dripless" latex paints. An uncommon phenomenon in actual practice has been defined as the increase in viscosity with time at constant shear rate or RHEOPEXY. These are shown as plots of viscosity (Fig. 13.3) with time at constant rate.

The experimental quantization of thixotropy of a latex paint has been detailed by Pierce [96] and Freling [97]. The plots of shear stress vs. shear rate show a hysteresis loop characteristic of the phenomenon when shear rate is decreased or increased.

The non-Newtonian flow of one fluid may not parallel that of another fluid. The viscosity-shear rate diagram in Fig. 13.4 shows one reason for complete viscosity characterization as opposed to determinations at a single shear rate. The comparisons of single-point viscosity measured at intercepts of A, B, or C

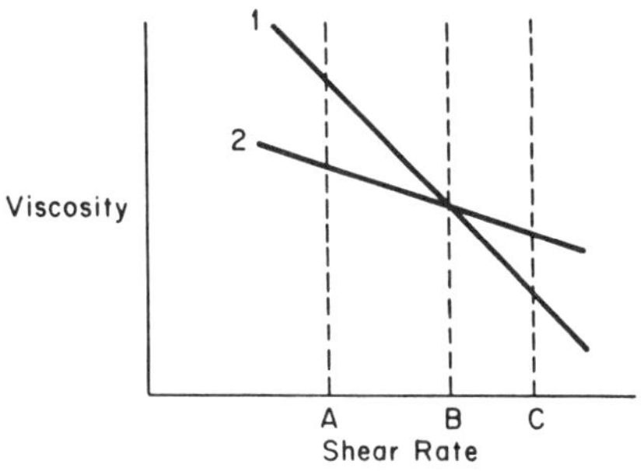

Note: $A_1 > A_2$, $B_1 = B_2$ and $C_1 < C_2$

Figure 13.4 Viscosity Comparison of Two Fluids.

on Fig. 13.4 are not likely to yield useful process-control information. Hence, a characterization at several shear rates and a physical concept of the in-process shear, at least to an order of magnitude, would probably lead to a more useful process control parameter. Of particular interest is the presentation made by Matsuda and Brendlay at the 1979 National convention of the Federation of Societies of Coatings Technology (FSCT) showing the variety of rheology characterizations possible in reverse roll-coating formulations and their relation to coating transfer and leveling. Several important points were made:

1. Measurement of a single point in the viscosity-shear rate plot could be misleading;
2. High shear viscosity proved an important control parameter; and
3. A predictive model equation was devised with constituent properties and operating parameters as factors.

The first point was demonstrated by their makeup of coatings to a 30-sec Ford cup flow time, while still showing orders of magnitude of viscosity difference at high shear. This phenomenon is also commonly found in extrusion of plastics.

This work has implications developed by Shaw [106] in a collaborative study by the Los Angeles Society for Coatings Technology Technical Committee. They sent samples of fluids designed to be Newtonian, thixotropic, pseudoplastic, and dilatant to about a dozen laboratories across the United States, asking that they measure viscosity on all samples by as many methods as available in each lab. The objective was to check comparability of methods (they *are not!*), precision among labs, and whether or not the chosen methods (as they were used) were capable of determining non-Newtonian flow. The presentation was made in the U.S. National Paint Show Voss Award competition, and was the first of that year's papers to be published [106] *though it did not receive an award*. The emphasis was on the myths of rheometry and the reality that is vastly different from the myths.

Another non-Newtonian flow behavior is described as Bingham plastic flow. Herein, no flow occurs until a flow energy barrier is overcome by the applied force. Common examples of Bingham plastic flow are putties, caulks, greases, and modeling clay. The representation of shear stress dependence upon shear rate is shown in Fig. 13.5. The value of the shear stress needed to induce flow is termed the yield value. The quantification by Bingham of the plastic flow behavior [97] is shown in Eq. (13.2). One may calculate the antisag characteristics

$$\eta = \frac{\tau - f}{\gamma} \tag{13.2}$$

of a paint from its yield value, according to Patton [98], and use this as a relative measure of the pumping problem encountered in emptying an adhesive tank or

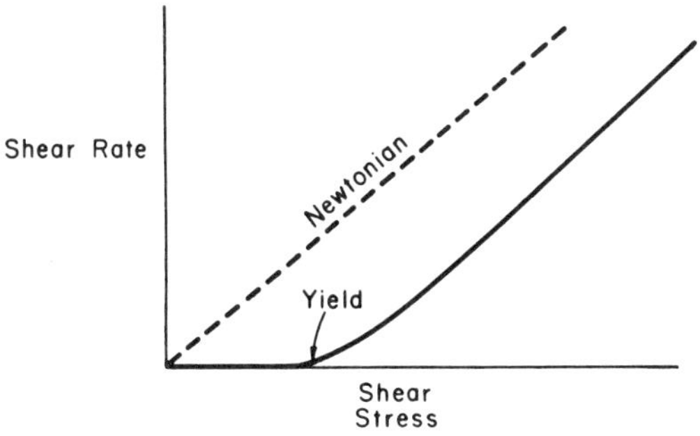

Figure 13.5 Bingham Plastic Flow Representation.

pipeline by simple gravity flow. In addition, the Bingham plastic yield is a measure of how thick a film one can paint on a wall without having sag or drip problems. Orchard [108] related film thickness to flow, and leveling with verification by profilometer testing on coatings.

The temperature effect on viscosity is important. Greet and Magill [99] characterized the temperature dependence of liquids viscosities and found group correlations based on intermolecular bonding similarities. Pierce [96] reported a nomograph for, and the practice of, the use of Andrade's equation, Eq. (13.3)

$$\eta = A \,.\, \exp(B/T) \tag{13.3}$$

for alkyd resin paints. He also reported the work of

Lewis and Squires, showing temperature dependence of viscosity for alkyd resins, solvents, and drying oils in a universal viscosity temperature curve.

Another interesting set of characterization examples may be derived using the terms "short" and "long". A "short" viscosity is one wherein the fluid breaks upon stress. A short high viscosity fluid is putty, whereas a short low viscosity fluid is mayonnaise. A "long" viscosity has a stringiness in character, giving an extensibility. A long high viscosity fluid is melted cheese, and a long low viscosity fluid is honey.

The components of viscosity in non-Newtonian flow may be separated into a plastic contribution and a viscoelastic contribution. The plastic component has little recovery, whereas the viscoelastic contribution is a measure of the energy stored, which may act as a restoring force when the stress is released. The plastic and viscoelastic components of viscosity are the realm of fluid mechanics dealt with in polymer mixing and extrusion rheology and in rubber tensile testing, but Greener and Middleman [100] relate them to the thickness attained in roller-coating operations as well. Excellent references on this aspect of rheology are the tests by Ferry [101] and Eirich [102].

Other viscosity measurement systems exist. Conversions from one system to another have risk, but tabulations may be obtained from Technical Measurement Inc., or Brookfield Engineering Laboratories (Table 13.1). In general, they relate an arbitrary unit (flow time from a cup, for instance) to centipoise units. Indeed, for a specific application, one may use tables to plot a calibration curve relating two viscosity systems (e.g., Ford-cup seconds to Brookfield centipoise).

Detailed discussions of viscometry, with the geometry involved as well as the engineering considerations, can be found in many references. The classic discussions of specific analyses of viscosity and applications were presented by Bird et al. [94]. The practices of viscometry and detailed instrumentation descriptions are provided by Van Wazer et al. [95] in a laboratory handbook. In addition, the suppliers of viscometers generally provide reprints or bibliographies on theory and applications of their instruments. The engineer would be well advised to have these or similar references available for aiding in practical approaches to problems involving viscosity.

Table 13.2 Commercial Viscometers

Manufacturer	Type	Features
ACE Scientific Supply 1420 E. Linden Avenue Linden, NJ 07036	Falling Ball	Viscosity Ranges 0.3–15 or 5–300 cP. For use with transparent or translucent fluids.
Automation Products Inc. 3030 Max Roy Street Houston, TX 77008	Dynatrol Detector (vibrational damping)	Electronic in-line monitor or control device with temperature-correction attachment available.
Baroid Petroleum Services Division NL Industires Inc. P.O. Box 1675 Houston, TX 77001 713-527-1100	Capillary Viscometer	Used to characterize drilling muds by pressurized extrusion through capillary; can be used for pigment or filler slurries.
	Rotational Rheometer	Hand-cranked model with two rotation speeds. Electrically driven model for five preset speeds or continuous variation of rotation speed.

Table 13.2 Commercial Viscometers (*Continued*)

Manufacturer	Type	Features
	FANN Viscometers (Rotor and Bob)	Hand-cranked or electrically driven models. Pressurized electronically monitored model available.
	FANN Consistimeter (recirculating pump)	Temperature and pressure monitors.
	FANN Model 70 HTHP Viscometer	Electrically driven high temperature/pressure coaxial cylinder device
Bohlin Reologi P. O. Box 6623 Edison, NJ 08818	VOR Rheometer	Computer-controlled device that measures viscosity and elasticity as a function of rate, time, and temperature.
	CS Rheometer Visco 88	Controlled stress rheology; portable viscometer.
C. W. Brabender Instruments 50 E. Wesley Street South Hackensack, NJ 07606 201-343-8425	Convimeter (rotational)	Couette device for continuous in-line viscometry up to 300° and 1000 psi for viscosities to 200,000 cP.
	Farinograph (rotational)	Dough- or food-mix rheometry by monitoring torque of mixing blades; can be heated to 50°C. Usually measures viscosity behavior upon addition of water to dry grain or food.
	VISCO-AMYLO-GRAPH (rotational)	Starch hydration by pin-stirrer torque monitor as swelling and dissolving occur. Variable and programmable temperature control during operation.

Table 13.2 Commercial Viscometers (*Continued*)

Manufacturer	Type	Features
	Rheotron (rotational)	Couette or cone and plate with electrical drive and data recording. Continuous shear rate variation is available, as is temperature control to 300°C.
	Viscocorder (rotational)	Paddles rotated in cup. Can be used for plastisols cements, paints, and other dispersions.
	Plasti-corder (rotational)	Temperature-controlled mixing torque recorder which may be used for themnoplastics, paints, plastisols, and other applications. Many rotor styles, including roller or sigma blade.
Brookfield Engineering Laboratories, Inc. 240 Cushing Street Stoughton, MA 02072 617-344-4310	Synchro-lectric Viscometer (rotational)	Many models of simple couette four- or eight-speed laboratory portable devices with electric drive, Bob geometry complex; an average shear stress gives good approximate measurement; commonly used in paper, textile, paint, and adhesives industries; Helipath device for plastisols; choice of dial or electronic readout.

Table 13.2 Commercial Viscometers (*Continued*)

Manufacturer	Type	Features
Brookfield Engineering	Wells-Brookfield Micro Viscometer (rotational)	Cone-and-plate viscometer for small sample volumes (0.5–2.0 mL). water jacket for temperature control; choice of dial or electronic readout.
	Models TT	Process monitor for viscometry fully flooded and under pressure.
	Models VTA	Process monitor for open-system viscometry.
Burrell Corporation 2223 Fifth Avenue Pittsburgh, PA 15219 412-471-2527	Burrell-Severs Extrusion Rheometer (capillary)	Room-temperature capillary-extrusion rheometer with or without gas pressure; requires separate purchase of timer and balance to time and weight extrusions.
Cambridge Applied Systems Inc. 23 East St. Cambridge, MA 02141 617-577-8808	Viscosity-Temperature Systems	Magnetically driven coaxial cylinders monitored electronically
Cannon Instrument P. O. Box 16 State Colelge, PA 16801 814-466-6232	HTHS High Temp. High Shear Viscometer (capillary)	Supplied with five cells for measurements, personal computer for calcuations, and temp.-press. indicators.
R. P. Cargille Laboratories 55 Commerce Road Cedar Grove, NJ 07009 201-239-6633	Viscosity Tubes (rising bubble)	Gardner Viscosity standards or ASTM Time-Test method.
EG&G Chandler Engr. 7706 E. 38th St. Tulsa, OK 74145 918-627-1740	Chan 35 Direct	Twelve speed rotating cylinder, dial indicating device for oil-well drilling muds.

Table 13.2 Commercial Viscometers (*Continued*)

Manufacturer	Type	Features
	RheoChan	Twelve speed rotational device for high pressure-temperature rheology on oil-well drilling muds; strip chart or IBM PC readout.
Eur-Control USA Inc. 2579 Park Central Blvd. Decatur, GA 30035 404-981-3998	In-Line Viscosity Transmitter (rotational)	Electric or Penumatic signal output for viscosity range of 18–1500 cP for paper, cement, or other industries; rotating and stationary disk torque sensor.
Ferranti Electric Inc. East Bethpage Road Plainview, NY 11803 516-293-8383	Ferranti-Shirley (rotational)	Automatic electrically driven servo-controlled continuous-shear variable-recording viscometer; Temperature, cycle time, and shear rate range controlled; several models available.
Ford Viscometer Inc. 1379 Dix Highway Lincoln Park, MI 48146 313-928-8853	Ford Cups (capillary)	Flow through an orifice in the bottom of the cup for comparisons of flow times for constant volumes.
P. N. Gardner Co. Inc. 316 NE 1st St. Pompano Beach, FL 33060	Ford and Zahn Cups (capillary)	Flow through an orifice in the bottom of the cup for comparison of flow times for constant volumes.
Haake, Inc. 244 Saddle River Road P. O. Box 549 Saddle Brook, NJ 07662 201-843-2320	Rotovisco (rotational)	Variable (continuous or single-point) multiple-shear couette device with temperature control.
	ViscoTester (rotational)	Portable-battery operated couette device.

Table 13.2 Commercial Viscometers (*Continued*)

Manufacturer	Type	Features
	Viscontrol (in-line rotational)	Continuous process monitor.
	Falling ball Viscometer	For transparent Newtonian fluids.
	Viscobalance (rising ball)	Ball drawn upward through tube by balance and weights, for opaque Newtonian fluids.
ISA Division Adamel Lhomargy d'instruments S.A. c/o Testing Machines Inc. 400 Bayview Avenue Amityville, NY 11701 516-842-5400	Nametre Viscometer (vibrational)	Single-shear-rate device for lab or in-line process measurements in six ranges of up to 10^5 cP.
J&L Instruments Corp. 2430 Blvd. of Generals Morristown, PA 19403 800-365-2635	Irvine-Park Falling Needle Viscometer	Falling-needle device for Newtonian or non-Newtonian fluids requiring no calibration.
Mitech Corporation 4399 Hamilton Parkway Willoughby, OH 44095 216-946-3130	Carri-Med CS Rheometer	Controlled-stress rotational device fitted for parallel plate, cone and plate, or couette at varying temperatures with electronic readout.
Nametre Company 101 Forrest St. Metuchen, NJ 08840 201-494-2422	Nametre Vicometers	Vibrating rod, sphere, or collet viscometers with electronic readout.
Norcross Corporation 255 Newtonville Avenue Newton, MA 02158	Norcross Ring Viscometer (falling rod)	Similar to falling-rod viscometer; may be used in-line for process or in lab work, electronic monitoring and air-driven cylinder return; environment Standard to 300°F and 1000 psi (higher if desired) for viscometry to 20000 cP.
	Shell Cups	Similar to Ford or Zahn cups.

Table 13.2 Commercial Viscometers (*Continued*)

Manufacturer	Type	Features
Paar USA 211 53rd St. Moline, IL 61265 309-762-7716	AMV200 Microviscometer	Requires only 0.15 mL; will measure between 0.5 and 70 cP; may vary angle of tube for measurements for extrapolation to zero shear rate.
Physica Messtechnik GmbH u. Co. KG Vor dem Lauch 6 D-7000 Stuttgart 80	Viscolab LC 1 (rotational)	Couette device with 27 speeds and temperature control for high shear measurements.
	Viscolab LC 10 (rotational)	Similar to Viscolab 1 with computer readout.
Rheometrics, Inc. 2438 U. S. Highway No. 22 Union, NJ 07083 201-687-4838	Sensitive transducer (rotational)	Cone-and-plate or couette adapters for the Rheometrics Mechanical spectrometers, separating viscous and elastic components of mechanics.
Ruska Instrument Corp. 3601 Dunvale Houston, TX 11063 713-975-0547	Model 1602 (rolling ball)	High temperature-pressure testing of drilling muds.
Schott (Jena Glaswerk Schott & Gen, Inc.) 11 East 26th Street New York, NY 10010 212-679-8535	KPG Viscometers (capillary)	Ubbelohde, Cannon-Fenske, and Ostwald capillary viscometers, temperature controllers, automatic flow timers and refillers for transparent or opaque fluid viscometry.
Sheen Instruments 8 Waldegrave Rd. Teddington, Middlesex England TW11 8LD 01-977-0051	Spindle and Paddle Instruments (rotational)	Brookfield and Stormer types.
	Flow Cups (capillary)	Zahn, Ford, DIN, and others.

Table 13.2 Commercial Viscometers (*Continued*)

Manufacturer	Type	Features
Shimadzu Scientific Instruments Inc. 7102 Riverwood Dr. Columbia, MD 21046 301-997-1227	CFT-500A Flowtester (capillary)	Heated-die extruder with multiple monitors and computer readout of temp., die conditions, and flow results.
Technical Measurements Inc. 19412 E. Ten Castle Road East Detroit, MI 48021 313-775-5290	Zahn Viscosity Testing Kit (Zahn Cups capillary)	Similar to Ford cups.
Tekmar Company P. O. Box 371856 Cincinnati, OH 45222-1856	Rheomat 108 (rotational)	QC viscometer with strip recorder.
513-761-0633	Rheomat 115 or 185 (rotational)	High shear devices with programmer and readout.
	Low Shear 30 Rheometer (rotational)	Couette device with elecronic readout.
	Ravenfield Viscometers (rotational)	Low shear Couette device.
	Covimat series (rotational)	In-line production controllers-monitors of viscosity.
Testing Machines Inc. 400 Bayview Ave. Amityville, NY 11701 516-842-5400	Laray Viscometer Model VM.01 (falling rod)	Thin-film viscosity by close fit of rod to cylinder, varying weight load on rod; Electric timing.
Theta Industries Inc. 26 Valley Rd. Port Washington, NY 11050 516-883-4088	Theotronic High Temp. Viscometers	Couette, parallel plate or bending beam (?) types.
Viscotek Corp. 1032 Russell Dr. Porter, TX 77365 713-359-5966	Model 100 Differential Viscometer	Paired capillary-flow device that balances the pressure needed to make two fluids flow at same rate, with pressure as readout.

Table 13.2 Commercial Viscometers (*Continued*)

Manufacturer	Type	Features
Wescan Instruments Inc. 3018 Scott Boulevard Santa Clara, CA 95050 408-248-3519	Automatic Viscosity Timers (capillary)	Cannon-Fenske, Ostwald or Ubbelohde capillary viscometers with automatic electronic timer.

2. Commercial Instrumentation

A wide variety of instruments are manufactured, which measure viscosity or some force related to it. A compilation of instruments is given in Table 1.7, listed alphabetically by supplier. They fall into basic categories of similarity due to the principles underlying their operation. This is substantially expanded over a table in an earlier publication [107].

A caveat regarding instrumentation in general should be noted. These devices have precision-machined surfaces to be contacted with the fluid. Hence, chemical corrosion or mechanical damage on these surfaces can destroy their accuracy. These surfaces must be scrupulously cleaned so that no particles or coatings deposit to alter the surface geometry.

The capillary viscometers and their automatic flowtime measurement equipped systems have been used to measure flow rates for many years. These may be the Zahm or Ford cups, or the Ostwald, Ubbelohde, or Cannon-Fenske capillary-tube glassware devices. They measure a flow rate by timing a known volume of liquid flowing past a known point in a standard orifice.

However, there are problems in using these devices. The temperature must be consistent for comparisons of one fluid or another. Viscosity standard fluids are required for calibration, as the volume perceived as flowing varies with the viscosity because of the residues clinging to the walls. The useful range of a particular viscometer may be narrowly limited. The devices must also be absolutely clean to ensure good wetting by the fluid. Precision in measurement of the time of flow is usually tested by several determinations, choosing the time values which agree within one or two seconds. Automatic timers usually require light and a photocell detector, and opaque liquids can not be investigated. The capillary-flow devices have some theoretical problems in their usage as well. The fluid passing through the orifice is not subject to constant hydrodynamic pressure as the hydrodynamic head pushing the fluid is decreasing. As the viscosity of samples increases, more fluid will cling to the walls (as noted above) and the volumes passing through the cup will not be comparable. Another concern is with regard to pressure to force the flow of these fluids, which changes the shearing force. Indeed, one company claims to "match" viscosities in paired comparisons of fluids by altering the pressure until their flows are equal. That is not strictly valid in my view, especially for non-Newtonian fluids. These capillary devices are generally not suitable for characterizing complex flow behavior (e.g., a non-Newtonian fluid).

The Bariod Capillary viscometer and its counterpart from Burrell Corporation is an unusual capillary flow device which utilizes a compressed-air pressure head. It is most frequently used for extremely high viscosity heterogeneous mixtures, such as oil-well drilling muds. It could also be used to estimate the flow through paint or adhesive plant piping or hoses. We recommend it to a carpet manufacturer.

Another class of viscometers employs a falling rod, needle or ball or a rising bubble as the rate-measurement device. Again, the time of rise or fall is measured in seconds. These instruments are normally not appropriate for opaque liquids. There are automated falling-rod devices which use nonoptical sensors that can make measurements on opaque fluids. The falling rod or ball devices usually have very small clearance between the containment wall and the mobile element, and several problems might be worth noting. The heterophase fluids (latex paints or adhesives, pigment slurries) may coagulate or flocculate because of shear. The instrument must be perpendicular to the ground (falling or rising force parallel to gravity). Cleanliness is again essential. The viscosity ranges may be limited, although shear stress may be varied in the Laray viscometer, a falling-rod device for which timing may be automated.

A variation on the falline-rod viscometer has purported utility for in-process measurement. The Norcross Ring Viscometer provides high temperature and pressure system resistance up to 2000 cP viscosities. The electric monitoring could be applied to a simple control logic system, as well as the current readout device.

Rotational viscometers have in common the cup (the fluid holder) and the rotating element upon which is measured the fluid resistance to flow. Some devices use a simple spring and pointer to measure torque flow resistance; others use transducers to generate an electrical signal. The variety of rotating elements can take almost any shape. Some systems have paddles or pins about the axis of the rotational element. The Couette devices have bobs or rotating cylinders, usually within a known small distance from the container wall, although one may assume an infinite distance for the smaller spindles used with the Brookfield Synchrolectric Viscometer. The design distance between bob and wall is a determining factor of the shear rate, while the rotational speed is another determinant. Most systems will have a variety of speeds and bobs which may be used to determine viscosity variation with shear rate. The simpler devices, such as the paddle, pin rotational devices, or the Brookfield Synchrolectric pose problems in calculating viscosity from mechanical principles, as the bob shape is complex.

The geometry of those bobs is such that at various points on the bob the shear may be quite different than at other points (linear speed at the base of a horizontal pin is slow, at an outer surface point it is fast). However, the use of calibration fluids may be recommended.

The variation of rotational speed for many instruments is a single-point step function, allowing measurement at a particular speed for as long as needed to attain the equilibrium thixotropic viscosity value. Other instruments are designed to continuously vary the shear rate and record the torque (yielding a curve as in Fig. 13.2b). These give good indicators of pseudoplasticity or dilatancy, but several successive trials must be made to determine the extent of thixotropic behavior.

Some unusual features were designed into a rotational viscometer which allow its use in high viscosity heterogeneous systems such as plastisols. The Brookfield Synchoelectric Viscometer, fitted with a horizontal pin-type bob on the vertical rotating element, can be equipped with a holder that automatically raises and lowers the viscometer in the fluid. The bob, rotating at a slow speed (e.g., 3 rpm), describes a helical arc in the cement, plastisol, or ink, and is always starting a new path through the fluid to reduce likelihood of thixotropy. These fluids commonly return very slowly to their original configuration after the passage of a pin, so this technique ensures that the path will have collapsed by the time the rotating pin has returned.

Temperature control is not common among the less expensive instruments. The more sophisticated devices have oil or water-jacketed cups to control temperature via a recirculating heater. Some specialized devices provide programmable heating and cooling to measure viscosity variation in the hydration of grain or food, or the swelling and solution of starch. These instruments may be useful to those who use starch-latex systems in paper coatings. Measurements of viscosity with rotational devices should also have a record of temperature for comparison qualification, a common practice in the paint industry.

In-line process-viscosity monitors based on rotational elements are commercially available. The Brabender Convimeter has high temperature, pressure, and viscosity specifications in the design. The Eur-Control instrument can give an electrical or pneumatic signal, based on a rotating-disk viscosity detector, which may be used in process control. Haake also makes a rotational device for in process monitoring. Cambridge Applied Systems has a sliding ring device to sense viscosity in their in-line monitor [109].

A special class of rotational viscometers utilizes cone-and-plate systems. These devices use much less fluid than the Couette instruments. The geometry is precisely known, so the physics of the fluid response to rotation may be rigorously described. Some instruments are fitted to detect the vertical component of force derived from horizontal rotation, which allows treatment of the fluid flow problem in terms of plastic and viscoelastic components. Unfortunately, the expense of these instruments and the high degree of skill needed for operation and interpretation of the data inhibit frequent application of this technique to practical problem solving. Larger companies (e.g., rubber companies) and research institutes have the devices and use them, but utility requires precise definition of the problem and design of the experiment series aimed at bracketing a range of solutions to the problem. This requires good understanding of the practical application by the rheologist, and of rheology by the engineer looking for a practicable answer. Both must be effective communicators to ensure this understanding.

A relatively new entry into the field of viscometry is the class of viscometers based on measurement of vibrational damping. These devices are electronic, and constructed to monitor or control viscosity. They measure at a single frequency, so non-Newtonian behavior may not be characterized. Temperature and pressure ranges of utility may be expected, and at least one device has a temperature correction setting which may be adjusted. These may be used for in-line correction of a coating process viscosity (e.g., in paper coating).

An unusual viscosity measurement device requires an understanding of viscoelastic flow thermodynamics. The Fann Consistimeter is a recirculating pump with monitors on pressure and temperature. Non-Newtonian flow with a pressure and temperature change will be detected by a plot of temperature and pressure versus time. Increases in temperature may be ascribed to Joule heating, in part, although the energy input necessary to convert the solution or dispersion to an equilibrium viscoelastic flow must also be accounted for.

One last comment is necessary on some sources of standards for viscometer calibration. The Cannon Instrument Co. has commercial oil and asphalt standards conforming to ASTM Standards. Brookfield markets white oils (silicones?) viscometers calibration.

3. Reprise

The literature of the adhesives, coatings, and paper industries abounds with the recognition that viscosity is a controlling factor in their processes. A rigorous mechanical or physical analysis of the controlling parameters, for instance, at a doctor blade, only gives indication that viscosity should be a controlling factor. The prime question then becomes "Which viscosity measurement is most meaningful?" The answer to this question requires a basic understanding of the devices used to measure viscosity. Examples cited herein show the variety of problems industry can face in defining the practical viscosity problem, and the variety of instruments for viscosity behavior measurement.

Although fluid mechanics is essentially a mechanical engineering discipline, the chemical engineer and even the chemist are often faced with the requirement that a process control parameter may be viscosity and may require a special measurement technique. The range of instruments available, and the extent of the problems wherein they may be used, suggest that an effective match may be found to derive a solution to the problem. However, experience shows there are still difficulties encountered, especially in heterogeneous systems, which reduce the practice to a search for the best answer available as an approximation. Engineering analysis and approximation then becomes the key, rather than a rigorous physical or mechanical definition.

Definitions of Mathematical Symbols

n	=	viscosity, poise = (dyne sec/cm)
δ	=	shear rate, dynes/cm?
τ	=	shear stress, sec $^{-1}$
F	=	Force, dynes
f	=	force required to induce flow, dynes
a	=	area covered by shear force, cm?
h	=	distance from wall to point in moving fluid, cm
v	=	velocity of fluid, cm/sec
T	=	temperature
A	=	Andrade equation constant
B	=	Andrade equation constant

REFERENCES

1. J. E. Carmichael, "The Effects in Paper and Board Coatings of Carboxylated S. B. R. Latex Particle Size With Protein Cobinder," TAPPI Coating Conference Preprint, p. 9, 1975
2. R. D. Athey Jr., TAPPI, *60(4),* 118 (1977).
3. R. J. Farris, "Prediction of the Viscosity of Multimodal Suspensions from Unimodal Viscosity Data," *Trans. Soc. Rheol., 12,* 281 (1968).
4. J. E. Smith and M. L. Jackson, "Mathematical and Graphical Interpretation of the Log-Normal Law for Particle Distribution Analysis," *J. Coll. Sci., 19,* 549 (1964).
5. R. D. Athey, Jr., et al., "Hydrodynamic Volume Evidence for Swellability of Styrene-Butadiene Carboxylated Latexes," *Coll. Polym. Sci., 255,* 1001 (1977).
6. E. A. Collins et al., "Review of Common Methods of Particle Size Measurement," *J. Paint Technol., 47,* 35 (May 1975). An excellent article!
7. T. M. Austin, "Techniques for Particle Measurement," *Indus. Res. Develop.* p. 129 (Feb. 1979).
8. J. G. Stockham and E. G. Fochtman, eds., *Particle Size Analysis,* Ann Arbor Science Publishers, Ann Arbor, (1977).
10. S. H. Maron et al., "Determination of Surface Area and Particle Size of Synthetic Latex by Adsorption," *J. Coll. Sci., 9,* 89 (1954).
11. *Ibid.*, p. 104

12. *Ibid.*, p. 263
13. *Ibid.*, p. 347
14. *Ibid.*, p. 382
15. J. Kloubeck et al, "Determination of Surface Area of Copolymer Latex Particles by Surface Tension Measurements," *J. Polym. Sci., 14,* 1451 (1976)
16. R. E. Nieman and G. A. Gorenkova, "Determination of the Adsorption Saturation of Synthetic Latices," *Kolloidny Zh., 37,* 300 (1975).
17. E. R. Meincke, unpublished work in 1975.
18. E. A. Collins et al., *op cit.*
19. P. Stenius, J. Kuortti and B. Kronberg, "competitive Adsorption Phenomena in Paper Coatings," *TAPPI, 67,* 56 (May 1984).
20. R. D. Athey Jr., unpublished work of 1969.
21. E. A. Collins et al., *op cit.*
22. M. Kerker, *The Scattering of Light and Other Electromagnetic Radiation,* Chap. 7, p. 32, New York, (1969).
23. H. Small, *J. Coll. Interf. Sci., 48,* 147 (1974).
24. R. F. Stoisits et al., *J. Interf. Sci., 57,* 337 (1976).
25. E. A. Collins et al., *op cit.*
26. T. L. Provder and R. M. Holsworth, *ACS Org. Ctgs. and Plas. Chem. Prepr., 36,* 150 (1976)
27. E. A. Collins, *op cit.*
28. S. Kinsman, "Some Particle Size Measuring Possibilities for Nonwoven Fabric Binders," TAPPI Paper Synthetics Conference Preprint, p. 249 (1978).
29. W. E. Franczek, "Floc Characterization with Coulter Counter," M. S. Thesis, SUNY College of Environ. Sci. and Forestry (Apr. 1977).
30. ASTM D1417-75, paragraph 12.
31. R. D. Athey Jr., unpublished work of 1969.
32. R. D. Athey Jr. and R. W. Ireland, unpublished work of 1983.
33. B. W. Greene and A. S. Reder, "Electrokinetic and Rheological Properties of Calcium Carbonate Dispersions Used in Paper Coatings," TAPPI Coating Conference Preprint, p. 53 (1974).
34. E. Matijevic and C. G. Force, "Colloidal Properties of Rubber Latex III. Interaction with Hydrolyzed Metal Ions," *Koll.-Z. A. Polym., 225,* 33 (1968).
35. B. W. Greene, *J. Coll. Interf. Sci., 43,* 149, 462 (1973).
36. H. Ono et al., "Studies on the Stability of Styrene-Acrylonitrile Copolymer Latex Dispersions. I. Stability and Electrophoretic Behavior Against Simple Inorganic Electrolyte," *Coll. Polym. Sci., 253,* 538 (1975).
37. P. E. Nieman et al., "Investigation of the Latex of the Copolymer of Alkyl Acrylates with Methacrylic Acid," *Kolloidny Ah., 37,* 574 (1975).
38. M. W. J. van den Esker and J. H. A. Pieper, "The Critical Coagulation Concentration of a Latex (Student Experiment)," Physical Chemistry: Enriching Topics from Colloid and Surface Science, H. van Olphen and K. J. Mysels, eds., Theorex (Publ.), La Jolla (1975).
39. J. Shute, Special Summer Student Report to G. Poehlein, Lehigh Emulsion Polymer Institute (2 Sep. 1976).

40. R. D. Athey Jr., M. M. Conrad and W. Kulhanek, unpublished work in 1976.
41. R. D. Athey Jr., unpublished work of 1979.
42. W. E. Franczek, *op. cit.*
43. P. A. Shaw, unpublished work of 1984.
44. R. D. Athey Jr. and R. W. Ireland, unpublished work of 1983.
45. P. A. Shaw and R. D. Athey, unpublished work of 1983.
46. R. D. Athey Jr. and K. V. Smith, unpublished work of 1985.
47. E. R. Meincke, unpublished.
48. L. A. O'Neill and G. Christensen, "Analysis of Thermosetting Resins Cooperative Study," *J. Paint Technol., 47,* 46 (1975).
49. F. Gaslini and L. Z. Nahum, "Conductometric Titrations of Very Weak Acids," *Anal. Chem., 31,* 989 (June 1959).
50. D. A. Kangas, "Characterization of Carboxylated Latexes by Potentiometric Titration," *ACS Div. Org. Ctgs. and Plas. Chem. Prepr., 36,* 353 (1976).
51. R. D. Athey and D. R. Campbell, unpublished.
52. D. H. Dicker, *Am. Laboratory,* p. 73 (Feb. 1969).
53. D. R. Campbell, unpublished work of 1976-1977.
54. Teijin Ltd., "Synthetic Latex Purification," Jap. Pat. 82 30705, (19 Feb. 1982); Cf: *CA 97,* 93343e (1982).
55. J. Hearn et al., "Problems Associated With The Preparation of Clean Polymer Latices", Presentation No. 79, ACS Div. of Coll. and Surf. Chem, 176th National ACS Meeting, Sep. 11–14, 1978.
56. R. D. Athey, A. K. Chatterjee, M. M. Conrad, and W. Kulhanek, unpublished work of 1976–1977.
57. G. D. McCann et al., "Effect of Ion Exchange on Latex Stability," *ACS Div. Polym. Chem. Prepr., 11(2),* 898 (1970).
58. A. A. Kamel et al., "Cleaning Latexes for Surface Characterization," *J. Coll. Interface Sci., 87,* 537 (1982).
59. D. H. Foster, R. S. Englebrecht, and V. L. Snoeyink, "Sorption of Proteinaceous Materials on Weak Base Ion Exchange Resins", *ACS Org. Ctgs, and Plas. Chem. Prepr., 34(2),* 43 (1974).
60. R. D. Athey, Jr., "Organic Polymer Dispersants for Pigments in Water," *Modern Paint Coatings, 72,* 31 (Feb. 1984); see Table III.
61. W. Obrecht et al., "Reactive Microgels: 5, About the Construction of Sulfate Group Containing Microgels by Emulsion Polymerization of 1,4-Divinylbenzene," *Makromol. Chem., 177,* 2235 (1976).
62. H. One and H. Saeki, "Preparation and Properties of Poly(methyl methacrylate) Latex Dispersions Having No Surfactant," *Coll. Polym. Sci., 253,* 744 (1976).
63. H. One and H. Saeki, "Preparation and Properties of Polymer Lattices(sic) Polymerised Without Surfactants," *Br. Polym. J., 7,* 21 (1975).
64. A. Homola et al., "Experiments with Soap-Free Polymerization of Styrene in the Presence of Alcohols," *J. Appl. Polym. Sci., 19,* 3077 (1975).
65. B. W. Greene, D. P. Sheetz and T. D. Filer, "In Situ Polymerization of Surface Active Agents on Latex Particles I. Preparation and Characterization of Styrene/Butadiene Latexes," *J. Coll. Interface Sci., 22,* 90 (1970).

66. *Ibid.*, p. 96.
67. K. Sakota and T. Okaya, "Preparation of Poly(styrene) Latexes in the Absence of Emulsifiers," *J. Appl. Polym. Sci., 20,* 1725 (1976).
68. D. C. Blackley, High Polymer Latices, Vol. I., p. 238, Palmerton Publishing Co., New York (1966).
70. P. Sennett and J. P. Oliver, "Colloidal Dispersions, Electrokinetic Effects, and the Concept of Zeta Potential," *Ind. Eng. Chem., 57,* 32 (1965).
71. P. Sennett et al., "Electrokinetic Effects in Paper Coating Colors," *TAPPI, 52,* 1152 (1969).
72. R. A. Stratton and J. W. Swanson, "Electrokinetics in Papermaking - A Position Paper," *TAPPI, 64,* 79 (1981).
73. A. L. Smith, "Electrical Phenomena," Chap. 7, *Dispersion of Powders in Liquids,* G. D. Parfitt, ed., Appl. Sci. Publ., London (1982).
74. V. M. Tereikovskii et al., "Temperature Coefficients of the Electrical Conductivity of Latex and Latex Mixes," *Internat. Polym. Sci. Technol., 3,* T/56 (1976).
75. A. A. Baran et al., "Effect of Temperature on the Electrical Charge of the Surface of Some Colloid Particles," *Kolloidny Zh., 38,* 328 (1976).
76. J. B. Kayes, "The Effect of Surface Active Agents on the Microelectrophoretic Properties of a Poly(styrene) Latex Dispersion — Microelectrophoretic studies," *J. Coll. Interf. Sci., 56,* 426 (1976).
77. E. Witt and D. S. Heller, unpublished work of 1965.
78. T. G. Bol'shakova et al., "Electrokinetic Properties of Poly(styrene) Latex in a Nonionic Emulsifier," *Zh. Prikl. Khim. USSR, 47,* 940 (1974).
79. B. V. Eremenko et al., "Adsorption of Poly(oxyethylene) and the Electrokinetic Potential of Antimony Sulfide Particle in Aqueous Solution," *Kolloidny Zh., 37,* 1083 (1975).
80. J. Lyklema, "Inference of Polymer Adsorption from Electrical Double Layer Measurements," *Pure Appl. Sci., 46,* 149 (1976).
81. A. K. Chatterjee, private communication.
82. P. Debye, *J. Chem. Phys., 1,* 13 (1933).
83. "Test Procedure for Detecting Enzymatic and Chemical Degradation of Viscosity in Latex Paint," Hercules, Inc. Bulletin VC-492B (1984).
84. R. D. Athey Jr., proprietary literature search report for client company, 1980.
85. ASTM D 1417, paragraph 5.
86. G. N. Fossick and A. J. Tompsett, *JOCCA, 49,* 477 (1960); cited in H. L. Gerhart and E. E. Parker, "Polymer Coatings," *Ind. Eng. Chem., 59,* 43 (1967).
87. L. K. Austin and G. P. Cunningham, *J. Ctgs. Technol., 52(665),* 49 (1980).
88. K. J. Rygle, "Trace Residual Monomer Analysis by Capillary Gas Chromatography," *J. Ctgs. Technol. 52(669),* 47 (1980).
89. W. H. Jordan, unpublished work of 1982.
90. R. J. Steichen, "Modified Solution Approach for the Gas Chromatigraphic Determination of Residual Monomers By Head-Space," *Anal. Chem., 48,* 1398 (1976).
91. D. Krockenberger and H. Gmerek, "Study of Styrene-Butadiene Latexes by Headspace Analysis," *Fresenius' Z. Anal. Chem., 311(5),* 485 (1982); Cf: *CA 97,* 56929e (1982).

92. J. B. Pausch and I. Sockis, "Application of Headspace Chromatography to Analysis of Residual Volatiles in Synthetic Rubbers," *Rubber Chem. Tech., 50,* 828 (1977).
93. R. P. Lattimer and J. B. Pausch, "Determination of Residual Volatile Chemicals in Polymers by Solid
94. R. B. Bird, W. E. Stewart, and E. N. Lightfoot, Transport Phenomena, Wiley, New York (1960).
95. J. R. Van Wazer, J. W. Lyons, and R. E. Colwell, *Viscosity and Flow Measurements, Interscience, New York, (1963).*
96. P. E. Pierce, *J. Paint Technol., 41(533),* 383 (1969).
97. J. Freling, *Instrumentation Technol.,* pp. 41–45, (June 1972).
98. T. C. Patton, *J. Paint Technol., 40(522),* 301 (1968).
99. R. J. Greet, and J. H. Magill, *J. Phys. Chem., 71,* 1746 (1
100. Y. Greener and S. Middleman, *Polym. Sci. Eng., 15(1),* 1 (Jan. 1975).
101. J. D. Ferry, *Viscoelastic Properties of Polymers,* Wiley, New York (1970).
102. F. R. Eirich, ed., *Rheology: Theory and Applications,* several Volumes, Academic Press, New York.
106. P. A. Shaw, *J. Ctgs. Technol., 56(718),* 59 (Nov. 1984).
107. R. D. Athey Jr., *Chemtech, 11,* 308 (1981).
108. S. E. Orchard, *J. Appl. Sci., All,* 451 (1962).
109. Anon., *Designfax,* p. 50 (May 1989).

14

Analyses

I. INTRODUCTION

Strictly speaking, analysis should yield compositional information from a material, whereas testing yields physical property information. Occasionally, the test can be related to some compositional factor, e.g., viscosity can be related to total solids of a polymer solution. In the main, a test is simply made to yield a number characterizing some physical fact (tensile strength, coefficient of friction, color). Many of the tests in the two preceding chapters are really analyses (residual monomer, pH) but they are characterizations commonly used for regulatory and process control purposes. Their inclusion in those chapters is based on an earlier presentation made to industry groups who needed or wanted the information.

The analysis of emulsion polymer systems is as complex as the formulations themselves. It is easiest to consider the problem in two phases, the fluid colloidal system and the basic polymer solid. We have already alluded to some of these techniques, but will repeat with amplification or emphasis when necessary.

II. COLLOIDAL FLUID ANALYSES

The colloidal system is complex, as already indicated in the preceding chapters. The analyses used to identify colloidal stabilization components are difficult. One may be able to identify the category of some surfactant or protective colloid, but not a brand name or grade. Strictly speaking, one is looking for clues as to how the formulation has been made and what it contributes to the product.

Analysis does not often identify processing steps, and one has to know the process well before being able to identify the steps.

Colloidal materials can be separated by thin layer chromatography, semi-permeable membrane washes, and column chromatography on ion exchange resin beads.

The first is used to identify surfactant blend compositions. However, one has to develop a library of Rf factors from known surfactants. The second technique uses polyethylene bags containing pure water, into which a soft semi-permeable membrane bag is placed which is partly filled with latex. This would extract smaller surfactant molecules and salts, but not protective colloids. The caution here is that the latex-filled membranes absorb water (by osmotic pressure) in the process; therefore they cannot be filled initially in order to leave space for the water that seeps in.

Ion-exchange chromatography was a good way for separating walter-soluble or dispersible oligomers from the caboxylated emulsion polymers, but salt and surfactant fractions were separated as well. One should use a mixed resin (acid and base), and conductivity water. Test the ion-exchange resin for extractable material, as some leach surface-active materials. Fractions collected were monitored for conductivity, indicating the salt or surfactant presence, but nonionic materials (surfactants, protective colloids) have to be isolated by evaporation. Turbidity shows the emulsion polymer fraction. One caution here is that the emulsion polymer needs to be diluted (5:1, or even 10:1) as the emulsion polymer phase gets very viscous upon stripping the salt and surfactant from the surface. We never tried the noncarboxylated latexes with this technique, but suspect that they would coagulate immediately.

One may use other analyses to find other components. Simply drying and ashing will give reasonable indications of inorganic material. Sulfate or phosphate may indicate salt or surfactant, and silica may come from the defoamer. The cations found may be related to rheological needs, as the cation used affects the performance of carboxylated thickeners. Later chapters will go into details of what these do in formulations.

Ashing may not be easy. The techniques vary from lab to lab, with some following a simple ASTM D-3723 test and others using a sodium peroxide fusion in a bomb. The ASTM D-3723 procedure is termed a low temperature ashing at 450 ± 25°C, but I prefer plasma ashing (using the PLASMOD device) for room temperature ashing of small samples. The PLASMOD is successful, except for certain aromatic polymers (for instance, it fails with polycarbonate according to IR evidence. We eliminated a styrene-acrylonitrile rheology modifier from an injection-molding grade polycarbonate by oxygen plasma [20]. We also showed that precoating oxygen plasma treatment had no effect on the surface [21]. The PLASMOD device is available from March Instruments.

Another analysis tool is gas chromatography. We have already discussed it in relation to residual monomer. But other chemical indicators for composition or process can show up in the gas chromatography trace. This presumes the gas stream can be analyzed (mass spectroscopy or isolation in a preparative stream for IR and NMR). Monomer derived indicators include acetaldehyde → vinyl acetate, acetic acid → vinyl acetate, benzaldehyde → styrene, methanol → methyl acrylate (or methyl methacrylate), vinylcyclohexene → butadiene, and phenylcyclohexene → butadiene and styrene.

Phenylcyclohexene was found to be in carpets at the US EPA offices in Washington, where it cause distress to some people [63]. It is a Diels-Alder addition product of butadiene and styrene, and is probably present in any hot-polymerized SBR latex. Others could be found, as the Diels-Alder reaction products of monomers are not uncommon in the literature. Indeed, some are offered commercially by specialty organic chemical companies, including Union Carbide. As soon as I smelled their vinylcyclohexene, I knew what the pervasive odor in hot butadiene polymerization facilities was. Many latexes smell of it and it can be easily identified by its odor.

Gas chromatography, if carefully done, can yield other process indicators. Typical examples include:

1. Isobutyronitrile, tetramethylsuccunonitrile → AIBN
2. Benzene, phenyl benzoate, biphenyl → benzoyl peroxide
3. Dimethyl dithioglycolate → methyl thioglycolate
4. Isopropanol → isopropanol solution polymer redissolved in water
   → isopropyl peroxycarbonate initiator

The last will also show up as isopropyl acrylate–methacrylate in the IR of the polymer film.

Some people who object to injecting a latex onto a gas chromatography column. They fear that the solids will deposit into the injector port and decompose, slowly but continuously bleeding an undesired organic into the column. These objections can be discounted. There are many examples of good lab practice using this technique in industry, and it is easy to clean an injection port; it can be done with a pocket knife, as noted earlier in Chapter 13.

III. SOLID POLYMER ANALYSES

A. Solid Polymer Isolation

When the form of the polymer is of no concern, the easiest method of isolation is the best. Coagulation, via salt, acid, or nonsolvent, was always the quickest. I prefer methanol as a coagulant, as it is easily evaporated and leaves no residue. I have seen HCl residues in carboxylated polymers, resulting in two neutralization curves in the conductometric titration. Calcium nitrate or chloride, or sodium chloride, will serve well for coagulations, but the isolated crumb needs to be washed thoroughly. Coagulants are discussed more thoroughly in a later chapter.

Powder or granular samples (e.g., for determinations of T_g or for salt-pellet transmission IR) do not have to be prepared from a coagulation. It is likely that the surfactants and other colloidal stabilizers are not retained with the polymer in a coagulation process. Hence, one may want to simply dry a film and grind it. A Waring blender or a Braun coffee grinder performs well for grinding to granular particle sizes. A Wiley mill may be preferred for smaller particles. Soft polymer film (e.g., an ethyl acrylate-based system), are placed in pieces in the Wiley mill with dry ice to yield a powder, which should be kept in a freezer to prevent the particles from combining again.

The preparation of free films for analysis or testing has always been fraught with problems. But it is a commonly used technique, especially for transmission spectroscopy or permeability or strength tests. We have seen or tried many forms of film casting, from glass plates, polymer films such as poly(ethylene) and TEFLON, mercury pools, and Pyrex baking dishes or Teflon-coated cake pans. ASTM has standard guides for preparing glass plates or steel panels for film casting or coating [1, 2]. Some films cast easily, but others crack, have bubbles, or are tacky.

Film thickness is important. For color measurement and hiding-power or corrosion-resistance studies, the film thickness must be known. The drawdown bar, the bird knife, or the wire-wound rod can be used, with recognition that the nominal space allotted for fluid to deposit a film is only a guideline. Supplies of wire-wound rods may be obtained from Consler Scientific, other film casting supplies from Paul Gardner Co.

Rheological factors are also important in the fluid movement in front of and under the bar, rod, knife, so you need to measure exactly to make sure the dry film thickness is that desired. In other instances, weigh the card or panel before and after coating, measure the area covered, and compute the film thickness from solids content, density, weight, and area measured.

Those films may yet be too thin for tensile or permeation tests or others. Multiple coating to build up a thickness may work, provided there are no bubbles within. Building up a film for permeation tests by spraying, gives results inferior to those of cast films. Porosity may be a consequence of sprayed coatings, as found in permeation testing of roof coatings.

Other polymers do not cast as films, as their minimum film formation temperature (MFFT) is too high for ambient casting; they have to be heated in an oven with all the attendant problems. The alternative is to add a so-called coalescent cosolvent to lower MFFT, but that can affect the physical properties [3]. Dissolving the polymer powder in an organic solvent for casting can also make a difference in the morphology of the polymer [4].

Some of these problems can be eliminated by so-called centrifugal casting.

We have used any number of variations on this technique, ranging from an expensive machined tube with a Teflon lining to a paint can with the closed end partly cut out. In the simplest case, we place the tube on the rollers used for jar mills for pigment grinding, and let the water evaporate. In one instance, we even used a Meker burner on the outer surface to heat the tube to cast a film of a hot melt adhesive. However, the soft or tacky film, upon centrifugal casting, may be difficult to handle. If tensile testing was the objective, a simple talc dusting made the soft film easy to remove and handle. If that has to be avoided for analytical reasons, place the casting device in a freezer, and the film can most likely be handled, at least while cold.

Centrifugal casting is not new. The rubber industry has been investigating it, especially for urethane-based systems, for tires for twenty or more years. They have achieved a modest success [5], and patents do exist on making products by rotational moldings [12]. Indeed, I toured a plant many years ago making dolls' heads by rotational casting with a plastisol. Progelhoff and Throne [13] examined the engineering modeling of such a system.

One may even obtain a desired thickness of film, by adjusting the fluid volume put in the rotational casting device. If the rollers are level, the film thickness will be consistent. A 6-in. diameter casting device will yield a film of about 6 in. × 18.5 in. in area. That is enough for six tensile test strips, three 2-in. permeation cup samples, 18 1-in. squares for solubility parameter determination, and a small amount left over for the files. (It is always advisable to retain a piece on a 3-in. × 5-in. card for later examination for discoloration, embrittlement, odor, development of tackiness, etc.). One does not have to use the jar mill roller. The biochemical industry offers a smaller quieter set of rollers (from New Brunswick Scientific). A speed controller on the rollers might be useful. High viscosity fluids can be rolled slowly, whereas a low viscosity latex may need higher rolling rate.

A free film or powder is not required for analysis. The film may be cast on a foil, film, or paper, and examined for detail by surface techniques. Attenuated total reflectance infrared spectroscopy gives good results, as does microscopy. Casting such a film on a flexible substrate requires some care. Again, the wire-wound rod or draw-down bar is the way to control film thickness, if desired. The flexible substrate needs to be held down, but there are vacuum plates that will keep it in place and provide the absolutely flat surface needed for the control of film thickness.

Film casting can have its problems. Croll [6] found internal stresses within films, whereas electron microscopy studies by Khodzhaeva et al. [7] found coalescence in anionic emulsified latex films although nonionic stabilized latexes maintained their globular nature upon drying. Yaseen and Ashton [8] found casting films on tin foil; amalgamating them off the foil with mercury gave more reliable moisture permeation data. Chow et al. [11] also found that solvent cast films on flexible substrates had strains induced. Eissler [10] recommends a fluorocarbon substituted silicone coating on glass for film casting to eliminate stresses. Blackley [9] has a discussion of film casting techniques.

B. Analyses of Solid Polymers

Although there are many analytical techniques available for solid polymer systems, only two of major significance are in use. They are infrared spectroscopy and gel-permeation chromatography (or, more correctly, molecular weight determination methods). Another technique for compounded systems, such as paint films, is based on x-ray fluorescence and related methods. Nuclear magnetic resonance (NMR), serves well for small organic molecules, but has difficulty with polymers due to peak broadening. Though there is a technique (called magic angle) to sharpen the spectra, one will have to find another source of information. Pyrolysis gas chromatography is being tried. Analyses of coatings are reviewed annually in Indust. Engr. Chem.; other reviews are by Greiser [73] and Johnson et al. [75]).

1. Pyrolysis Gas Chromatography

Pyrolysis gas chromatography is a simple technique. Take the film sample into a closed chamber at the inlet of a GC, pyrolyze the polymer, and analyze the off gases for organic components. Simple systems have been studied fairly thoroughly, and the results seem as rational as many of the other fragmentation methods for organic materials (e.g., mass spectroscopy). A reference book on chromatographic methods gives examples [14]. Haken [62] reports that pyrolysis does not have to precede gas chromatography, as lab chemical degradation schemes can be devised.

The problem with unknowns lies in the mixtures of materials present in polymer emulsions, or their downstream formulations as paints, inks, adhesives, and other applications. In an instance of paint comparisons for a trade-secret theft case, five formulations (two ostensiblly alike) were tested, and no two were identical. The case did not depend on the analysis scheme, but the original and the copy were judged the most alike by several experts. Analysis, however, should be more precise than those judgments.

If analytical techniques are used, there are several preliminary steps needed to make the results worthwhile. The formulation components should be subjected to the analysis technique individually before the whole formulation is analyzed. This allows relating a fragment to an individual component. It also allows search for the fragment reactions that occur to make an expected fragment disappear. This level of basic research will need a substantial justification, as it is expensive but valuable.

2. Infrared Spectroscopy

We will presume that the reader is familiar with the basics of infrared (IR) spectroscopy and with the fact that the spectra shown are based on intramolecular motions of specific atom combinations (C-H stretching, O-H bending, and others). The fundamental problem with polymer spectroscopy is the mixture of materials contributing to the spectrum. One can obtain identical spectral from two substantially different paint formulations because the components had the same absorptions and intensities. One might suspect that one formulation was developed to match the infrared spectrum of the other. That has happened, but a simple chloride analysis showed the two formulations were drastically different.

For review, we note that there are two main techniques of IR spectroscopy. They are transmission (shining the light through the sample) and reflectance, with the latter subdivided into multiple internal reflectance (MIR) and attenuated total reflectance (ATR). The early devices used primarily transmission, using a moving diffraction grating set to control the transmitted wavelength variation with time. As computer capabilities became more readily available, a total spectrum could be sorted out by the computer through Fourier transform mathematics; this technique is labelled FTIR (Fourier transform infrared). Pattacini and Anacreon [40] described the computer techniques for subtracting one spectrum from another. We were able to use the computer subtraction technique to show that only 2–5% of the paper coating binder is visible, with the clay pigment being the remainder [41]. Reviews of IR spectroscopy technique are available from a variety of sources [22–25, 45]. References sources for spectra of monomers and polymers are also available [26, 27].

Uses of IR in polymer film analysis have covered a wide range. Graf et al. [30] compared reflectance and transmission spectra of polymers, while Carlson et al. [31] characterized coating cure reactions by FTIR and mechanical techniques. Ishtani [32] recommended FTIR for characterizing industrial mixtures, while we recommended a standard low-cost IR transmission device for quality analysis (QA) in an injection molding facility [33]. Frazee [34] preferred reflectance to transmission IR for characterization of highway department coatings and plastic films, while a Canadian group used reflectance IR to monitor water loss in adhesives on papers [35]. Fowkes and McCarthy [36] used FTIR to help assessment of interfacial bonding of polymer to pigment, and Baier and Zisman [37] used reference IR to argue that wetability differences of solvent-cast films were based on an intramolecular hydrogen bonding differences. Luongo [38] argued that many physical processes changed the structure of the polymer, as evidenced by his IR studies. Aronson [39] described IR absorption and emittance for camouflage coating formulations.

Polymer degradation or defects have been described after IR analyses. Webb et al. [42] used FTIR to characterize degradation of solar-mirror coatings by

UV. Isakson [43] used IR reflectance to characterize the degradations of melamine-cured resins in artificial weathering devices using UV, or in chemical degradative environments (ozone chamber, outdoor weathering). Skrovanek [44] looked at coating defects (craters) with IR to see if a specific formulation component, such as silicone, was responsible. Shaw, faced with a lot of latex blend that did not conform to MFFT specification, showed that this lot had a different IR spectrum than past lots from thin films, using the simplest possible low-cost grating spectrograph [77]. The instrument need not be expensive, complicated, or computerized to be useful.

Unusual applications of infrared spectroscopy have been reported. Compton et al. [15] used FTIR attached to a thermal gravimetric device to identify components evolved from a variety of polymers, including a poly(vinyl acetate). Harrington et al. [16] make the point that FTIR may now be used in GC applications because of design changes and improvements in optics, so all the uses of GC we have discussed previously (monomer, by-products, and others) allow FTIR identification on-line. Williams [17], used ATR to determine both total solids content and monomer composition of latexes. Nguyen et al. [18] used ATR-FTIR to establish the degree of surface oxidation on poly(ethylene) after glow-discharge processing, while Nowlin and Smith [19] used reflected and transmitted IR to evaluate the oxygen plasma effect on poly(p-xylylene), both studies being aimed at enhancing wetting for subsequent adhesive or coating application. As noted earlier, this is not successful on all polymers, as polycarbonate is unaffected by oxygen plasma treatment [20, 21]. Mirabella [28] detailed a variety of IR techniques, including computer manipulation of spectra for multilayer polymer films. Infrared microscopy is available, and a newsletter describing its applications is published monthly [29].

Sample preparations need not be complicated. The powdered coating or polymer may be mixed into KBr granules and pelletized for transmission spectra, and the free film of the unpigmented polymer may be used as is for transmission spectra. If the intensity is too high, make a thinner film or use less KBR, as the beam interuption techniques to lessen intensity also damp out fine structure that may be needed. Stretching the film to make it thinner may modify the spectrum [38]. In some instances, the film of a thermoplastic can be melted onto a salt plate [33]. This may also give a modified version of the spectrum, as compared to a completely relaxed film [38], but newer technology in spectral correlation may help interpret such differences [64]. Preparations of samples for reflectance may even be simpler, as pigmentation does not interfere as much. Simply flatten the chip, film, or plaque of coating, plastic, or rubber, and push it (gently) onto the reflecting crystal surface to be used.

Sample preparation requires smooth film surfaces to avoid scattering in transmission spectra and ensure close surface mating in the ATR or other reflectance method. Pigmentation may cause scattering or strong absorptions that mask structural details; therefore the film surface should be free of pigment. Ensure a spectrum record between 10 and 90% transmission. Results should be quantified by recording at least three spectra, making the appropriate peak height measurements, and comparing at least two peaks (the reference and the desired peaks) on each spectrum. The film must be well dried, as absorbed water distorts and broadens polar segment IR absorptions.

My experience with IR is that it can be used in many of the most difficult cases. It simply requires attention to detail in sample preparation and operation of the spectrograph. With those aids, one can get enough detail in the spectrum to identify the observed structures. Comparisons are the best way to confirm identity; verification by other analytical methods may be useful.

3. Molecular Weight Determinations

The molecular weight of the resin greatly affects its physical properties. The strength in bulk or the viscosity (in solution or in melt) are among the important properties that depend on the molecular weight.

Molecular weight is, however, not a single number. The polymer molecules are constrained statistically by the synthesis technique to yield a distribution, some small and some large. The distribution is important, as some properties (e.g., solution vapor pressure) are governed by the number average molecular weight, whereas other properties are governed by other averages. Review your standard introductory texts on polymers for a further discussion of the types of molecular weight averages, their mathematical formulae, and their impact on physical properties.

The concept of molecular weight of resins for polymer systems containing appreciable amounts of *gel* is different. Here we use the term "gel" as in the Flory statistics for cross-linking of condensation polymers or step-growth polymers. But those gels also exist in the chain-growth or addition polymers we emphasize herein. For instance, any polymerization using butadiene, isoprene, or chloroprene that is not short-stopped at about 60% conversion will contain gel particles, as these monomers are prone to chain transfer to polymer at high conversions. That chain-transfer point becomes a branch, and when two such radicals meet, a cross-link forms. There are appreciable amounts of crosslinks in crylate, styrene, and other copolymers. Hence the first step to take, if molecular weight seems to be a factor in performance, is to determine the gel content, usually gravimetrically by filtration through a fine mesh (400+) or micropore filter.

If the gel content is less than 5%, you can use any one of a myriad of molecular weight determinations.

We summarize the common molecular weight determinations in Table 14.1.

Table 14.1 Molecular Weight Measurement Techniques

Method	Capital Expense	Technologist Expense	Time To Run	Results	Limits on MW
Light scattering	Medium to high	High	High	Distribution, Wt., Z avg.	Normally above 20000[a]
Gel-permeation chromatography	Medium to low	Medium to low	Medium to low	Distribution, all averages	None
Ultracentrifuge	Medium	Medium	Low to medium	Distribution Z avg.	20000 Minimum
Viscometry	Low	Low to medium	Low	Viscosity avg. wt. avg. in special cases	Varies
Membrane osmometry	Medium	Medium	Medium	Number avg.	20000 Minimum
Vapor-phase osmometry	Low	Medium	Medium	Number avg.	30000 Maximum
End-group analysis	Low	Medium	Medium	Number avg.	20000 Maximum
Cryoscopy	Low	Low	Medium	Number avg.	20000 Maximum
Ebulliometry	Low	Low	Medium	Number avg.	20000 Maximum

[a]One purveyor of small angle light scattering (SALS) instruments claims 5000 MW is their lower limit.

The majority of the methods cited therein are of classical interest, and will not be discussed further. However, we will touch on three methods that are relatively easy and can be used in quality assurance situations for special cases. These methods are vapor phase osmometry (VPO); gel permeation chromatography (GPC); and, intrinsic viscometry (IV).

We choose these from our experience of their ease of operation, and the volume of literature on their use. The instruments are commercially available, and may be operated by technicians, although a professional interpretation will likely be needed for many situations.

a. Vapor phase osmometry. Vapor phase osmometry is a misnomer. Osmometry involves a membrane separating two solutions of unequal concentrations (one concentration may equal zero). The migration of solvent through the membrane from the more concentrated solution occurs until the two cells have opposing equal pressures, osmotic from one side and hydrostatic from the other. Vapor phase osmometry does not act like this, but the misnomer has remained with the method. These devices are available from Mechrolab, Hewlett-Packard, Wescan, and other sources.

The VPO operating principle is a comparison of the vapor-evaporation and condensation rates from two thermistor beads in a thermostatted compartment saturated with solvent vapor, one bead coated with solvent and the other with solution. One monitors the difference in temperature between the two beads by monitoring the resistance of the thermistor circuit, which changes as the solvent evaporates from each bead. Remember that the vapor pressure of a solution is lower than that of the pure solvent, and the molecules leave the solvent-coated bead faster than they leave the solution-coated bead. The solvent-coated bead cools faster than the solution-coated bead, because of the enthalpy of evaporation (a measure of heat loss due to evaporation *per mole*).

The procedure for the determination of molecular weight by VPO is relatively simple. The instrument is turned on to raise the thermostatted chamber up to the required temperature and the solvent therein to evaporation equilibrium. Several solutions of polymer and a calibration standard (benzil or tristearin are often recommended) at different weight concentrations are made up in a suitable solvent. Most instruments have fittings for syringes for several solutions ready for coating the appropriate thermistor, so these may be loaded. The solvent syringe is at a different position and is aimed at its thermistor. Coat the solvent, and immediately coat the other thermistor with solution. Start with a timer and monitor the resistance change with time. Two minutes later, one records the resistance (R) for that concentration of solution. Meeks and Goldfarb [46] found that the resistance measured with time first rises and then can fall off, so choosing a standard time to measure the resistance after the initial rise is an important aid to precision. Their choice of two minutes from experiments with benzil was borne out in my own work with poly(styrene) samples [47]. As soon as the resistance is recorded at the correct time, the solution thermistor bead is thoroughly rinsed off with solvent, and the whole cycle is repeated with new solvent on the solvent thermistor and solution on its thermistor.

The calculations involved in finding the molecular weight are straightforward. A plot of resistance, R, versus concentration (c, in grams per liter) is drawn, and the R/c intercept at zero concentration determined (by extrapolation or least squares straight line fit). The plot for the calibration compound yields an instrument and system constant (K) as shown in Eq. (14.1).

$$[(R/c)\lim\rightarrow 0]\overline{Mn} = K \tag{14.1}$$

The same type of plot with the polymer unknowns yields molecular Mn (this is a number average) by the straightforward rearrangement of Eq. (14.1) to Eq. (14.2).

$$K/[(R/c)\lim\rightarrow 0] = \overline{Mn} \tag{14.2}$$

A Fortran IV computer program is available for these calculations, complete with an error estimate for the molecular weight calculated [48]. Expected error estimate may be 10–20%, but precision of about 10% is a reasonable expectation.

Some precautions have to be taken: The coatings on each bead must be about the same thickness [46]. The Sickfeld cautions on thermistors must be observed [49].

Determine the change in resistance with concentration for at least five concentrations to obtain a reasonable least squares line fit for the molecular weight determination. The calibration standard for molecular weight should be as high as reasonably possible. I have used a 4000 MW poly(styrene) from Pressure Chemical with success. Since resistance changes with time, use a stopwatch to ensure that the resistances recorded are all taken at the same interval from the start of the determinations.

Kuma and Kobayashi [50] discuss the error inherent in VPO measurements and recommend the Mechrolab Model 301 VPO with a limit of about 10000 MW units, although Hewlett-Packard claims their Model 302 will measure up to 25000 MW [51].

b. Gel permeation chromatography. Gel permeation chromatography (GPC), now more correctly called size-exclusion chromatography, has developed rapidly over the past 25 years. Commercial computer-controlled and interpreted systems are available with broad spectrum capability.

Gel-permeation chromatography is another hydrodynamic volume-separation technique (like the hydromanic chromatography we discussed in particle size determinations, Chap. 13). These techniques are now combined in a category called size-exclusion chromatography. An excellent primer on GPC was presented by Cray Valley Products in 1977 [52]; more extensive and formal reviews were prepared by Johnson et al. [53] and Cazes [54]. The premise underlying GPC is simple: a polymer solution percolating through a bed of porous beads will separate by molecular weight because the larger molecules do not migrate into the pores as deeply as the smaller molecules. Hence the larger molecules have a smaller hydrodynamic volume to pass through faster than the smaller molecules. Separations are enhanced by several columns having gel packings of differing porosities. The method will give an indication of the complete distribution of molecular weights.

Many liquid chromatographs may be modified to act as GPCs, but Waters Associates designed specific GPC instruments. Perkin-Elmer, Dupont, and Chromatix also sell GPC instruments. The separation of the solution effluent can be monitored by refractive index (provided polymer and solvent have appropriately different refractive indices) or by UV (again, provided solvent or polymer has a high absorption, preferably the polymer). Waters Associates recently introduced a viscometric detector for their systems.

Samples can be collected from the effluent for further characterization, with care. Indeed, one may isolate the additives in a compounded polymer formulation by GPC (antioxidants, plasticizers). Pharmacia markets a suitable fraction collector. The sample must be completely soluble, as the column can clog quickly with a hot melt formulation based on an oligomer of acrylate or vinyl acetate with a hydrocarbon wax additive.

The actual determination of molecular weight is made from the strip chart record of the detector. The strip chart will often have time blips (called counts) at intervals on the trace. Mark the chart at the time of sample injection and note the polymer effluent time range. Calibration samples are normally chosen to have a very narrow ratio of weight average to number average, and you can plot a calibration curve of number average versus count. Calibration materials for solvent-based chromatography are available from Pressure Chemical and Polymer Laboratories Ltd.; Mann Research Laboratories offers a variety of purified proteins as "Molecular Weight Markers" for aqueous determinations. On the unknown, you can write at each count the number average of that fraction. Given the distribution, one can calculate the number, weight, "Z", or any other average. Laboratory Micro Systems has introduced a microcomputer to monitor and calculate all needed values, and Waters Associates have complete packages of chromatographs, computers, CRTs, and all the necessary connections and software.

Caveats for GPC include: the best solubility parameter match must be chosen for solvent and polymer, to ensure the largest possible radius of gyration of the dissolved polymer. This will enhance elution rate and separation. Filter samples through microporous filters to remove dust, bacteria, microgels, or other particles to avoid the column plugging. The choice of column packing material is critical, not only from the expected standpoints of particle size and porosity (size and amount), but also for surface polarity. The polymer may adsorb on the packing surface, slowing and spreading the separations and invalidating the universal calibration. Ted Provder has published many discussions of controlling GPC separations by polymer functionality as well as molecular weight. Calibration runs with polymers of known molecular weight should be done every time the instrument is turned on after a period of disuse, or once a week if in continuous use. You may find that once a week may be extended according to your own experience. The commercial copolymers of interest to you do not often have molecular weight standards available for calibration. Hence, you will have to assume that the hydrodynamic volumes of unknown and calibrator are similar, or make use of some theoretical universal calibration scheme, which has been published [55].

The GPC system will perform well for a long time, with care. Gel-permeation chromatography has been used to separate species involved in electrophoretic paint deposition [56], to guide polymer syntheses [57], and to monitor polymer degradation [58].

c. Intrinsic viscometry. Intrinsic viscosity (IV) is a comparatively old technology in molecular weight measurement. It involves measuring viscosity of several polymer solutions at different concentrations and extrapolating a plot of viscosity variable versus concentration to a zero concentration intercept. Strictly speaking, for quality assurance or control, a single viscosity measurement at a known solution concentration will suffice, as long as the control limits are well known and documented with a precision statement.

The equipment and instrumentation for molecular weight measurements are normally the capillary viscometers used in undergraduate chemistry courses, the Ostwald viscometer, for instance. Jena Glaswerk Schott & Gen., Inc has a system that monitors the flow through these viscometers with a photocell, and normal stopwatch timing can be eliminated. Schurz and Hochberger [59] suggest that a rotational viscometer may be adequate for the determination, as they were able to determine critical molecular weight for the Bueche entanglement point in concentrated solutions.

The calculations involved are relatively straightforward, if the background work has been done. The IV $[n]$ is the y-axis intercept of a plot of either $(\ln n)/c$ or $(nsp)/c$ versus c. Eq. (14.3), attributed to Mark and Houwink, relates molecular weight and IV. The presumption herein is that the solvent is a theta solvent, minimizing the hydrodynamic volume, so $a = 0.5$ and M weight average. Otherwise, the equation has to have the constants determined experimentally with independent molecular weight verification (GPC?)

$$[n] = kM^a \tag{14.3}$$

Your supplier may have such information if you cannot find it in the chemical literature.

Again, pitfalls may exist in the use of IV for molecular weight determinations: The glassware must be rigorously cleaned. Polymer solutions must be filtered to remove any gels, bacteria, or dust. Aqueous solutions must be strictly controlled for salt concentration and pH, as the results only yield a straight line when the ionic strength is high enough. The sample-filled viscometer must be equilibrated in a strictly controlled thermostat to ensure that the temperature is the same for all determinations.

The thermostat may be filled with water, but some algicide and slimicide may be necessary to reduce the number of cleanings and fillings one could expect.

The IV method has a long and respected history of usage in the rubber industry. It can be tedious for the practitioner, but is easy to set up, and gives relatively fast and reliable results.

4. Chemical Analyses of Films

Many analytical techniques of films are available to identify the composition of the surface, primarily for use in compounded, pigmented, and filled formulations.

These techniques includes ESCA (electron spectroscopy for chemical analysis) and the like. Verbanic [60] summarized more than a dozen of these currently in use for coatings, ceramics, and other industrial applications; reviews of such techniques are available [61]. Wilkinson [65] reviewed the development of electron microscopy and Clark [66] ESCA for polymers.

These techniques can be used many ways. Brennan and Bienenstock [67] used grazing angle x-rays on a surface to determine orientation of atom arrays, and LeGrand et al. [74] argue that the small-angle x-ray scattering (SAXS) shows the development of microcracks in fatigued surfaces. Herglotz [68] showed the difference between an integral coating and islands of a coating with ESCA signal response, among other features in his review of the technique. Sparrow [69] used ion-scattering spectroscopy (ISS) and secondary ion mass spectroscopy (SIMS) to characterize aluminum prepared for coating. Gardella and Hercules [70] compared ESCA, ISS, and SIMS for surface analysis of methacrylate polymers, identifying the alkyl ester fragments. They also noted four reviews of ESCA technology on polymers. Dwight [71] demonstrated sodium etch on Teflon to improve water wetting with ESCA. Everhart and Reilly [72] used ESCA to show surface inhomogeneity of low-density poly(ethylene) films treated with a nitrogen plasma. Chilcoat [76] found that cast acrylic sheet had much more hydrocarbon on their surface than would have been calculated from the chemical structure of poly(methyl methacrylate). We advised him that he simply had proof of backbone and backbone methyl groups at the surface, with the oxygens buried.

X-ray fluorescence (XRF) is another technique in common use. Jenkins [78] describes several techniques in use with XRF. This is not the x-ray diffraction that observes crystal sizes and shapes, another useful tool in polymer science, which is discussed by Stout and Jensen [79]. X-ray analysis of paints is the topic of a task group within ASTM, and one method is already a standard determination of ratios of anatase to rutile in titania [80]. A subset of the x-ray analyses is energy dispersive x-rays (EDAX), a fluorescence technique used for quality control on cements [81]. Henry [82] described a combination of EDAX with electron microscopy, which I have seen used in the FBI labs in Washington for paint pigment analysis. McCarthy et al. [83] Compare two such techniques. Puumalainen et al. [84] point out that XRF analysis of coating weight on a paper coating machine is insensitive to the moisture of the coating or paper.

These techniques use a high energy particle or radiation to impact the polymer film or coating. The incident beam may be reflected or scattered, which happens at an angle that can yield information on that surface, if you know how to interpret it. In other cases, the incident beam is absorbed, and a fragment or energy wave different from the incident beam is emitted (fluorescence or similar phenomenon). Again, capture and analysis (mass spectroscopy or IR for fragments, wavelength identity for the beam emission) yields information on the structure of the surface. These techniques can be reduced to quality control analyses, if the need arises. Typical examples in my experience are the XRF analyses for titania or lead in paints for certification purposes, a common practice by the suppliers of government-specification paints. In another instance, we found that the mixture of titania and clay in a coating was not as well dispersed as we had supposed, as the titania had flocculated to inhibit thorough mixing throughout the clay and binder.

Needless to say, these are techniques to be used by experts. Those of us who are not experts need to be well enough acquainted with the usefulness of these tools. Then, faced with the potential need, we must query the expert in the usage of the technique, to see if they can be of assistance. We need to give them enough information to ensure that their expensive analysis is worthwhile.

REFERENCES

1. ASTM D-3891.
2. ASTM D-609.
3. J. Malac, E. Simunkova and J. Zelinger, *J. Polym. Sci., A-1, 7,* 1893 (1969).
4. N. Minora, *J. Appl. Polym. Sci., 27,* 1007 (1982).
5. *Designfax,* 22 (Oct. 1988).
6. S. G. Croll, *J. Ctgs. Technol., 50(638),* 33 (Mar. 1978).
7. I. D. Khodzhaeva, Z. M. Ustinov, Z. N. Tarasova, S. S. Voyutski, L. I. Sedakova, and I. A. Gritskova, *Koll. Zh., 38(2),* 403 (Mar.–April. 1976); Cf: *RAPRA Abstr.* 45931L (13 Sep. 1976).
8. M. Yaseen and H.E. Ashton, *J. Ctgs. Technol., 49(629),* 57 (June 1977).
9. D. C. Blackley, High Polymer Latices, Vol. 2, pp. 514–516, Palmerton Publishing Co., New York (1966).
10. R. L. Eissler, *J. Appl. Polym. Sci., 12,* 1983 (1968).
11. T. S. Chow, C. A. Liu, and R. C. Penwell, *J. Polym. Sci., Polym. Phys. Ed., 14,* 1311 (1976).
12. P. Merriman and F. Farnsworth, Br. Pat. 1139643 (8 Jan. 1969).
13. R. C. Progelhof and J. L. Throne, *Polym. Eng. Sci., 16(10),* 680 (1976).
14. C. G. Smith, N. E. Skelly, R. A. Solomon, and C. D. Chow, *Handbook of Chromatography,* Vol. 1, *Polymers,* CRC Press, Boca Raton, FL (1982).
15. D. A. C. Compton, D. J. Johnson, and M. L. Mittleman, *Res. Dev.,* 68 (Apr. 1989).
16. H. W. Harrington, R. J. Leibrand, M. A. Hart, and W. P. Duncan, *Res. Dev.* 88 (Sep. 1988).
17. R. C. Williams, Paper No. 8, 1983 Pittsburgh Conference.
18. L. T. Nguyen, N. H. Sung, and N. P. Suh, *J. Polym. Sci., Polym. Lett. Ed., 18,* 541 (1980).
19. T. E. Nowlin and D. F. Smith, *J. Appl. Polym. Sci., 25,* 1619 (1980).
20. R. D. Athey and P. A. Brown, unpublished study.
21. R. D. Athey and P. A. Shaw, unpublished study.
22. A. Dilks, *Anal. Chem., 53(7),* 802A (June 1981).

23. J. K. Barr and P. A. Flournoy, *Physical Methods in Macromolecular Chemistry,* Vol. 1, p. 109, B. Carrol, ed., M. Dekker, New York (1969).
24. "Infrared Spectrograms ar Film-Former Fingerprints," Technical Bulletin M-332A, Hercules Inc., Wilmington.
25. M. D. Low and H. Mark, *J. Appl. Polym. Symposia, 10,* 145 (1969).
26. *Infrared Spectra Atlas of Monomers and Polymers,* Sadtler Research Lab. Inc., Philadelphia.
27. Chicago Soc. for Coatings Technol., *Infrared Spectroscopy, Its Use in the Coatings Industry,* Federation of Societies of Coating Technology, Philadelphia (1969) and later editions.
28. F. M. Mirabella, Jr., *Spectroscopy, 1,* 49 (1986).
29. "Scan Time," Spectra-Tech Inc., Stamford, CT.
30. R. T. Graf, J. L. Koenig, and H. Ishida, *ACS Polym. Div. Prepr., 25(2),* 188 (1984).
31. G. M. Carlson, C. M. Neeg, C. Kuo, and T. Provder, *ACS Polym. Div. Prepr. 25(2),* 171 (1984).
32. A. Ishtani, *ACS Polymer Div. Prepr., 25(2),* 186 (1984).
33. R. D. Athey Jr., P. A. Brown, and H. L. Jones, *The Enigma of the Eighties: Environment, Economics, Energy, SAMPE National Meeting Prepr., 24(1),* 553 (1979).
34. J. D. Frazee, *JOCCA, 57,* 300 (1974).
35. H. K. Huynh, P. E. Lancaster, P. LePoutre, and A. A. Robertson, *TAPPI, 61(12),* 63 (1978).
36. F. M. Fowkes and D. C. McCarthy, *ACS Polym. Div. Prepr., 24(1),* 228 (1983).
37. R. E. Baier and W. A. Zisman, *Macromolecules, 3(4),* 462 (1970).
38. J. P. Luongo, *J. Appl. Polym. Symposia, 10,* 121 (1969).
39. J. R. Aronson, DTIC Technical Report No. 14076.2-0: 17451.1-0. (Jan. 1982).
40. S. C. Pattacini and R. W. Anacreaon, "Applications of Computer Difference Spectroscopy in Polymer Analysis," Perkin-Elmer Infrared Bulletin No. IRB-71 (Dec. 1979).
41. R. D. Athey Jr., unpublished work at Mellon Institute, 1979.
42. J. D. Webb, P. Schissel, and A. W. Czandema, *ACS Org. Ct. Appl. Polym. Sci. Proceedings, 46,* 718 (1982).
43. K. E. Isakson, *J. Paint Technol., 44(573),* 41 (Oct. 1972).
44. D. J. Skrovanek, *J. Ctgs. Technol., 61(769),* 31 (Feb. 1989).
45. U. Schernau, *Europ. Coat. J., 6,* 510 (1989).
46. A. C. Meeks and I. J. Goldfarb, "Time Dependance and Drop Size Effects in Determination of Number Average Molecular Weight by VPO," *Anal. Chem., 39,* 908 (1967).
47. R. D. Athey Jr., "Telechelic Polymers from Dimethyl Dithioglycolate," Ph.D. Dissertation, University of Delaware (1974).
48. Contact the author. Copy available for $5.
49. J. Sickfeld and B. Heinz, "Application of Thermoanalysis for the Investigation of Coating Deficiencies," *Farbe Lack, 85,* 366 (1979).
50. S. Kuma and H. Kobayashi, "Vapor Phase Osmometry," *Makromol. Chemie, 79,* 1 (1964).

51. "Vapor Pressure Osmometers," Hewlett-Packard Data Sheet 3020 (1/67).
52. Cray Valley Products "Gel Permeation Chromatography," *Pigment Resin* Technol., p. 5 (Sept. 1977).
53. J. F. Johnson et al., "Gel Permeation Chromatography with Organic Solvents," *Rev. Macromol. Chem., 1,* 393 (1966).
54. J. Cazes, "Gel Permeation Chromatography, Part I," Section XXIX in *Topics in Chemical Instrumentation,* S. Z. Lewin, ed., *J. Chem. Educ., 43,* A567 (July 1966).
55. For instance, Z. Grubisic et al., *J. Polym. Sci., 5B,* 753 (1967).
56. R. A. Ellis, *Pigment Resin Technol., 7,* p. 4 (Apr. 1978).
57. W. A. Pavelich and R. A. Livigni, "The Use of GPC in Guiding Polymer Syntheses," *J. Polym. Sci., 21C,* 215 (1968).
58. J. G. Hendrickson, "Basic GPC Studies. I. Polymer Degradation," *J. Appl. Polym. Sci., 11,* 1419 (1967).
59. J. Schurz and H. Hochberger, "Untersuchungen über den Lösungzustand von massig konzentrierten Polyisobutylen-Lösungen," *Makromol. Chem., 96,* 141 (1966).
60. C. Verbanic, "Surfaces: Where the Action Is," *Chemical Business,* p. 14 (Feb. 1989).
61. P. F. Kane and G. B. Larrabee, *Anal. Chem., 51(5),* 308R (Apr. 1979).
62. J. K. Haken,*Prog. Org. Ctgs., 7,* 209 (1979).
63. A. C. Nixon, *The Vortex, 51(1),* 11 (Jan. 1990).
64. S. Stinson, *Chem. Eng. News,* p. 21 (Jan 1., 1990).
65. G. Wilkinson, *Today's Chemist,* p. 24 (Dec. 1989).
66. D. T. Clark, *Critical Reviews in Solid State and Materials Science,* 1 (Dec. 1978).
67. S. Brennan and A. Bienenstock, *Res. Dev.,* p. 52 (Dec. 1989).
68. H. K. Herglotz, AFP SME (Soc. of Mfg. Eng.) Technical Paper FC77-673 (1977).
69. G. R. Sparrow, SME Technical Paper IQ77-442.
70. J. A. Gardella and D. M. Hercules, *Anal. Chem., 51,* 1879 (1951).
71. D. Dwight, *Chemtech,* p. 166 (March. 1982).
72. D. S. Everhart and C. N. Reilly, *Anal. Chem., 53,* 665 (1981).
73. R. H. Grieser, *Prog. Org. Ctgs., 3,* 1 (1975).
74. D. G. LeGrand, W. V. Olszewski, and G. R. Tryson, *ACS Org. Ctg. Plas. Chem. Div. Prepr., 45,* 628 (1981).
75. S. Johnsen, K. V. Poulsen, and H. K. Raaschou Nielsen, presentation to the 21st FATIPEC Congress (1972).
76. R. Chilcoat, unpublished.
77. P. A. Shaw, unpublished.
78. R. Jenkins, *X-ray Fluorescence Spectroscopy,* Wiley, New York, (1988).
79. G. H. Stout and L. H. Jensen, *X-Ray Structure Determination—A Practical Guide,* 2nd ed., Wiley, New York (1989).
80. ASTM D3720-84.
81. D. A. Gedcke, B. D. Wheeler, and N. Jacobus, *Indust. R & D,* p. 111 (Apr. 1979).
82. P. Henry, *Indust. R & D,* p. 59 (Jan. 1978).
83. J. J. McCarthy, J. J. Christenson, and J. F. Friel, *Amer. Lab.,* p. 67 (Sept. 1977).
84. P. Puumalainen, H. Venalainen, and R. Rantanen, *TAPPI, 63(7),* 55 (July 1980).

IV

Additives for Postpolymerization Compounding

Many industries are based on a formulator's art which comes into play to attain all the desired properties. Sometimes there is what Bill Coder of General Tire calls the OEI—"the one essential ingredient—that makes the thing work." The key, however, is proving the need for each ingredient. I once worked for an organization where everybody that touched a formulation had to add his own OEI. After spending a lot of my time trying to prove that propylene glycol in an adhesive formulation did nothing, it ended up staying in the formulation because the pilot plant people "needed it." That's when I decided to learn statistical treatment of experimental data for hard verification of data comparisons.

The basis for this section will be to describe classes of additives and their functions. I will give some chemical structures and explain how the additive actually works. However, I will not give brand names or suppliers, since the experienced chemist (or your local expert consultant) knows where to get the supplies needed. They are useful to the coater, when an adjustment needs to be made on line to make the final applied film look or perform just right. Indeed, the TAPPI Paper Coatings Division has a committee on additives to advise the mill coating engineers and chemists on what materials are available and from whom; the committee publications would be good references for the latex user for textiles, paints, inks, and other materials.

15

Colloidal Stabilizers

I. MATERIALS AND THEIR FUNCTIONS

The waterborne coating or adhesive is composed of colloidal materials, that is, polymer molecules or particles too small to be seen with a standard optical microscope. There may also be pigment particles that are likely to be colloidal in size (if the grind is correct). Ideally, the coating or adhesive is supplied in a ready-to-use form, just put it in the application equipment and run the product to be coated through. However, the problems of foam, substrate wetting, gelling, viscosity rise, and coagulation appear now and again *during* application. Surely it is the responsibility of the formulation manufacturer, but the production line may be interrupted while you ask his advice. You may be able to solve it yourself, with an understanding of colloidal systems and the chemicals you add.

A. Surfactants

Surface-active agents (surfactants for short) are indeed surface active. They will deposit on any available surface, including the piping, tank, mixer blades, latex particles, pigment, dirt, cigarette butts, monkey wrenches, and anything else. Surface-active agents are water soluble, but orient themselves on any surface, including the air–water interface. Strictly speaking, the surface-active agents are indeed water soluble, but they are constantly engaged in an adsorption–desorption equilibrium with any available surface.

Actually, surfactants are a combination of hydrophilic and hydrophobic blocks that cause this orientation to surfaces. An example of an anionic surfactant mixed-block structure is:

$$\underbrace{CH_3CH_2CH_2CH_2CH_2CH_2CH_2CH_2CH_2CH_2CH_2CH_2(C_6H_6)}_{\text{hydrophobe}} \Big| \underbrace{SO_4^- \; Na^+}_{\text{hydrophile}}$$

The surfactants reduce the surface tension substantially, by orienting their hydrophilic ends into the water phase and their hydrophobic ends away from the water phase—to any surface.

The term surfactant is commonly used as a catch-all, often applied to surfactants, dispersants, wetting agents (that I would call hydrotropes), and to protective colloid materials. However, I prefer to more rigorously define these terms. Table 15.1 shows this differentiation for the first three terms. You can consider a protective colloid as equivalent to the dispersants in this table. A good annual review of commercially available surfactants is published [1]

In latex formulations surfactants have a twofold function. The first is colloidal stabilization. That is, they protect the colloid against premature mechanical or chemical coagulation. Indeed, the surfactant is used as the primary stabilizer during emulsion polymerization. However, it should be remembered that the other components also contribute to stabilization.

The surfactant can also be used to make an emulsion of a liquid additive (such as a plasticizer) for incorporation at some time later in the formulation. One may prepare the surfactant solution in water and stir in the liquid additive until there is no phase separation. If the liquid additive is of reasonable viscosity, one may mix the surfactant into the liquid and add the mixture to water with vigorous stirring. Both techniques give stable emulsions which may be added to the latex when needed. These techniques are necessary when the addition of the liquid additive directly to the latex will not yield a smooth emulsion, or when the additive floats to the top as an oily layer.

One may be sure to incorporate the organic liquid into the polymer (rather than having a mixture of emulsions) by using a cationic surfactant for the organic additive preemulsification when the latex contains an anionic stabilizer. An excess of anionic stabilizers in the latex is necessary, as cationic and anionic emulsifiers are antagonistic. This technique is patented (General Tire) [6].

The standard charged surfactant species are best for stabilization against mechanical coagulation. There are several chemical types of charged surfactant species: anionic, cationic, and amphoteric. However, the anionic sulfates, sulfonates, and carboxylates are the only materials of interest used in substantial quantities. These charged molecules impart a charge that covers the surface of the particles after adsorbtion, building a so-called Helmholtz layer. This charged layer around the particle causes the particles to repel each other by electrostatic forces. The electrostatic charges also guard against mechanical coagulations,

Table 15.1 Colloidal Stabilizer Comparisons

Property	Surfactant	Hydrotrope	Dispersant
Disperses pigments	Yes	Yes	Yes
Emulsifies oils	Yes	Metastable, at best	No
Reduces surface tension	Yes	Slightly	No
Critical micelle concentration	Ca. 0.01%	About 1%, if any	None
Typical chemicals	Dioctyl sulfosuccinate Lauryl sulfate Docecylbenzenesulfonate Octylphenoxypoly (ethylene oxide)	Dibutyl Sulfosuccinate Octyl Sulfate	Poly(acrylic acid) Poly(isobutylene-*co*-maleic acid)

since the surfactants cover all the shear surfaces (e.g., of pump impellers, stirrer blades, and other equipment).

The best protection against chemical coagulation is provided by nonionic or a combination of nonionic–anionic surfactants. The mechanism by which this class of surfactants works is called steric stabilization. The water-soluble segments of the surfactant molecule extend into the water phase to form a viscous barrier layer that inhibits the close approach of another particle. This is especially of interest to us since we intend to compound with a variety of additional materials, which are essentially salts, which we will discuss later.

The second function performed by surfactants is to reduce the surface tension for a variety of reasons. For example, one may want very good rewetting of the final product, or also want to foam the product prior to application. Either of these needs will require a lowering of surface tension.

B. Wetting Agents

I prefer to call these materials hydrotropes. However, they are particularly useful to assure rewettability of a substrate that requires absorbency or other wetted properties. The hydrotrope may be too polar to be compatible with the bulk polymer after the latex particles coalesce, and thus exude to the surface to aid the rewetting.

This hydrotrope exudation is a mixed blessing. In exterior house paints, it results in the first rain washing away the hydrophilic exudate, making the coating more moisture resistant, which is desirable. However, in roof coatings, this exudate carried by water is concentrated into low spots on the roof, and remains wetted much longer than the other parts of the roof. This damp exudate attracts dirt, and eventually develops a brown coloration. Roofers and roof inspectors complain of coatings that form "tobacco juice." Some analyses even show protective colloids (from the section below) in this staining hydrophilic mixture.

On occasion the hydrotropes form mixed micelles with standard surfactants to support emulsion polymerization. Thus, often a low surfactant recipe will consist of a combination of a small proportion of a good surfactant and a larger amount of hydrotrope (e.g., 0.5 and 2.5%, respectively). The prime reason to use this technique may be to stabilize the latex during the polymerization, and increase its salability by claiming it is a high surface tension fluid (perhaps to inhibit migration into porous substrates like a cloth). That argument may be specious, as the real control of the dynamics of wetting on a fabric in the standard continuous processing equipment may be the viscosity, normally high enough to inhibit rapid flow at low shear.

Since the hydrotrope is usually an anionic material, its ionic nature contributes to the total salt load of the system.

C. Dispersants

Although most coating systems contain particulate materials such as pigments or flame retardants in the final coating formulation, most also contain a small amount of organic chemicals known as dispersants. It is common practice in the emulsion polymerization synthesis of a latex to include a small amount of a dispersant, commonly 0.1–0.5% by weight on the polymer. Dispersants aid in the mechanical stabilization of the latex and subsequent formulation. They facilitate the addition of other particulate colloids, such as pigments, antioxidants, UV stabilizers, or flame-retardant fillers.

Another source of polyions in the binder may be the polymerization "recipe" itself. If the binder system is a copolymer which contains relatively hydrophobic monomers and some charged (or chargable) monomer (such as a carboxylic acid monomer, e.g., acrylic acid), the hydrophobic monomers generally polymerize first with little charged monomer incorporated. The charged monomer builds up in relative concentration until the latter stages of the polymerization, and then penetrates into the copolymer. These latter-formed copolymers are highly charged, and may be water soluble, similar to the carboxylated copolymers used as commercial dispersants. Indeed, some years ago, I was involved in the development of a chromatographic technique for isolating the charged polymer fractions in a latex; some of these were low molecular weight water soluble polymers, which make it easier to disperse pigments in the waterborne formulations used for coatings.

Since dispersants are ionic in nature, they add to the total salt load of the system. It can be reasonably assumed that part of the polymeric organic dispersant function is that of a charged protective colloid.

The dispersants are also useful in preparation of the particulate additives in a high solids preliminary dispersion alone. Thus, one may prepare a dispersion of a solid additive in a Cowles dissolver by emptying the additive in powder form into water which contains 1–2% of dispersant with stirring. The evaluation techniques and types of dispersants available were reviewed in the Sept. and Oct. 1975 issues of TAPPI [2].

These dispersants, however, affect the final properties of the coating. Corrosion is enhanced by using dispersants neutralized with sodium hydroxide. Certainly any permanent salt will add to the hydrophilicity of a coating, reducing its water-barrier properties.

D. Protective Colloids

The discovery over 100 years ago that fine particle colloid dispersions could be stabilized by organic hydrocolloids (such as starch, gelatin, and others) led to the naming of these water-soluble polymers as "protective colloids." This technology is a very important part of formulating emulsion polymerization recipes for poly(vinyl acetate), poly(ethylene-*co*-vinyl acetate), and certain other vinyl acetate copolymers.

The protective colloids give the polymer dispersions considerable mechanical and chemical stability because they are absorbed on the particle surface and, in some instances, are even grafted to the polymer. Current hydrocolloids used as protective colloids include a variety of poly(vinyl alcohol)s, dextrins, and other starch-based materials. The mechanism by which they work is again steric stabilization. However, protective colloids are a mixed blessing, because their structure is mostly substituted with hydroxy functionality. Thus, they will impart water sensitivity to the bulk polymer that could reduce wet strength. Of course these hydroxy-functional polymers can be cured into the system with the help of the appropriate cross-linking mechanism, such as the methylolamide monomers, or melamine formaldehyde-urea formaldehyde (MF/UF) resins. The highly hydroxylated polymers impart excellent adhesion to metals and fibers of high polarity (e.g., cotton, wood pulp, and nylon). These highly polar protective colloid polymers also act as stiffeners whenever flexibility is required.

E. Defoamers and Antifoams

Although these materials are usually sold as separate entities, they are formulated mixtures which may contain oils, silica, and surfactants. Defoamers and antifoams containing surfactant, participate in the dynamics of adsorption–desorption of all the surface-active species, and a new equilibrium in time will be established.

Since the nonsurfactant components may not stay dispersed (due to the reequilibration), one may occasionally see an oily layer on the surface of the fluid coating; these components may also deposit on the surface of the dried coating film (reducing gloss, making the surface slick, and prohibiting recoating). Assuring compatibility of the antifoam–defoamer within the system is another problem facing the formulator. It has to be solved with an Edisonian approach because the suppliers of these products are unlikely to provide information on components or formulation.

F. Salts

Certain standard water-soluble materials, called salts, are constituents of the formulation. They originate from many sources; some are added as pH buffers or as initiators for emulsion polymerization. Others act as cure catalysts, flame retardants, or pigments (use a solubility-product table to calculate how much zinc oxide or calcium carbonate dissolves in your formulation). If you do not use deionized water, your water supply may be an accidental source. There are by-product salts in some of the raw materials you use in your formulation, such as the sulfates in sulfate–sulfonate surfactants and dispersants, or the peroxidisulfate used to initiate polymerizations. Those sulfates can be corrosion catalysts in metal coatings.

In small amounts, the salts contribute to colloidal stability. The earliest synthetic colloids were dispersions of silver chloride or barium sulfate stabilized by the counterions of the salts present in excess. You would think, intuitively, that the salt ions would remain in the water phase, keeping the ionic concentration equivalent throughout. That is the case, but there is no charge within the pigment or latex particle. There is a driving force that has the ions perceive that uncharged volume and more over into the uncharged volume. The ions cannot move through the surface of the latex or pigment particle, remain at the surface adsorbed. Since these salt ions are charged, they make a measurable contribution to the charge on the surface. This model works for me.

Salts are the source of chemical coagulations, and we are reminded of the Schulz-Hardy rule. A hundred years ago, colloid chemists observed that colloids were coagulated to varying degrees by salts. The Schulz-Hardy rule is a semi-quantitative statement of empirical observations that trivalent ions were tenfold more effective in coagulating a colloid than divalent ions, and that divalent ions were about tenfold more effective than monovalent, as we noted in Chapter 12. But the point here is that *any salt* contributes to coagulation.

Strictly speaking, when we consider salt-base coagulation, we should include all contributions to the ionic strength or to what I call the "salt load" of the system. I have seen instances where the simple salt effect increasing the ionic strength renders a good surfactant no longer compatible with water by simply "salting out" the surfactant. The salting out followed a Scatchard plot in some formulations [3].

An adhesive formulation problem, where our statistically designed experiments led to a model, showed good additive dependence of viscosity on associative thickener and surfactant levels charged, but an unexplained adverse (negative) term based on the interaction of those two components. We tested this by preparing the same formulations without latex or filler to hide any results by turbidity; the system mixture contained only salts, surfactants, and a surface-active thickener. In this formulation, we found substantial evidence for the adverse interaction, namely, an increase in viscosity and development of turbidity. The total ionic strength, counting the contribution of surfactant and thickener acrylate anion moieties, was the controlling factor, that is, a simple "salting out" of all the surface-active species.

G. Colloidal Destabilizers and Gelling Agents

There are good reasons for gelling the waterborne formulation before it can move away from where it should deposit. The examples of dipped goods coagulated by polyvalent cations cited above are typical of gelling systems commonly used in industry, but there are others.

Some surfactants are specifically designed to degrade upon heating (perhaps by losing the charged portion of their structure) and thus eliminating the colloidal protection that ordinarily prohibits coagulation. Some protective colloids and nonionic surfactants are no longer water soluble at the elevated drying temperatures, and the system containing them will gel. Some additives are relatively innocuous in the formulation, but break down upon heating to a strong acid, which is a coagulant by converting charged anionic colloidal stabilizers to nonionic materials, having substantially less effect on the colloid stability.

The choice of stabilizers (surfactants, salts, thickeners, or dispersants) should be made carefully, in order to facilitate the task ahead. A carboxylic surfactant is more readily inactivated by a drop in pH than the sulfate–sulfonate surfactant, which is readily understandable, since pKa governs the percent of ionization as pH varies. Since the acrylates are sensitive to pH drop (they precipitate at pHz), these are preferred as dispersants or thickeners when gellation is desired. The maleate-derived dispersants and thickeners are insensitive to pH drop, and would not promote pH gellation.

This sort of technology to help gel the coating on a porous substrate (rather than have it drain off or absorb) will promote drying and curing and result in the film thickness and coverage needed. This way you can design the colloidal destabilization needed for your application. Chapter 20 will cover some coagulant and gellant technology.

II. TESTING OF SURFACTANTS AND PROTECTIVE COLLOIDS

The main reason for employing surfactants and protective colloids is to make the aqueous formulation more stable to shear, heat, or chemical additives (salts, cosolvents, plasticizers, fillers, pigments). The effectiveness of the colloidal stabilizers may be evaluated by using some of the performance tests on compounded latexes, discussed in Chapters 12 and 13, such as:

1. filtration for coagulum determination after standing some reasonable time period;
2. electrophoretic mobility;
3. foam tests;
4. critical coagulation concentration;
5. mechanical stability; and,
6. aging tests.

These tests may be done on a series of formulations, to which the colloidal stabilizer is added incrementally, in order to find the minimum needed to provide protection. Compare the material of several suppliers, not only on the basis of price, but in the formulation in web or coating applications. There was a case where nominal triethanolamine lauryl sulfate from six suppliers was compared, and only one did not discolor in a standard textile exposure testing method [5].

Surfactants may also be added to nonaqueous liquid systems to improve wetting of the substrate, in order to act as release agents where blocking occurs, and to modify sprayability [4]. Surface-tension measurement may correlate with effectiveness of such additions.

When critical, quality analysis on surfactants may include total solids content, pH, conductivity (if salt by-products are found important), and ash content (another indicator of salt by-products). Infrared spectroscopy, and GLC analysis for organic by-products may also be useful; e.g., unreacted alcohol may increase the foaminess of a luaryl sulfate. High pressure-liquid chromatography may be used to look at the distribution of molecular weights of the nonionic surfactants.

III. SUMMARY

The many components of the colloidal protection system are water soluble and adsorb onto surfaces to varying degrees. There is competition between each of these species adsorption–desorption equilibria that can require two days to become equilibrated. In addition, occasionally adding a stable dispersion to a stable emulsion gives a product that resembles cement or cottage cheese. The problem is not due to the individual components, but due to the fact that the two colloidal systems had to reestablish adsorption–desorption equilibria which produced an unstable colloid mixture. Probably a third to a quarter of the time spent in laboratory investigations of colloidal systems should be devoted to the stability of mixed colloids and their adsorption–desorption reequilibration. You will find that if you follow this advice, new products or developments will easily pass through pilot plant and production trials.

REFERENCES

1. *McCutcheon's Detergents and Emulsifiers Annual,* McCutcheon, Ridgewood, NJ.
2. R. D. Athey Jr., *J. Water Borne Coatings* (Aug. 1983).
3. R. D. Athey Jr. and M. M. Conrad, unpublished work, 1976.
4. See surface tension section for reference to work of A. H. Heitcamp.
5. R. D. Athey Jr. and J. W. Booth, unpublished work of 1964.
6. R. W. Kreider, US 3479313 (18 Nov. 1969).

16

Rheology Modifiers

I. RHEOLOGY MODIFIERS

A. Polycarboxylates

There are many categories of carboxylated water-soluble polymers that act as thickening agents for general use in coatings and adhesives. They range from the simple high molecular weight copolymers of maleic anhydride and ethylene, to methyl vinyl ether–maleic copolymers, or poly(acrylic acid) copolymers and homopolymers. The earliest versions of the polycarboxylic thickeners were hydrolyzed poly(acrylonitrile)s and poly(acrylamide)s. These may adsorb to some degree, but they mainly tend to thicken the water phase.

However, in the late 1950s and early 1960s, a class of modified copolymers similar in structure to those above, but also containing hydrophobic species, was developed. These copolymers might contain a small amount of a nonionic emulsifier converted to the acrylate or maleic ester, so that these unusual monomers could be incorporated into the copolymer. The thickening efficiencies of these polymers, that is, the number of centipoise viscosity rise per percent of thickener polymer added, were substantially improved over those of the homopolymers. There is argument as to whether the improvement is due to these hydrophobized thickeners adsorbing completely around the one particle to raise the hydrodynamic volume, or whether the polymers form tenuous particle-to-particle bridges that are easily broken upon shear. Regardless of the mechanism, these materials perform as desired. The increased efficiency yields several benefits to the user, such as reduction in the amount (cost, ionic load) of polycarboxylate present with attendant reduction in water sensitivity of the final product. A discussion for a TAPPI conference of the chemical factors governing effectiveness of these thickeners was published in German [1].

A class of carboxylated rheology modifiers is supplied as a latex on the acid side of the pH range. These may or may not be called the "alkali-swellable" latexes, but they perform by swelling and solubilizing to a degree as the pH is raised. They can also act as cobinders. The main feature of this class of thickeners is that as they swell. The surfactants used in their synthesis are eliminated to move about and find another surface to adhere to. Indeed, we showed that the degree of swelling in any carboxylated latex was considerable and should be accounted for rheologically. The surfactant reequilibration with pH changes, especially with any carboxylated latex species present, is also an important factor to check with rheological, foaming, and surface-tension measurement.

There are some caveats to using these carboxylated polymer systems in formulations. Polycarboxylates will coagulate the latexes containing poly(vinyl alcohol) or other hydroxylic protective colloid stabilization. Some thickeners of this class are susceptible to precipitation by polyvalent cations (zinc from zinc oxide, calcium or lead from the carbonate, etc.) [2, 3]. This precipitation is not a solitary occurrence, as nonionic surfactant may be desolubilized by calcium poly(acrylate) [4]. The contribution of the polycarboxylate to the total ionic strength may result in "salting out" some surfactants, with some even following the Scatchard [5] mathematical model for salting out phenomena. The *whole* formulation must be checked for compatibility, rather than just the latex.

B. Cellulosic Types

A variety of cellulosic derivatives may be used as viscosity modifiers. The simplest possible systems would be the cooked starches which are inexpensive, but require some sort of treatment to make them water soluble (cooking, enzyme degradation). Modified starches and celluloses are commercially available that do not require this treatment. They may simply be dissolved as received. They include the celluloses or starches that have been chemically reacted to produce methyl, ethyl, hydroxyethyl, or hydroxypropyl ethers, and even mixed ethers having a selection of methyl–hydroxylic species.

Cellulose can also be converted to the sodium alcoholate form and reacted with chloroacetic acid to give sodium carboxymethylcellulose (CMC). This product is contaminated with sodium chloride to some degree (which varies from grade to grade) that your supplier can characterize. This salt can adversely affect the emulsion stability, as noted earlier and can act as a corrosion catalyst for some metals. The CMC materials may gel or precipitate in the presence of some polyvalent cations.

Certain naturally occurring polymers qualify as cellulosic thickeners. These include the natural gums arabic, guar, and tragacanth. Natural kelp can be treated to give a series of alginate polymers that are very popular for thickening formulations. These polymers (especially the alginates, which contain some carboxylation) may be sensitive to polyvalent cations.

The advantage of cellulosic-derived thickeners is that they are not as sensitive to polyvalent cation precipitation or the salt load of the system as the polycarboxylates. However, that is not to say that there is no viscosity loss with increasing salt load. The hydrodynamic volume occupied by any water-soluble polymer is reduced with increasing salt concentration. This is the reason that the intrinsic viscosity of water-soluble polymers is measured in salt solutions. You have to determine the degree to which you can accept viscosity reduction by your salt load, and determine which thickeners will give the minimum viscosity reduction.

Another benefit of the cellulosic systems is their viscosity response to pH variation. Many carboxylated thickeners have a relatively narrow pH range for their maximum viscosity, perhaps 7 to 9. The cellulosic systems can be capable of viscosity control through the broader range of 5 to 9 in the short term, but they may be subject to hydrolytic degradation on long-term storage with attendant viscosity reduction. Long-term storage can also present potential problems in biological degradation. Glass [6] showed advantages in usages of cellulosic ethers in reducing spatter of coatings applied from a roller.

The pitfalls in using hydroxylic water-soluble polymers as thickeners are, in part, already described above. However, these cellulosic ethers will coagulate highly carboxylated polymer latexes [7]. Recent publications suggest that combinations of hydroxylic-type thickeners and carboxylated thickeners yield desirable rheology, but finding a compatible combination may be time consuming. Additions of alcohols (isopropanol, for instance) may "shield" the carboxylate and minimized interaction.

Another pitfall to usage of cellulosic (or other hydroxylic) thickeners lies in their interaction with surfactants. Saito [8] showed that the Kraft point (the temperature at which surfactant solubilities start increasing drastically) was lowered by methyl cellulose, and poly(vinyl alcohol) along with other water-soluble polymers. This could change the apparent critical micelle concentration, with attendant disturbance of the adsorption equilibria.

C. Proteins

A certain class of thickeners was used in the past, but is no longer employed extensively. These are the natural proteins, which have to be cooked to dissolve. Work we did on coatings about ten years ago identified three functions of casein (dispersancy, rheology control, and binding strength). Stabilization of clay dispersions with casein showed that viscosity and minimum viscosity dosage of casein both were pH dependent [9]. Other proteins, available at reasonable prices, can perform the same functions. These would include collagen and soya protein.

These thickener proteins will yield good rheology control, and will act as cobinders as well. They can be cross-linked into the polymer matrix of the binder, and have the capability of viscosity stability at a wider range pH. We also showed that the casein and soya proteins were effective thickeners for highly carboxylated latexes that would coagulate with hydroxyethyl cellulose [10].

However, the proteins are subject to the potential of biological degradation on long-term storage in solution, as are the cellulosics. Anyone having used an old casein-thickened latex paint from years past will remember the sour milk odor when applying the paint, though the odor disppeared after a few days. Some of these paints may discolor on aging. Another storage problem arises because the proteins are edible. A paper company using casein in its coatings had a substantial rat problem in the warehouse. There are reports of surfactant association with dissolved proteins, a potential reducer for the stabilization of the latex and any other particulates in the formulation [11].

D. Nonionic "Associative" Thickeners

Associative thickeners are relatively new introductions to this market. They are nonionic hydrophilic surfactants of high molecular weight with a number of hydrophobic "arms" to adsorb onto any particulate (pigment or latex). Kossmann [12] described the "diurethane" thickeners as early as 1978. Nonionic surfactants (which are mono-alcohol functional) can be reacted with polyisocyanates to give a "star" or "comb" like structure, for another approach. Glancy [13, 14] described some of the development findings that require several grades of the polyurethane-based thickeners, simply because of differences in adsorption characteristics of different latexes.

Another approach, probably the one used by Hercules, is to treat a hydrophobic reactive chemical (such as a "Ketene-ized" fatty acid) with a cellulosic thickener. The hydrophobe may be derived from a fatty acid-derived isocyanate. Shaw and Leipold [15] describe the performance of hydrophobized cellulosic thickeners in latex paint formulations.

These materials associate like the surfactant-modified carboxylate thickeners (and the "associative" characteristic as a "new" item is somewhat specious), but the viscosity control is over a wider pH range, which can be an advantage on the acid side of that range. Glass [16] offered his perspective of the thickener developments for water borne coatings.

E. Analysis and Testing of Thickeners

Viscosity control agents are available in a variety of forms. Solid polymers, such as the cellulosic, protein, or alginate thickeners, should be checked for absorbed water (by simple hygroscopic uptake) in an effort to have the best control of an effective formulation weight addition factor. The synthetic thickeners supplied as solutions should be tested for total solids content. Cover the aluminum pan bottom with a poly(ethylene) circle, as the alkalinity will attack the aluminum, and the alumina generated will give a false result.

Acid precipitations can isolate all but the maleic copolymers for solid polymer cleaning and characterization. We have been able to identify thickeners made by hydrolysis of methyl acrylate (co-?)polymers by the methanol found in gas chromatographic analysis of the water after the polymer was precipitated. Surface tension measurements on the much diluted solutions of high efficiency thickener systems may give a clue as to the reproducibility of the incorporation of the surface-active monomers. Thin layer chromatography has been used to identify the nonionic surfactant basis for thickeners made with polymerizable nonionic esters.

These analysis and testing techniques may be more of interest to the manufacturers of thickeners in a competitive market. In cases where the thickener may be the cause of a formulation problem, these tests may show why one material is superior to another.

F. Recapitulation

There are three theories as to why a water-soluble polymer will act as a thickener to increase viscosity of a waterborne formulation:

1. The component raises the viscosity of only the water phase;
2. The component adsorbs to all particles, increasing their hydrodynamic volumes
3. The component adsorbs to several particles to bridge them, making them act like a substantially larger particle.

All of these factors may be operating in a formulation, but in cases of adsorption, one has to consider the dynamics of surfactant adsorption–desorption as well. Indeed, I have seen formulations where foaming problems occurred only after the thickener was added. In another case, the dynamic adsorption competition required addition of a poststabilization surfactant and stirring for 20 minutes before adding the adsorbing thickener. Otherwise an unusually large amount of thickener would be required to attain the specified viscosity, and the viscosity would increase to unreasonable high values for days thereafter.

The caveats about inadvertent coagulation–precipitation by other formulation should be reiterated. The adverse interaction of a hydroxylic polymer with a polycarboxylate (as soluble polymer or latex) has ample precedent. Many polyvalent cations will adversely affect the solubility of the thickener polymer. All water-soluble polymers have their radius of gyration reduced as the ionic strength increases so that the higher the salt load, the less efficient the thickener.

REFERENCES

1. R. D. Athey, Jr., "Viskosität wasserverdünnbarer Beschichtungsstoffe: Chemische Einflussfacktoren", (transl. "Chemical Factors Which Affect Water Based Coating Viscosities"), *Farbe Lack, 85,* 757 (Sept. 1979), English translation available for $5 from author.
2. R. D. Athey Jr., Ind. Eng. Chem. *Prod. R & D, 14(4),* 319 (1975).
3. R. D. Athey Jr., *J. Water Borne Coatings, 5(3),* 25 (Aug. 1982).
4. Jap. Pat. 50035-034, assigned to Momotani Junten Ltd.; Cf: Derwent 60317x/32.
5. G. Scatchard, *Chem. Rev., 4* 383 (1927).
6. J. E. Glass, *J. Ctgs. Technol., 50(641),* 72 (June 1978) and preceding numbers in that series.
7. R. D. Athey Jr. and R. W. Ireland, unpublished work of 1983.
8. S. Saito, *Kolloid-Zeitschrift, 275(1),* 16 (1967).
9. R. D. Athey Jr., *Modern Paint Coatings,* p. 31 (Feb. 1984), see Fig. 2.
10. R. D. Athey Jr. and R. W. Ireland, unpublished work of 1983.
11. O. Laurie and J. Oakes, *J. Chem. Soc. 5, Faraday Trans. I,* 1324 (1976).
12. H. H. Kossmann, *Farbe Lack, 84(12),* 955 (Dec. 1978).
13. C. W. Glancy, *Amer. Paint Ctgs. J.,* p. 48 (Aug. 6, 1984).
14. J. W. Glancy and D. R. Bassett, *ACS Org. Ctgs. Plas. Chem. Prepr., 51(2),* 348 (1984).
15. K. G. Shaw and D. P. Leipold, *J. Ctgs. Technol., 57(727),* 38 (Aug. 1985).
16. J. E. Glass, *Amer. Paint Ctgs. J.,* p. 54 (Aug. 6, 1984).

17

Plasticizers, Cosolvents, and Coalescents

I. PLASTICIZERS

Any additive that produces a reduction of the glass transition temperature, a lower strength, or an increased elongation can be considered a plasticizer. These performance measures show that the bulk polymer molecules are better able to move around and rearrange themselves because the additive is acting as a solvent for the polymer molecules with which it is compatible.

Years ago, there was some controversy as to whether the plasticizing moieties were just slippery entities in the bulk polymer matrix that lubricated the movement of the larger molecules without regard for compatibility, or whether the plasticizer was a true solvent for the polymer molecule, creating a high solids solution. The controversy was resolved by considering the efficiency of the plasticizing component along the lines of solubility parameter theory; this hypothesis has taken over.

Thus, the water in a poly(vinyl alcohol) film or woodpulp fiber, the unreacted epoxy resin curing agent, the surfactant that can dissolve in the polymer matrix, or the oily liquid or greasy semisolid added purposely can all act as plasticizers. Some years ago, Sam Daroowalla of General Tire and I performed some experiments aimed at the glass transition temperature contribution of acrylamide in polymers. We found that we could only get reproducible results when we dried the samples rigorously, because the acrylamide is hydrophilic enough to adsorb water from the air. Hence, all formulation components that may be reasonable candidates as possible plasticizers for your binder should be tested at temperature–humidity ranges that are suitable for your system in its end use.

The commercial plasticizers are sold in several classes. The first are the low molecular weight polymers or oligomers in the range of 2000–7000 MW. These are viscous molasses-like liquids, may be preemulsified for addition to the formulation; hot emulsification can help. Their structures can vary from condensation-derived polyesters to free-radical-derived polyacrylates or a poly(butadiene-*co*-acrylonitrile). There are also polymers derived by free-radical or other addition polymerization with end-group functionality; these are called "telechelic" polymers; they were reviewed with a listing of some commercially available [9]. Whichever of these polymeric plasticizers you use, the benefits derived are a longer-lasting property; plasticizers are not volatile and are not extracted as easily as the even lower molecular weight materials.

The second class of plasticizers, the simple organic esters, has several subcategories based on their structures. They all have a central polyfunctional component at which the oily subgroups have reacted, resulting in the plasticizer action. The polarity of the central segment and the oiliness of the substituents determine the Solubility Parameter of the molecule and the materials with which it will be compatible. Examples of the diversity of the central components include phosphates, aromatic polyacids (phthalates, mellitates), and aliphatic polyacids (sebacates, adipates).

The phosphate-based plasticizers are reported to add a measure of flame retardancy. The alkyl or aromatic side chains of the plasticizer molecule are derived from alcohols or phenols that have been used to esterify the acid center segments listed above; butanol, 2-ethylhexanol, benzyl alcohol, or cresol may be used, for instance. As the organic ester products are lower in molecular weight than the polymeric plasticizers, they are more likely to be volatile or extractable. In addition, they are less expensive.

Addition of plasticizer to the formulation may not be as simple as just stirring it in. The plasticizer can be stirred if there is enough surfactant in the formulation to emulsify it. However, the plasticizer may just float to the surface, looking like the oily layer it is. One may add emulsifier to the plasticizer (like stirring one oil into another, a nonionic or a fatty acid like oleic will serve) at about 5% by weight; the mixture is then stirred into the waterborne formulation. If you use the oleic acid, check the pH of the final formulation to ensure that it is high enough to meet your needs and neutralize the oleate as well. If that does not succeed, you can preemulsifiy the plasticizer at 50–65% solids in water with surfactant at about 5% by weight of plasticizer. The assumption is that the surfactant will be compatible with the other surface-active materials in the formulation, and that the dynamic exchange among all the particles will not cause any difficulty.

There is another assumption underlying the two preemulsification schemes:

that the plasticizer will migrate into and coalesce with the polymer upon drying (which requires heat). General Tire has a patented process wherein they pre-emulsify the plasticizer with a cationic surfactant, and add this to a polymer emulsion stabilized with anionic surfactant. The cationic plasticizer particles migrate to and merge (coalesce) with the anionic polymer particles [13].

There is a technology wherein the plasticizer is a solvent for the polymer only at elevated temperatures. That is the basis for *plastisols*. Therein, a PVC is dispersed in the plasticizer of choice and does not dissolve (or even swell). The mix is fluid to some degree (some are almost like putty), and can be coated on a mold or can fill a mold cavity. Upon heating the mass becomes transparent and appears to solidify, a process that is incorrectly called gelation. Since true gelation is the formation of a cross-linked matrix, this process is more correctly the viscosity increase of several orders of magnitude to yield a Bingham plastic fluid. It is sometimes correctly called a fusion process. Upon cooling, the molded mass is rubberlike, and regarded as a solid in the tactile sense.

There are plastisol emulsions wherein a PVC emulsion and a plasticizer emulsion are mixed, with knowledge that they will fuse in the heating-drying stage of the process. Typically, the plasticizers used are emulsion poly(butadiene-*co*-acrylonitrile) polymers, phosphate triesters, or phthalate dialkyl ester emulsions. The phosphate triester emulsion mixtures with emulsion PVC polymers are mainly sold as flame-retarding coatings for fabrics or structures.

The most efficient plasticizers are those that closely match the Solubility Parameter of the polymer. The simplest way to test this is to dissolve 1 g of polymer in 10 g of plasticizer and measure the viscosity (if it dissolves). Comparing many such solutions of the polymer (in many plasticizers) will show as best those solutions with the highest viscosity. They are the solutions where the polymer molecules are in their largest radius of gyration form. These preferred plasticizers are the least likely to bloom out of the polymer, and the best for Tg reduction. There was a problem with the molded rubber label from the maker of a famous tennis shoe, where the rubber sole of the shoe was plasticized and the label became gumlike during storage. The problem was that the label rubber more closely matched the plasticizer Solubility Parameter than the sole rubber did, and the plasticizer migrated.

A more rigorous approach to match polymer to plasticizer would be to use one of the mathematical modeling techniques for solvency. One such technique is called the Solubility Parameter concept. A recent book by Barton reviews that concept and its many extensions [1]. The original concept, as formulated by Hildebrand and others at the University of California at Berkeley, was that the evaporation of solvents, and the interaction of solvent with solute (penetration of solvent between solute molecules to dissolve the solute), depending upon their cohesive energy densities. Burrell applied the concept to paint and ink technology [2]. Hansen devised a three-dimensional modeling technique, which

Table 17.1 Solubility Parameters for Plasticizers [7]

Material	Sigma d	Sigma p	Sigma h
Tricresyl phosphate	9.30	6.00	2.20
Butyl benzyl phthalate	8.80	6.00	2.00
Dibutyl phthalate	8.50	4.80	1.60 [3]
	8.10	3.40	1.50
Dioctyl phthalate	8.30	3.30	1.50 [3]
Dioctyl adipate	8.15	3.00	1.70 [3]
Diethyl pimelate	7.02	4.15	4.05 [1]

refined the system to eliminate some of the problems with the earlier (linear) model [3]. This work confirmed the polymer–plasticizer interactions and the optimum choices.

The technique has been used successfully to describe many polymer and solvent interactions. Members of the ASTM F23 Committee on Chemical Protective Clothing have collected information on permeation of chemicals through glove or other protective materials. The Matrecon Lab in Oakland has tried to model the permeation of hazardous waste pond liners by organic chemicals using the technique. We developed a statistical technique for determination of the 3D Solubility Parameter [10] with the Golden Gate Society for Coatings Technology and compared it to the Matrecon [11] technique in a subsequent article [12]. Swedlow workers modeled the solvent crazing of aircraft window substrates. The technique has been used to choose the best solvents for HPLC of bioactive materials (medicines and drugs). Table 17.1 gives Solubility Parameters for representative plasticizers.

However, the method has flaws. Among them is the need to discard methylene chloride data from the Swedlow study. Alternative mathematical solvent modeling schemes have been proposed by Fowkes (of Lehigh University) based on acid–base interactions [4], and by Taft (University of California Irvine) et al. [5] combining the acid-base theory with other phiscochemical factors. However, these flaws are not serious when considering the interaction of plasticizer with polymer, as the plasticizers are generally fairly close in Solubility Parameter to the polymer they plasticize and there is little need for the correction terms.

The basic reference to commercially available plasticizers is the *Red Book* for various industries, giving the plastics industry reference a broad coverage of materials [6]. Other such *Red Books* may be found for your industry as well.

II. COSOLVENTS AND COALESCENTS

These are liquid chemicals incorporated in the coating formulation to solve problems with film formation. For skinning problems in storage, or on drying

Table 17.2 Differences between Cosolvents and Coalescents

Property	Coalescent	Cosolvent
Turbidity change	More turbid or no change	Opacity reduced or eliminated
Viscosity change	None	Increased
Evaporation rate to dryness	None, for the water phase, though the polymer may remain tacky	Generally slowed

of thick films the appropriate cosolvent may be the answer. If you want to slow the drying to give better recoating, add high boiling water-miscible cosolvent.

The cosolvent has two functions: it *changes drying or open time* by speeding up the evaporation of water with an azeotrope, or by slowing the drying by leaving the film later than the water does. The main point is that the cosolvent keeps the binder particles from merging completely until the system is almost dry.

The other function of the cosolvent *may* be to help extend the water-swollen molecule of the binder. We have all seen examples of opaque polymer dispersions made transparent or translucent by the addition of cosolvent. This may have taken place because the cosolvent changed the refractive index of the water–cosolvent mixture or because the polymer molecule became more swollen by the mixture (with an attendant rise in viscosity).

The coalescent cosolvent (an anomalous and imprecise term) should actually be called *only* coalescent. It has only *one* function, to be absorbed by the polymer molecule and to plasticize it (temporarily) during drying to form a more continuous (and possibly more glossy) film. It lowers the minimum film-forming temperature (MFT) of the latex or polymer dispersion particles. Table 17.2 summarizes the differences between cosolvents and coalescents.

Some cosolvents seep into the polymer, and some coalescents are water soluble. However, the idealized definition is for the guidance of the formulator, to what the added organic liquid is *supposed* to do for the film, rather than how many functions it actually can contribute.

Cautions on coalescents and cosolvents exist. The favorites of mine and others over the years were derived from ethylene glycol (ethers and esters), but toxicity problems are making these no longer viable in the market. These organic liquids are generally flammable, as well.

We also need some caution with regard to the formulation. All cosolvents and coalescents (at least those coalescents that are water soluble) effect a change in the dipole moment of the water used as the dispersion medium. This is particularly important in systems using an electrostatic charge to stabilize the colloid(s). You can destroy the Helmholtz double layer of charges surrounding the colloidal particles with too much organic solvent. One of my favorite techniques for isolating a polymer or a colloidal mixture from the dispersion is to add an excess of a water-miscible solvent that is a nonsolvent for the polymer. I would use methanol or acetone for poly(styrene) latexes, for instance, and obtain a good yield of an easily dried product.

The choice of cosolvent or coalescent may be aided by the Solubility Parameter concept (see discussion in preceding section). Table 17.3 contains a listing of three-dimensional solubility parameter data gathered over several years from the literature. We have divided this Table into six sections, corresponding to the six planes in Fig. 17.1. This makes choosing a solvent easier, especially when you want to chose seven from different spatial segments for a modeling study. Table 17.4 shows typical polymer solubility parameter values. However, many of the newer solvents being sold by suppliers trying to replace ethylene oxide derivatives with propylene oxide derivatives are not yet characterized. They have to be tried without much guidance. Some of the data in Table 17.3 have been contributed by the suppliers of the newer solvents, but not all are represented. One may also try to calculate the solubility parameter from structural parameters. The Barton book [7] gives some guidance.

Determination of solubility parameter of a polymer requires more effort than one might expect. The polymer must be divided into small samples, each of which will be treated with a different solvent. Swelling determinations are convenient for cross-linked polymers, though any solvent-dependent physical property will suffice as an indicator of the polymer solvent interaction, be the property a strength or some other. In earlier years, prior to the introduction of the three-dimensional Solubility Parameter by Hansen [3], one could plot the percent swell on the y-axis and the one-dimensional Solubility Parameter on the x-axis. The peak swelling was a good approximation of the solubility parameter. However, some polymers gave two peaks, which were not easy to explain (though some argued that they were indications of block-copolymer structure.)

The determination of the three-dimensional Solubility Parameter is more difficult. One must use *many* solvents, and a statistical computer program capable of multiple regression must be used to calculate the model equation. You then find the maximum swell point in that model, which is the Solubility Parameter set of three dimensions for the polymer. How many solvents must be used? Querying many statistics experts on a six-plane representation of the three-dimensional space of the Solubility Parameters, I have received no answers. For my own satisfaction, I would use solvents from the segments of the six planes marked X in Fig. 17.1. One may not find a solvent for each of the 27 marked plane segments, but one may have sampled the space *thoroughly* enough to have reasonable estimates in the statistical model equation. Indeed, most modeling computer programs give some idea of the "goodness of fit" through an F-statistic

Table 17.3A Solubility Parameter Values for Common Materials

Solvent Solubility Parameter					SIGMA d Range = 7.0 to 7.5 Actual Values			
SIGMA p = from	to	SIGMA h = from	to	Solvent	Refs.[a]	σ_d	σ_p	σ_h
0	2.0	0	2.0	hexane (1), heptane	(2)	7.24	0	0
				FREON 113	(2)	7.20	0.80	0
				dipropylamine	(1)	7.50	1.30	1.70
		2.0	4.0	isoamyl acetate	(2)	7.50	1.50	3.40
				isobutyl acetate	(2)	7.40	1.80	3.10
				diethyl ether	(2)	7.10	1.40	2.50
				diethylamine	(1)	7.50	2.30	2.30
		4.0	6.0	methylisobutylcarbinol	(1)	7.47	1.00	6.00
				butyric acid	(2)	7.30	2.00	5.20
		6.0	8.0					
		8.0	10.0					
		10.0	12.0					
		12.0	up					
2.0	4.0	0	2.0	diethyl ether	(1)	7.05	2.45	1.00
				isobutyl butyrate	(1)	7.38	2.65	1.80
				methyl chloride	(2)	7.50	3.00	1.90
				EXXATE 900	(5)	7.20	2.90	1.80
				EXXATE 1000	(5)	7.30	2.80	1.50
				EXXATE 1300	(5)	7.40	2.50	0.80
		2.0	4.0	butyraldehyde	(2)	7.20	2.60	3.40
				methylisobutyl ketone	(1)	7.49	3.00	2.80
				EXXATE 600	(5)	7.10	3.40	2.60
				EXXATE 700	(5)	7.10	3.30	2.30
				EXXATE 800	(5)	7.20	3.20	2.00
		4.0	6.0	acetaldehyde	(2)	7.20	3.90	5.50
		6.0	8.0	acetic acid	(2)	7.10	3.90	6.60
		8.0	10.0					
		10.0	12.0					
		12.0	up					
4.0	6.0	0	2.0	methylal	(1)	7.35	4.10	1.50
		2.0	4.0	ethyl acetate	(1)	7.44	4.60	2.50
		4.0	6.0	diethyl pimelate	(3)	7.02	4.15	4.05
		6.0	8.0	l-pentanol	(3)	7.22	4.44	7.17
		8.0	10.0	formic acid	(1)	7.47	4.50	8.50
				formic acid	(2)	7.00	5.80	8.10
		10.0	12.0	methanol	(1)	7.42	5.50	11.20
				methanol	(2)	7.40	6.00	10.90
6.0	8.0	0	2.0					
		2.0	4.0	propionitrile	(2)	7.50	7.00	2.70
		4.0	6.0					
		6.0	8.0					
		8.0	10.0					
		10.0	12.0					
		12.0	up	water	(1)	7.00	8.00	20.90
8.0	10.0	0	2.0					
		2.0	4.0	acetonitrile	(2)	7.50	8.80	3.00
		4.0	6.0					
		6.0	8.0					
		8.0	10.0					
		10.0	12.0					
		12.0	up					

[a]References in Table 17.3F.

Table 17.3B

Solvent Solubility Parameter					SIGMA d Range = 7.5 to 8.0 Actual Values			
SIGMA p = from	to	SIGMA h = from	to	Solvent	Refs.	σ_d	σ_p	σ_h
0	2.0	0	2.0	octane	(2)	7.60	0	0
				dodecane	(2)	7.80	0	0
		2.0	4.0	diisobutyl ketone	(2)	7.80	1.80	2.00
				butylCELLOSOLVE acetate	(4)	7.80	1.90	3.90
		4.0	6.0	butyl acetate	(2)	7.70	1.80	3.10
				butyl lactacte	(7)	7.95	1.80	4.50
		6.0	8.0	2-ethylhexanol	(2)	7.80	1.60	5.80
		8.0	10.0					
		10.0	12.0					
		12.0	up					
2.0	4.0	0	2.0	methyl isoamyl ketone	(2)	7.80	2.80	2.00
		2.0	4.0	ethyl acetate	(2)	7.60	3.50	3.70
				mesityl oxide	(2)	8.00	3.50	3.00
				diethyl ketone	(2)	7.70	3.70	2.30
				butyl CARBITOL	(2)	7.80	3.40	3.40
				butyl lactate	(1)	7.65	3.20	3.80
		4.0	6.0	diacetone alcohol	(2)	7.70	4.00	5.30
				i-propyl lactate	(7)	7.80	2.50	5.00
				ethyl lactate	(7)	7.95	3.40	5.40
				methoxypropoxypropanol	(4)	7.80	3.40	5.20
				CARBITOL acetate	(4)	7.90	2.00	4.00
				methoxypropyl acetate	(4)	7.70	2.30	4.20
				dibutoxyethylene	(4)	7.80	3.40	5.20
				butoxyethanol	(2)	7.80	2.50	6.00
				methyl salicylate	(2)	7.80	3.90	6.00
				ethyl lactate	(1)	7.80	4.00	5.80
		6.0	8.0	*n*-butanol	(1)	7.81	2.50	7.80
				methoxypropanol	(4)	7.60	3.60	6.80
		8.0	10.0					
		10.0	12.0					
		12.0	up					
4.0	6.0	0	2.0	nitropropane	(2)	7.90	5.90	2.00
		2.0	4.0	methyl ethyl ketone	(2)	7.80	4.40	2.50
				CELLOSOLVE acetate	(1)	7.78	5.40	2.50
				acetone	(2)	7.60	5.10	3.40
		4.0	6.0	ethyl formate	(2)	7.60	4.10	4.10
				diethylene glycol monomethyl ether	(2)	7.90	4.50	6.00
				diethoxyethylene	(4)	7.90	4.50	6.00
				diacetone alcohol	(1)	7.65	4.90	4.50
		6.0	8.0	CELLOSOLVE	(2)	7.90	4.50	7.00
				methyl CELLOSOLVE	(1)	7.90	4.50	7.90
				methoxyethanol	(2)	7.90	4.80	8.00
		8.0	10.0	ethanol	(1)	7.73	4.00	9.70
				ethanol	(2)	7.70	4.30	9.50
		10.0	12.0					
		12.0	up					
6.0	8.0	0	2.0					
		2.0	4.0	nitroethane	(2)	7.80	7.60	2.20
		4.0	6.0	epichlorohydrin	(3)	7.76	6.00	5.12
		6.0	8.0	phenylhydrazine	(3)	7.61	7.27	6.68
		8.0	10.0	triethylene glycol	(2)	7.80	6.10	9.10
				diethylene glycol	(1)	7.86	7.50	9.70
				diethylene glycol	(2)	7.90	7.20	10.00
		10.0	12.0	dipropylene glycol	(1)	7.77	6.50	11.70
		12.0	up	water	(2)	7.60	7.80	20.70
8.0	10.0	0	2.0					
		2.0	4.0					
		4.0	6.0					
		6.0	8.0					
		8.0	10.0					
		10.0	12.0					
		12.0	up					

Table 17.3C

Solvent Solubility Parameter						SIGMA d Range = 8.0 to 8.5 Actual Values		
SIGMA p = from	to	SIGMA h = from	to	Solvent		σ_d	σ_p	σ_h
0	2.0	0	2.0	cyclohexane	(2)	8.18	0	0
				xylene	(1)	8.50	1.20	2.00
				1,1,1-trichloroethane	(1)	8.13	2.00	1.80
		2.0	4.0	furan	(1)	8.43	1.50	3.00
		4.0	6.0	1-octanol	(1)	8.30	1.60	5.80
		6.0	8.0					
		8.0	10.0					
		10.0	12.0					
		12.0	up					
2.0	4.0	0	2.0	cyclohexyl chloride	(1)	8.45	3.00	1.00
				dioctyl phthalate	(1)	8.30	3.30	1.50
				dioctyl adipate	(1)	8.15	3.00	1.70
		2.0	4.0	1,1-dichloroethylene	(2)	8.30	3.30	2.20
				tetrahydrofuran	(2)	8.20	2.80	3.90
				tetrahaydrofuran	(1)	8.22	3.25	3.50
				isophorone	(1,2)	8.10	4.00	3.60
				di-*n*-butyl phthalate	(2)	8.10	3.40	1.50
		4.0	6.0	cyclohexanol	(1)	8.50	2.00	6.60
				mixed diester DBE	(4)	8.10	3.40	4.1
		6.0	8.0					
		8.0	10.0					
		10.0	12.0	propylene glycol	(1)	8.24	3.50	11.80
		12.0	up					
4.0	6.0	0	2.0	benzonitrile	(2)	8.50	4.40	1.60
				dibutyl phthalate	(1)	8.40	4.80	1.60
				2-nitropropane	(1)	8.15	5.50	2.00
		2.0	4.0	dimethyl phthalate	(1)	8.40	5.50	2.80
		4.0	6.0					
		6.0	8.0					
		8.0	10.0					
		10.0	12.0					
		12.0	up	ethylene glycol	(1)	8.25	4.50	13.30
6.0	8.0	0	2.0	nitroethane	(1)	8.11	7.30	2.00
		2.0	4.0					
		4.0	6.0	*N,N*-dimethylacetamide	(2)	8.50	6.70	5.50
		6.0	8.0					
		8.0	10.0	ethanolamine	(1)	8.35	8.50	9.80
		10.0	12.0					
		12.0	up					
8.0	10.0	0	2.0					
		2.0	4.0	formamide	(2)	8.40	9.00	2.50
		4.0	6.0					
		6.0	8.0					
		8.0	10.0					
		10.0	12.0					
		12.0	up					

Table 17.3D

Solvent Solubility Parameter					SIGMA d Range = 8.5 to 9.0 Actual Values			
SIGMA p = from	to	SIGMA h = from	to	Solvent	Refs.	σ_d	σ_p	σ_h
0	2.0	0	2.0	mesitylene	(2)	8.80	0	0.30
				carbon tetrachloride	(1)	8.65	0	0
				decalin	(2)	8.80	0	0
				benzene	(2)	9.00	0	0
				toluene	(1)	8.67	1.00	2.00
				toluene	(2)	8.80	0.70	1.00
				xylene	(2)	8.70	0.50	1.50
		2.0	4.0	trichlorethylene	(1,2)	8.78	1.50	2.60
				chlorform	(1)	8.65	1.20	3.00
		4.0	6.0					
		6.0	8.0					
		8.0	10.0					
		10.0	12.0					
		12.0	up					
2.0	4.0	0	2.0	ethylene dichloride	(1)	8.97	3.30	2.00
		2.0	4.0	acetophenone	(1)	8.55	3.80	2.50
				methylene chloride	(1)	8.91	3.00	3.10
				methylene chloride	(2)	8.90	3.10	3.00
				cyclohexanone	(1)	8.65	3.40	3.40
				cyclohexanone	(2)	8.70	3.10	2.50
		4.0	6.0					
		6.0	8.0	*m*-cresol	(1)	8.82	3.00	6.10
				benzyl alcohol	(2)	9.00	3.10	6.70
		8.0	10.0					
		10.0	12.0					
		12.0	up					
4.0	6.0	0	2.0	butyl benzyl phthalate	(1)	8.80	6.00	2.00
		2.0	4.0	nitrobenzene	(1)	8.95	4.90	3.00
		4.0	6.0					
		6.0	8.0					
		8.0	10.0					
		10.0	12.0					
		12.0	up					
6.0	8.0	0	2.0					
		2.0	4.0	*N*-methyl-2-pyrrolidone	(2)	8.80	6.00	3.50
		4.0	6.0	dimethylsulfoxide	(2)	9.00	8.00	5.00
		4.0	6.0	dimethylformamide	(4)	8.5	6.70	5.50
		6.0	8.0					
		8.0	10.0					
		10.0	12.0					
		12.0	up					
8.0	10.0	0	2.0					
		2.0	4.0					
		4.0	6.0					
		6.0	8.0					
		8.0	10.0					
		10.0	12.0					
		12.0	up					

Table 17.3E

Solvent Solubility Parameter					SIGMA *d* Range = 9.0 to 9.5 Actual Values			
SIGMA *p* = from	to	SIGMA *h* = from	to	Solvent	Refs.	σ_d	σ_p	σ_h
0	2.0	0	2.0	tetralin	(1)	9.40	0.50	1.40
				benzene	(1)	9.03	0.50	1.40
		2.0	4.0	chlorobenzene	(1)	9.20	1.90	2.00
		4.0	6.0					
		6.0	8.0					
		8.0	10.0					
		10.0	12.0					
		12.0	up					
2.0	4.0	0	2.0					
		2.0	4.0	*o*-dichlorobenzene	(1)	9.43	2.10	2.10
				benzaldehyde	(2)	9.50	3.60	2.60
				pyridine	(1)	9.25	3.70	3.60
				quinoline	(2)	9.50	3.40	3.70
		4.0	6.0					
		6.0	8.0					
		8.0	10.0					
		10.0	12.0					
		12.0	up					
4.0	6.0	0	2.0					
		2.0	4.0	dimethyl phthalate	(2)	9.10	5.30	2.40
		4.0	6.0					
		6.0	8.0					
		8.0	10.0					
		10.0	12.0					
		12.0	up					
6.0	8.0	0	2.0					
		2.0	4.0	tricresyl phosphate	(2)	9.30	6.00	2.20
				furfural	(2)	9.10	7.30	2.50
		4.0	6.0	dimethyl sulfoxide	(1)	9.42	6.50	5.90
				butyrolactone	(1)	9.26	7.60	4.50
		6.0	8.0					
		8.0	10.0					
		10.0	12.0					
		12.0	up					
8.0	10.0	0	2.0					
		2.0	4.0	butyrolactone	(2)	9.30	8.10	3.60
		4.0	6.0	2-pyrrolidone	(2)	9.50	8.50	5.50
		6.0	8.0					
		8.0	10.0	dimethyl sulfone	(2)	9.50	9.50	8.60
		10.0	12.0					
		12.0	up					
10.0	12.0	0	2.0					
		2.0	4.0	ethylene carbonate	(2)	9.50	10.60	2.50
		4.0	6.0					
		6.0	8.0					
		8.0	10.0					
		10.0	12.0					
		12.0	up					

Table 17.3F

Solvent Solubility Parameter					SIGMA *d* Range = 9.5 to 10.0 Actual Values			
SIGMA *p* = from	to	SIGMA *h* = from	to	Solvent	Refs.[a]	σ_d	σ_p	σ_h
0	2.0	0	2.0	carbon disulfide	(2)	10.00	0	0.30
				tetralin	(2)	9.60	1.00	1.40
		2.0	4.0	1-bromonaphthalene	(2)	9.90	1.50	2.00
		4.0	6.0					
		6.0	8.0					
		8.0	10.0					
		10.0	12.0					
		12.0	up					
2.0	4.0	0	2.0	1-bromonaphthalene	(1)	9.74	2.50	2.00
				bromobenzene	(2)	10.00	2.70	2.00
		2.0	4.0					
		4.0	6.0	aniline	(1)	9.53	3.70	4.20
				ethylene dibromide	(2)	9.60	3.30	5.90
		6.0	8.0					
		8.0	10.0					
		10.0	12.0					
		12.0	up					
4.0	6.0	0	2.0	acetophenone	(2)	9.60	4.20	1.80
		2.0	4.0	nitrobenzene	(2)	9.80	4.20	2.00
		4.0	6.0					
		6.0	8.0					
		8.0	10.0					
		10.0	12.0					
		12.0	up					
6.0	8.0	0	2.0					
		2.0	4.0					
		4.0	6.0					
		6.0	8.0					
		8.0	10.0					
		10.0	12.0					
		12.0	up					
8.0	10.0	0	2.0	propylene carbonate	(1)	9.83	8.80	2.00
		2.0	4.0					
		4.0	6.0					
		6.0	8.0					
		8.0	10.0					
		10.0	12.0					
		12.0	up					

[a](1) C. M. Hansen, *J. Paint Technol.*, *39*, pp. 104 ff (Feb. 1967).
(2) A. F. M. Barton, *Chem. Rev.*, *75*, pp. 731 ff (1975).
(3) A. F. M. Barton, Handbook of Solubility Parameters and Other Cohesion Parameters, Table 2, pp. 94 ff, CRC Press (1983).
(4) DuPont Company Brochure E88491, "Solvent Properties Comparator."
(5) EXXON Chemicals Brochure SOL 87-1720, "Exxate Solvents, Coalescing Agents Industrial Acrylic Latex Paints."
(6) H. E. Haxo Jr., T. P. Lahey, and M. L. Rosenberg, "Factors in Assessing the Compatibility of FMLs and Waste Liquids," EPA Contract No. 68-03-3213 Final Report (1988).

or some such estimate. The solvents to be chosen may be found in the Hansen and Barton references [1,3] or in Table 17.3.

The Technical Committee of the Golden Gate Society presented an evaluation of a scheme for using seven solvents chosen to represent the corner points and the center of a Hansen 3D Solubility Parameter space. The found that three of four polymers chosen at random (a butyl rubber, a silicone, an aliphatic urethane, and an aromatic urethane) could be mathematically modeled, using replicate swellings, the square of the Solubility Parameter, and the molar volume of the solvent molecules. Their results were presented at the 1987 National Paint Show Voss Award competition [10].

There are many complicating factors in experimental work done to determine the Solubility Parameter of a polymer. Again, remember the possible toxicity of some of these solvents (benzene, butyrolactone, ethylene glycol derivatives, dimethyl sulfoxide, chlorinated solvents). In addition, some react with other materials, such as the polymer. The interactions of pyridine or quinoline with a chlorinated polymer may well fit the Fowkes modeling scheme, although not the Hansen scheme. The Golden Gate Society Technical Committee study showed that solvent blends did not yield the same results as a single solvent, because of the preferential absorption of one solvent over another [10].

The real problems are due to the three dimensions. The polymer model yields a three-dimensional figure somewhat like an oblate spheroid in shape. The centroid of that figure is given the Solubility Parameter designation, although the whole figure should be so designated. The size of the figure depends upon the molecular weight of the polymer, becoming *smaller* as the molecular weight increases. Hence, one can rationalize the solution polymer fractionation of molecular weights by additions of nonsolvents, with the higher molecular weights being eliminated as coacervates or even precipitates as the nonsolvent concentration increases.

Another three-dimension problem appears in modeling certain solvent blends. Federal specifications for certain kinds of thinners and cosolvents give the supplier a range of concentrations for the solvents in the formulation. If one uses the equations for calculating the solubility parameter of the blends at each of their formulation limits, one finds that the solvent blends allowed construct a solid figure in the three-dimensional space. When this solvent formulation is supposed to dissolve a certain polymer, the figures of the polymer and the solvent blend may be plotted to see the extent of overlap of the solid figure. In one instance, we found the solvent figure to be a torpedo shape that penetrated the disk of the polymer figure, with the ends outside the disk. The implication is that the hypothetical solvent blend is not always a solvent for the polymer.

The determination of Solubility Parameter for a soluble polymer (or a liquid, like a plasticizer or potential coalescant) is considerable easier than the determination of swell on a cross-linked polymer. The determination *procedure* starts

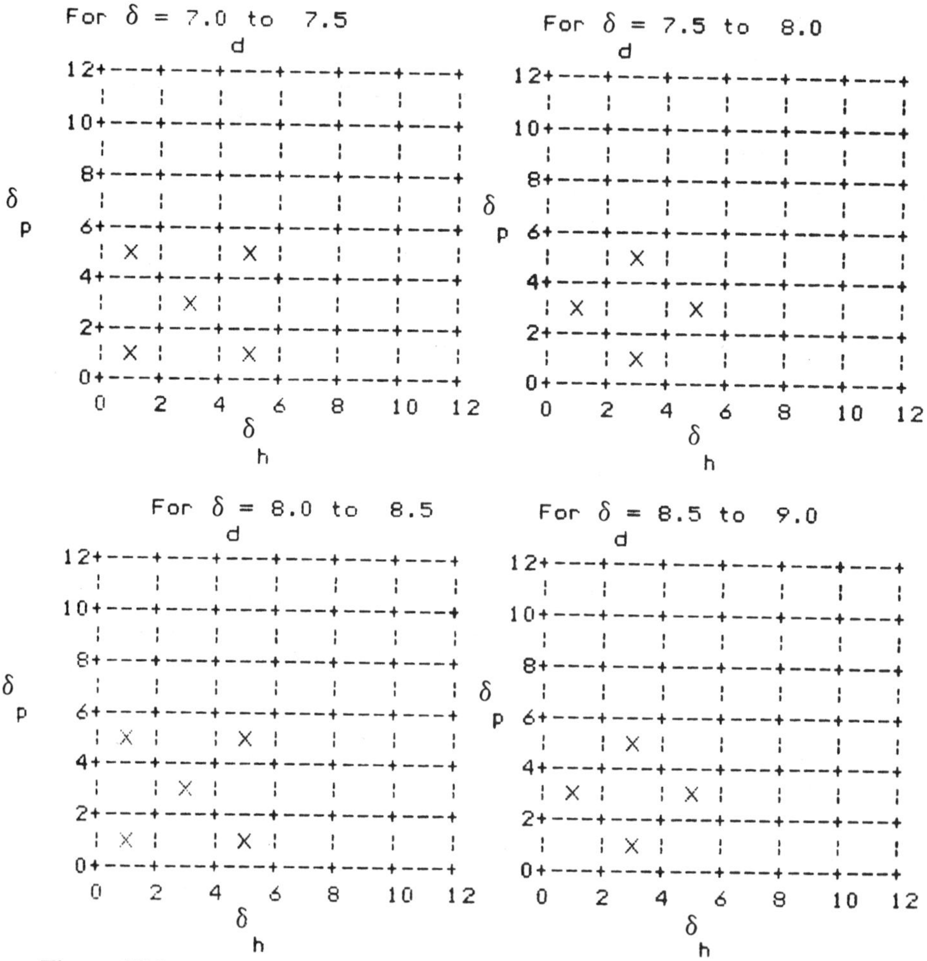

Figure 17.1 Planes in Three Dimensional Space of Hansen Solubility Parameters

in the same way by adding 10% by weight of the unknown material to a series of solvents of known Solubility Parameter. Use a range of solvents, such as the 27 suggested in Fig. 17.1. You simply look for phase separation, turbidity, or complete transparency, as indication of miscibility or solution. In instances where a range of solvents dissolve the polymer, determine the viscosity of the solution, with the highest viscosity being the best match for Solubility Parameter, as we noted earlier.

The solubility parameter concept may be extended to relate directly to coalescense and cosolvency. Some interesting work on evaluating the effectiveness

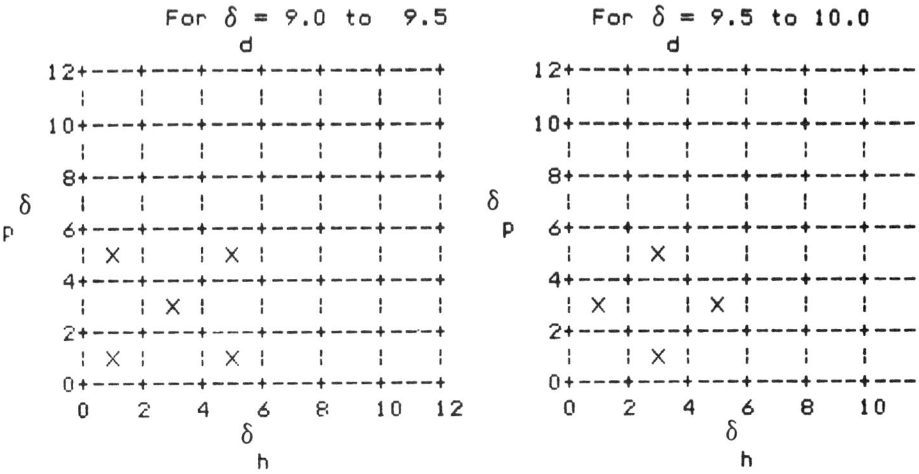

Figure 17.1 Continued

of the newer propylene oxide-based solvents suggests that such evaluations, using polymers of known solubility parameter, would allow assignment of solubility parameters to the new solvents. The work used minimum film formation temperature determinations to show type and amount of solvent needed for several polymers. It was originally presented at the 1987 Southern Society for Coatings Technology Annual Symposium by Guthrie and Czerepinski [8].

This is an active area of research which will make the formulator's job easier. Even now, the formulator is using these solvent-characterization tools to work on volatile organic compounds (VOC) reductions, and other regulatory-driven reformulations. We look forward to seeing the public results.

Table 17.4 Typical Polymer Film Solubility Parameters [7]

Polymer	σ_d	σ_p	σ_h
Poly(vinyl chloride)	7.99	5.39	3.91
Epichlorohidrin rubber	9.23	5.54	4.56
Poly(ethylene-*co*-propylene)	9.07	0.64	0.65
Poly(ethylene-*co*-vinyl acetate)	8.96	0.88	0.98
Neoprene	9.29	1.72	1.95
Nitrile rubber	9.02	2.50	3.58

REFERENCES

1. A. F. M. Barton, *CRC Handbook of Solubility Parameters and Other Cohesion Parameters,* CRC Press, Boca Raton, FL (1983).
2. H. Burrell, *InterChemical Review,* p. 3 (Spring 1955); *J. Paint Technol., 40(520),* 197 (1968).
3. C. M. Hansen, *Ind. Eng. Chem., Proc. R&D, 8(1),* 1 (1969); and *J. Paint Technol., 39(505),* 104 (1967).
4. F. M. Fowkes, 1983 ACS Rubber Div. Meeting, Houston; *ACS Polym. Mater. Sci. Eng. Div. Prepr., 51,* 522 (1984).
5. R. W. Taft et al., *Nature, 313(31),* 384 (1985); M. J. Kamlet et al., *Chemtech, 16(9),* 566 (1986).
6. *Plastics Compounding 1986/87 Redbook,* Harcourt Brace Janovich Publications, Denver.
7. A. F. M. Barton, *Chem. Rev., 75,* 731 (1975).
8. D. H. Guthrie and R. Czerepinski, *Amer. Paint Ctgs. J.,* p. 37 (July 27, 1987) and p. 41 (Aug. 3, 1987).
9. R. D. Athey Jr., *Prog. Org. Ctgs., 7(3),* 289 (1979).
10. R. D. Athey Jr., *Europ. Coatings J., 89(2),* 122; *89(3),* 216, and *89(4),* 287 (1989).
11. H. E. Haxo, T. P. Lahey, and M. L. Rosenberg, "Factors in Assessing the Compatibility of FMLs and Waste Liquids," EPA Contract No. 68-03-3213 Final Report (1988).
12. R. D. Athey Jr., *Europ. Coatings J., 89(5),* 400 (1989).
13. R. W. Kreider, US 3479313 (18 Nov. 1969).

18

Curatives

I. INTRODUCTION

Classically, the cure of the first rubbers used latex coating binders was the standard rubber vulcanization (with sulfur, bismercaptobenzothiazolyl disulfide (MBTS), and other thermal sulfur activators) which is a cross-linking reaction bridging between macromolecules. The first acrylic ester rubber cures were done with diamines. However, we have come a long way since then, and color, odor, and aging brittleness are nowhere near the problems they were 50 years ago. We will deal with specific agents, called curatives, used to bridge two macromolecules and separately with those agents that promote cure, called catalysts.

Some commentary on the curing reactions is essential. There are three points to be made concerning the basics of curing reaction chemistry. They are based on:

1. thermodynamics of curing reactions;
2. kinetics of curing reactions; and,
3. rheology control of cure.

Indeed, recognizing these scientific controls will aid research and development in avoiding "dead-end" projects.

The presentation by Peter Pappas at the New Orleans Water Borne and Higher Solids Coating Symposium, February 1983, clearly gives the thermodynamic rationale why formulation stable at room temperature cannot give a cure in a second at 100°C. Of course, those of you who remember what an Arrhenius plot is can figure it out. It must be remembered that the cure rate doubles with every increase in temperature of 10°C (four times with 20°C, eight times with 30 and so on). So the 70°C difference between boiling water temperature and room temperature means that the 1-min cure at 100°C gives a pot life at room temperature of 2^7 min, a little more than 2 hr. While considering the thermodynamics of cure reactions, one should remember the equivalency of the various forms of energy used to induce the cures (especially thermal, IR, and microwave) [7].

The second point is that amine cures or melamine formaldehyde-urea formaldehyde (MF-UF) resin cures are condensation reactions. They are *equilibrium* reactions and cures are inhibited by the competition of the reverse reaction; the water by-product of the condensation reaction must be removed quickly to eliminate the competition, for instance. You may have to dry the equilibrium curing system completely (or heat it to remove other by-products of the cure reaction) for a full cure. The reaction can also be reversed by an excess of one of the resulting products (high humidity may retard the condensation cure). Indeed, the urethane cure reaction has no inhibiting by-products (unless you are using a blocked isocyanate), but it is still a condensation reaction, and is therefore reversible and vulnerable to hydrolytic attack.

Other equilibria exist as well. Thermal input may induce thermoplastic flow of ionic cures with ionomers like SURLYN, even though they act like thermosets at lower temperatures. The association of carboxylate anion with curative cation or the hydrogen bond-induced apparent "cure," will both be less stable at higher temperatures.

The third point is that until the cure is complete, the cure agent may weaken the system because it acts as a plasticizer. That has advantages because the plasticization promotes the movement of the molecules around to the correct geometry to reach the curing reactants. Indeed, adding a "temporary plasticizer" like a coalescent aid, can speed up the cure reaction by lowering the viscosity of the reactants at the final stages of cure. The higher the molecular weight of the polymer, the slower it moves to approach the next cure-reaction site, to reemphasize the first point on curing rates. This cure dependence on flexibility was demonstrated in a presentation by an acrylic copolymer manufacturer in the 1987 Western Coating Symposium. The cure was essentially stopped as the T_g of the copolymer and curative passed room temperature, because the chains were no longer mobile enough to continue the reaction.

II. CURING AGENTS

The curing agents are the chemicals that convert the thermoplastic coating binder into a thermoset. The curing agent is bonded to two different molecules of binder polymer and acts as a cross-link bridge. Examples are:

Table 18.1 Base Materials for Methylolamide Curatives

Compounds	Structure
Urea	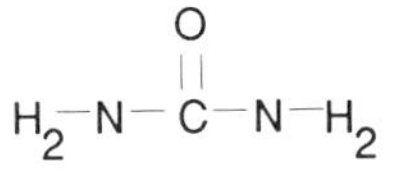
Glycouril	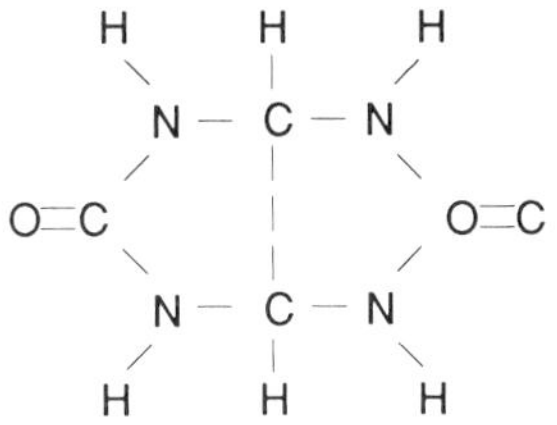
Melamine	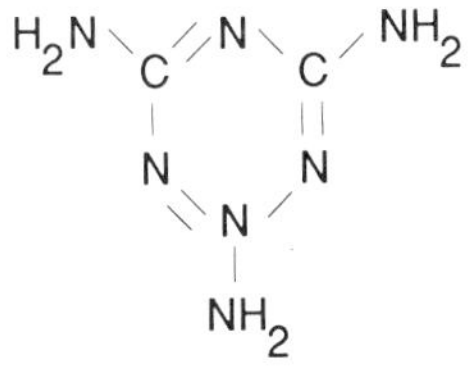
Benzoguanimine	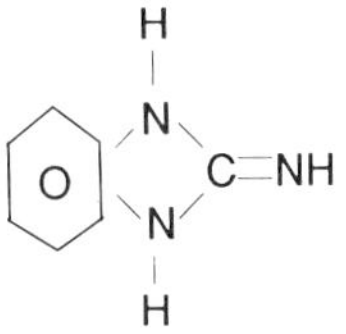

1. melamine (or other precursor)/formaldehyde resins to cross-link hydroxy, amide, carboxy, or amine functional polymers;
2. epoxy resins to cure the same types of functional polymers;
3. polyvalent cations (zinc, zirconium, calcium) to cure carboxylates;
4. carbodiimides to cure carboxylates; and,
5. aziridines to cure carboxylates.

The last group is not approved for many applications requiring FDA clearance.

The curatives listed above have some unique properties, which are a guide to the choice of a particular curative as the *one* desired for the particular waterborne formulation. We have already alluded to that in mentioning FDA considerations for can coatings.

A. Methylolamide Curatives

Commercial methylolamide curatives fall into three major groups. The resins are made with the materials shown in Table 18.1 and are methylolated with an excess of formaldehyde. The reaction of formaldehyde with the pendant amino groups is a simple addition across the carbonyl double bond. In water, it will slowly continue with a reaction between two methylolated amino groups to form an ether with elimination of a water molecule, or a similar reaction to form a methylene bridge between the two amino groups with elimination of a formaldehyde molecule and then a water molecule. This is also the curing reaction upon drying. The formaldehyde elimination reaction is a factor in not choosing this type of curative for certain disposable articles (e.g., diapers), because consumers fear of formaldehyde.

If the formylation (another term for methylolation) takes place in a hydroxylic solvent, the cure reaction is blocked by etherification of the methylol group by the alcohol. Commonly available methylol ethers are based on methanol and butanol. During the cure of polymers with pendant carboxylic or hydroxylic groups, the alcohols evolve as by-products, adding to VOC (see 1987 Northwest Society for Coatings Technology, Voss Presentation) or temporarily plasticizing some polymers [16]. Ester formation of methylolamides with hydroxylic polyesters based on trimellitate baking enamels was shown by Stevens [13].

These cures are acid catalyzed. Unpublished work we did with James Wonnell of Scott Paper confirmed that pKa of the acid is an important consideration. Hence, among latent acid catalysts, ammonium bromide was better than ammonium chloride, which in turn was better than oxalic acid. Many coating systems use *p*-toluenesulfonic acid, though methanesulfonic acid and trifluoromethane sulfonic acid (triflic acid [8]) have been introduced as effective catalysts. A real concern is the fate of the catalyst after the cure, as a residue of the strong halide or sulfur acids could promote corrosion. Emphasis in the textile industry on "fugitive" catalysts (ammonium chloride) was based on their desire to avoid acid weakening of fibers.

Incorporation in the waterborne formulation is a concern for methylolamide curatives. They can be supplied as water-soluble or dispersible materials, or they may need preemulsification. As one might expect, pH is an important consideration in the stability of the formulation, of the final mixture, or of the preemulsion. Indeed, pH is reputed to control the type of cure reaction, methylene bridge versus methylol ether. The long-term storage stability of methylolamide curatives in emulsion polymers may be suspect. We observed a possible surface reaction over time of methylolmelamine with carboxylated SBR latex (with subsequent agglomerations, charge reversals?) in some hydrodynamic chromatography work in the mid-1970s. This might be an interesting area for a more thorough investigation.

The methylolamide-cured material may need a balance of the type and amount of methylolamide source with the carboxylic–hydroxylic polymer, because of the rate of methylolamide homopolymerization compared to the rate of curing reaction. Homopolymerization of the methylolamide source may add a desirable reinforcing and opacifying element as a hard phase, but will require much higher than a stoichiometric cure ratio. Solvents or cosolvents can affect this homopolymerization vs. cure ratio [11]. Bauer and Dickle [12] showed that solvent resistance was a function of cross-link density with methylolated melamine–acrylic (hydroxylic or carboxylic), and that variation of type of methylolation and catalysis was necessary to optimize cures in solvents or waterborne acrylic enamels.

B. Epoxy Curatives

Epoxy resins cure with a variety of reactants, such as amines, carboxylic acids, mercaptans, and others. But we are interested in the incorporation of epoxy resins into a latex to act as a curative. The presence of the water imposes several restrictions on epoxy-based curative. Ideally, it should be an oil that can be dispersed, emulsified, or dissolved in the aqueous medium. Many commercial epoxy resins fit this category, and may be preemulsified or dispersed before incorporation into the latex. The technique may be that often used with plasticizers or oils, wherein an emulsifier (nonionic or carboxylic, such as oleic acid) is mixed with the bulk 100%-active epoxy resin before emptying the mix into stirred water (with amine or ammonia neutralizatin for the carboxylate, if necessary). The alternative is to prepare the surfactant–dispersant mixture in the water, and add the oil with high shear agitation.

The preemulsified epoxy resin may be added to the latex or aqueous polymer dispersion. A caution is added to the BFG Chemical literature, in that their aqueous hydrosol acrylic carboxylate resins have limited "pot life," that is, they are concentration dependent [1]. For instance, a blend of Shell's EPON 828 with Carboset 514H is stable for two to three days at 34% solids, but for about 30 days at 20% solids. I have also seen other formulations that were stable (no viscosity rise or gelation) for 30 days.

Epoxy resin formulations can benefit from catalysis in their cure. Chromium salts of carboxylic acids are sometimes recommended; the octoate was used in the Carboset literature [11], almost halving the room-temperature cure time. However, a formulation of vinylcyclohexane diepoxide with a chromium catalyst did not succeed in curing a carboxylate-functional polymer in my own work; the epoxide may have been too volatile in the heating stage [2].

There is literature available, suggesting that epoxy-carboxylate systems give good coatings. Bauer [3] gives a formulation of about 80/20 CMD-979 (a Celanese carboxylate resin)/EPONEX 1513 (a Shell epoxy resin) for a white pigmented waterborne enamel. He also reports that clear films did not yield the blushing (from water) that noncarboxylate epoxy cures gave. We have published the Golden Gate Society for Coatings Technology Technical Committee's work on room temperature cures of carboxylated latexes with epoxy or other curatives [14].

One additional caution is that any dispersed epoxy may be degraded (undesirably reacted with or homopolymerized by) acid or alkaline materials. Hence, one must observe the pH and ensure that the amine or ammonia used is not causing a problem.

C. Polyvalent Cations

The polyvalent cation cures are usually based on zirconium or zinc ions. Both have stable ammonia coordination complexes, as shown below. Commercial solutions of these complexes (zinc from S. C. Johnson or Ultra-Adhesives) and zirconium from Magnesium-Elektron) may be readily mixed into the waterborne formulations of carboxylated polymers for cure upon drying. The basic principle is that the ammonia-coordination complex disappears on drying, leaving the bare metal ion to react with the carboxylate polymer. As the ammine complexes, these divalent cations do not threaten the coagulation or precipitation that we so frequently warn of in our discussions of colloid stability.

$Zn(NH_3)_4^{+2}$	$Zr(NH_3)_6^{+4}$
zinc ammine complex	zirconium ammine complex

Other unpublished work suggests that even calcium carbonate can act as a curative. Russ Meincke of General Tire showed that films of a carboxylated SBR had a higher modulus on baking than the previously air-dried version with as little as 10% calcium carbonate (limestone) in the formulation. This was a lab confirmation of suspicions aroused when carboxylated SBR carpet coatings filled with limestone were not launderable upon drying, but were launderable after an additional 5-min "cure" in the oven after dryness was gravimetrically verified.

The effectiveness of the polyvalent metal cation cure has been discussed in the literature. Peter Moles' presentation to the 1987 Water Borne and Higher Solids Symposium described several properties improved by addition of zirconium ammonium carbonate solutions to carboxylated polymer coatings or inks, including adhesion improvement and solvent resistance [6]. His interpretation of the zirconium structure as an inorganic polymer oxide (similar to some representations of silica) which forms an ester with free carboxylates on drying is not the salt neutralization I envision, but they both function to yield the important property improvement expected by curing. It may be that their product line contains both the polyvalent cation as such (their AZC, ammonium zirconium carbonate) and the polymeric form (their Bacote 20). Other precedents for polyvalent cation-derived hydroxylic polymer materials exist; for example, the alum flocculants for waste water are considered inorganic polymers. The titanate literature of DuPont's TYZOR and the Tioxide equivalent cause viscosity increases in aqueous formulations containing cellulosic polymers, hypothetically through hydrogen bonding of inorganic polymer to cellulosic polymer. John Hall, of Tioxide, gave a presentation at the 1986 National Paint Show on the viscosity increases induced by titanates on latexes having some hydroxylic polymer grafted to their particle surfaces [15].

D. Polycarbodiimides

A new technology has recently appeared using curatives which promises to be effective. Indeed, it could prove to be the best way to cross-link latexes with carboxylic functional groups on the surface of their particles, or the way to cross-link colloidal dispersions of "solubilized" acrylics or alkyds. The polycarbodiimides are also recommended as cross-linkers for the soluble or emulsion polyurethane polymers recently developed for coating formulations.

The new technology employs a series of commercial products with many *carbodiimide* functional groups pendant from a central chemical structure. The suppliers do not reveal the details of the structure, but you can safely assume that it is based on polyisocyanate chemistry. The carbodiimides are dehydrated ureas, and ureas are the reaction products of amines and isocyanates. Only a primary amine can be converted to a carbodiimide. Commercial polycarbodiimides are sold as solutions in an oxygenated solvent of high polarity. The solvent will have some degree of water solubility, but will contribute to VOC. It will also help to confer gloss, coalescence, and final drying-stage plasticization.

The carbodiimide must be treated carefully in the water solution, as it may be unstable unless the pH is kept within the range of 8–9. Therefore the carboxylic polymer emulsion or dispersion must be preneutralized with an amine or ammonia to about pH 8.5, the highest attainable with ammonia.

Some latex or dispersion systems can accept the polycarbodiimide as it is supplied in solution. However, some formulations will not accept the carbodiimides as supplied; in this case carbodiimide must be preemulsified with a cellulosic ether–surfactant combination, again, the pH must be close to 8.5. The penalty here is that hydrophobicity of the final coating may suffer from these added colloidal stabilizers. In addition, more biocide may be needed for maintaining the inhibition of fungi or bacteria.

The reaction of the carbodiimide is interesting, as it produces an acyl-substituted urea. This may be subject to a hydrolytic attack in extremely alkaline exposures, but the advantage of the reaction is that there are no by-products to evaporate, contaminate, exude, contribute to VOC, or yield extractables into food. The reaction can be catalyzed to go faster, but few details have found their way into the literature to date. One could expect that acid catalysis and the use of amines or ammonia as the neutralizers for the carboxylates results in driving them off on drying, leaving the carboxylic acid as an autocatalyst. One may expect some transition metal catalysis as with urethane syntheses.

The cure rate varies with the carbodiimides supplied. Some will cure to excellent films at room temperature, while others need 30 min at up to 100°C. The commercial brochures show 200+ MEK rubs at the end of the good cures. (A rag is wetted with MEK (methyl ethyl [Ketone]) and wrapped around a finger which is then rubbed over the coating surface until the coating is broken through.) Materials used in examples of these cures include styrene–acrylic latexes and polyurethane latexes or dispersions.

E. Aziridines

Although these are frequently avoided in product applications, they do have utility. One must be sure to handle them safely, according to the supplier directions. They have found some use in aircraft coatings, but may have been replaced by urethanes and epoxies by now. Tyskwicz and Tsirovasiles [4] used a polyfunctional aziridines to cross-link urethane carboxylate dispersions for a variety of demonstration formulations. Supplier literature shows the solvent resistance of carboxylated acrylic latexes and lacquers and urethanes improved by multifunctional aziridines [5], while others stress new polymeric forms with lower toxicity (?) [9].

III. CURE CATALYST

A catalyst participates in the reaction, but is not incorporated into the cured product. It may therefore may extracted by water or solvent, or rise to the surface to be exposed to the air. This may affect uses in can or bottle coatings which require FDA approval, and the surface properties of the coating.

There are four classes of catalysts. The *acid* catalysts aid in formation of ether, ester, or amide linkages with MF/UF resin cures. Most latexes contain some latent catalyst acid as a consequence of the peroxidisulfate initiator decomposition, producing acid sulfate as by-product. *Alkaline* or *basic* catalysts help epoxy or solvent-borne urethane cures, and can help homopolymerize phenolic or MF/UF cures. Both acid and alkaline catalysts may show sensitivity to pH. *Transition* metals can speed the cure of unsaturated polymers and alkyds as oxidation promoters [10]. There are a series of chromium salts that are reported to speed up epoxy cures (possibly by some *coordination complex* formation), and there is certainly evidence for transition metal catalysis of urethane cure reactions. Remember that transition metals are polyvalent cations, and will possibly be destabilizers for any polycarboxylated binder or additive in the waterborne formulation.

Choice of a catalyst can be critical, as it may impede the curing reaction, resulting in a phenomenon called reversion by rubber chemists, that is, the reduction in strength upon overcure. One will commonly see "reversion" as a parabolic arc on the plot of strength (tensile, for instance) versus cure time (at some temperature). The cure reaction and some other unlinking reaction are competing; the latter may simply be the reverse of the cure reaction which is accelerated by the catalyst just as the cure reaction is. This is the reason that "fugitive" catalysts are frequently used for the self-crosslinking polymers based on N-methylolacrylamide. The catalysts contribute to the ionic strength or total "salt load" of the system, a point that we will deal with later.

Actually, more research is needed on the "reversion" phenomenon. The point that a cure reaction proceeding one way and a cleavage reaction of a completely different sort are competing, needs to be clearly identified, so that inhibitors for the cleavage reaction may be incorporated into the formulation to give a "cure plateau" for the formulator.

REFERENCES

1. "Reacting Carboset Resins With Epoxy Resins," TDS 216, BF Goodrich Chemical, Cleveland, OH.
2. R. D. Athey Jr., unpublished work of 1975.
3. R. S. Bauer, Presentation to the 1982 Water-Borne and Higher Solids Coating Symposium, Shell Chemical Technical Bulletin CS:729-73.
4. A. S. Tyskwicz and J. Tsirovasiles, 1987 Water-Borne and Higher Solids Symposium Preprint, p. 443.
5. Polyvinyl Chemicals Inc. Technical Bulletins CX-100E and CX-101A, Wilmington, MA.
6. P. J. Moles, 1987 Water-Borne and Higher Solids Symposium Preprint, p. 314.
7. D. L. Hertz, *Elastomerics,* p. 30 (July 1987).
8. R. R. Alm, *Mod. Paint Ctgs.,* p. 88 (Oct. 1980).
9. XAMA-2 and XAMA-7 Technical Bulletins CC-17076 and CC 10076, Cordova Chemical Co.,
10. "Driers and Drying," Manchem Inc., Princeton, NJ.
11. M. I. Karyakina et al., *1974 FATIPEC Conress Preprint,* p. 599.
12. D. R. Bauer, and R. A. Dickle, *ACS Org. Ctg. Plas. Chem. Preprint, 41,* 451, 457 (1979).
13. J. R. Stevens, *Off. Digest,* p. 380 (Apr. 1963).
14. R. D. Athey Jr., *Euro. Coat. J.,* 490 (June 1989).
15. J. E. Hall, Annual FSCT Meeting and Paint Show, Atlanta, GA, 5–7 Nov. 1986.
16. E. C. Ferlauto (for the Northwestern Soc. for Ctg. Technol.), Annual FSCT Meeting and Paint Show, Dallas, TX, 5–7 Oct. 1987.

19

Heat and UV Stabilizers

I. ANTIOXIDANTS AND HEAT-AGING STABILIZERS

Some systems of formulated binder compositions deteriorate upon heating or ambient aging by an oxidation mechanism. Hence, stabilizers against heat-aging problems are generally antioxidants; we will also discuss a few special cases where heat aging is not an oxidation problem. The oxygen of the air may add to the polymer (e.g., to an allylic hydrogen to form a hydroperoxide) and react to form colored bodies and new cross-links resulting in brittleness, or to break polymer chains, thus weakening the system. Years ago, we tested a commercial poly(butadiene) based nonwoven fabric binder and found excellent properties, flexibility, and strength combination. When we checked the samples we had kept for hand-feel and other tests, we found that the soft webs we had spray-bonded had hardened in a few months' storage in a closed box. This resinification could have been prevented by an antioxidant, but there were other problems with those webs, so we did not pursue the resinification problem.

Antioxidants generally perform by capturing and destroying free radicals. They can capture the oxygen or attack the free radical product of the oxygen addition to the polymer. The chemical structures used as antioxidants may include aromatic amines, thioethers, and hindered phenols. These act essentially as free-radical traps by chain transfer. These materials form very stable free radicals with a very slow decomposition mechanism that does not involve reinitiation of a polymer radical. You can understand why these radicals are added to the polymer after the synthesis reaction. A wide variety of commercial materials are available for evaluation. There are even food-grade antioxidants used in cooking oils. Hence, you can surely find one to suit your needs. Antioxidants are frequently available as waterborne dispersions to be added in the latter stages of formulation. You may find the commercial materials in the standard *Redbook* [1].

Poly (vinyl chloride) and other chlorinated polymers tend to dehydrohalogenate upon aging or heating. A flame-retardant (poly(vinylidene chloride) latex that we used as an adhesive for a laminated nonwoven exhibited orange stripes if dried too long. We found a less colorful product in the final composition but did not learn why the first color was orange. The problem is very simple, if you remember the conjugated double bond structure of carotene and its color. If you dehydrohalogenate a sequence of five or more vinyl chloride units, you approximate the conjugated carotene structure, and the color deepens as more conjugation is created. Equation (19.1) shows the dehydrohalogenation to the double bond in polymer chains (denoted by P in the equation).

$$\mathrm{P{-}\underset{\displaystyle H}{\overset{}{C}}H{-}CH{-}P} \quad \text{(H on the first C, Cl on the second C)}$$

$$\text{P—}\overset{\text{H}}{\overset{|}{\text{CH}}}\text{—}\overset{\text{Cl}}{\overset{|}{\text{CH}}}\text{—P} \longrightarrow \text{HCl} + \text{P—CH=CH—P} \qquad (19.1)$$

A variety of stabilizers can reduce color in PVC. As the dehydrohalogenation is acid catalyzed and the by-product is HCl, the reaction is autocatalytic. Fillers and pigments based on carbonates or oxides provided early solutions to the problem. More recent work recommends cadmium thio-organic stabilizers or mixed metal stabilizers [2]. Some work is under way to correlate metal-stabilizer effectiveness with a common chemical characteristics (ion diameter, hard acid-soft base concept) and we look forward to the publication of their findings. These commercial stabilizers may be in the form of powders, liquids, or pastes.

Dehydrohalogenation stabilizers are most commonly added to the dry powdered PVC in compounding mixers in preparation for sheeting or extrusion. The supplier will give advice on incorporation into emulsion forms of the polymer. However, these stabilizers merely delay the inevitable, as they cannot consume *all* the acid generated, and will eventually be spent.

Some heat-initiated discolorations are not due to dehydrohalogenations. Nitrile-based polymers, containing acrylonitrile or methacrylonitrile, will form reddish aging products by other conjugated double-bond formation. This color is due to the reaction of adjacent nitrile groups to form a nitrogen-containing ring in the polymer chain. Heating poly(acrylonitrile) (PAN) in caustic causes the dispersion to become brick red before the hydrolysis of the nitrile groups results in water solubility; the solution retains an orange tinge, perhaps because not all the nitrogen-containing rings are destroyed by hydrolysis. One can envision a pair of nitrile groups reacting to form a 2-iminoazine with 3,5-chain linkages, and air oxidation giving four double bonds in the structure, as shown in Eq. (19.2) (where P again denotes the rest of the chain). Two of these iminoazine structures in a row would yield the dark red absorption.

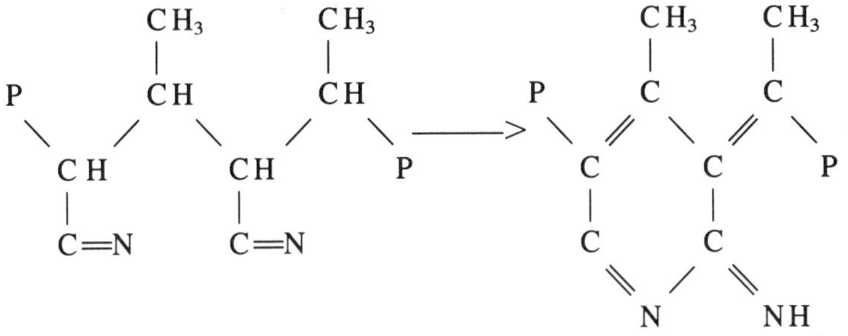

(19.2)

This discoloration is only harmful if color is important, as with the acrylic or butadiene copolymers used in textiles or the acrylic copolymers used in coatings. A PAN fiber polymer can be cyclized by heating to form "black acrylic" fibers (e.g., ORLON) which are much less likely to hydrolyze, and are stiffer than the uncyclized PAN.

Testing for discoloration is not difficult. It is a simple matter of subjecting free films or coatings to a standard heating routine, and observing when and if discoloration occurs after a reasonable time. We used saturated filter paper (pure cellulose) for heat-aging studies of a variety of nonwoven binder latexes, some of which had been treated with antioxidants [3]. We recommended color determinations (HunterLab colorimeter) as a way to characterize the transitions from yellow to red to brown. We were able to mathematically model the discoloration determined by a GE Brightness meter reading on the webs; by far the best performing material was a commercial ethylene–vinyl acetate copolymer (EVA) latex [4]. However, I would prefer to use the "Delta E" of the colorimeters.

II. UV STABILIZERS

The energy in ultraviolet radiation is on the order of the energy levels of the electrons moving about in a chemical compound. There are especially sensitive segments in chemical compounds that absorb UV radiation; energy is also involved. A good portion of the energy will be harmlessly emitted at a different wavelength, but some harmful things can happen. The molecule can rearrange to form a linkage that is easily cleaved by oxidation or hydrolysis. The electrons at the new, more energetic level may generate a free radical to cross-link or degrade the polymer or split off a small molecule, leaving a color-inducing double bond behind. These possibilities reduce the useful lifetime of the binder polymer.

A variety of additives act as UV screens to trap the UV radiation before it reaches the polymer. The protective molecule operates on the same principle as sun-block lotions. The important structural features of the UV stabilizers usually include an aromatic moiety and a carbonyl substituent. Your supplier may provide structural details, and help design a testing scheme to find cost-effective usage level for these relatively expensive materials.

Some pigments are UV absorbers. Pigments such as zinc oxide can absorb UV and provide corrosion protection, whereas titanium dioxide will whiten as well as absorb UV. Lithopone is an even better UV absorber.

These UV materials (the pigments as well as the organic stabilizers) are generally powdered solids. One may disperse them in water, like any pigment. The material may need grinding to satisfy the particle size requirements of the coating formulation. It may then be added to the coating formulation, provided that the pH and the colloidal stabilizers are compatible. Again, the commercial purveyors may be found in the *Redbooks* [1].

Testing the effectiveness of the UV stabilizer(s) with laboratory instruments in a polymer application has a substantial history in textiles, plastics, paints, and other materials. There are standards for the application of lab instrument UV exposure alone or with heat and humidity to test materials. However, a control standard material *must* be included along with the experimental materials to be tested. Even then, the correlation of the lab exposure instrument data to normal weathering exposures may not be possible or realistic [6]. Indeed, Ford Motor Company researchers report that exposures' attacks on some systems can be rationalized by IR analysis of peaks appearing in the degrading surface, and will guide your choice of appropriate stabilizers [7]. The examples included clear acrylic topcoats, where the melamine-cured material needed antioxidant and UV stabilizer on a weathering that correlated Florida exposure to a UV-humidity cabinet where the IR identified the formaldehyde and a peroxide derivative of it in the degrading coating. Another clear acrylic topcoat with an isocyanate-based cure could not correlate between those exposures, and antioxidant did not prevent degradation.

REFERENCES

1. *Plastics Compounding 1986/87 Redbook,* Harcourt Brace Janovich Publications, Denver.
2. *Plastics Compounding 1986/87 Redbook,* Harcourt Brace Janovich Publications, Denver.
3. D. Anderson and R. D. Athey Jr., *1977 TAPPI Paper Synthetics Conference Preprints,* p. 31.
4. R. D. Athey Jr. and M. M. Conrad, unpublished work of 1976.
5. ASTM G23, G26, and G53.
6. R. D. Athey Jr., et al., *J. Ctgs. Technol., 57(726),* 71 (July 1985).
7. J. Gerlock, Presentation to the 1987 FSCT Technical Advisory Committee Dinner Meeting, 17 Nov. 1987.

20

Biocides

I. BACTERICIDES

These materials, sometimes called bacteristats, are toxic to microorganisms. This suggests a potential problems with OSHA- or EPA- regulations, when you need this sort of additive. Fortunately, your supplier has been faced with these problems before, and can advise you as to safe usage and disposal. Although a *bactericide* and *bacteristat* may be the same chemicals, the meaning of the words is not the same. A bactericide is used in large enough quantity to *kill all* the bacteria, whereas the bacteristat is used in smaller quantity, just to kill excess bacteria. Table 20.1 lists bactericides and their manufacturers.

Varieties of these agents are available. They may be used to protect a fluid coating formulation and its components before formulation, or the composite structure of substrate, binder, and additives in the dried product. Aldehydes (formaldehyde and others) are probably most frequently used for latex systems, and have the advantage of low cost and volatility (they will be absent from the final product). Other organic chemicals that find application are amines (similar in structure to the cationic surfactants), phenols, and heterocyclics (triazoles and thiazoles). Some salts (barium metaborate, for instance) may be used in special applications that do not involve skin or food contact. However, there are inexpensive effective materials and there is no need to use large doses or expensive materials.

The best advice for protection of the fluid components of the formulation is to use a minimum of a bacteristat (for no longer than six months) and then change to another bactericide of different chemical structure. The pH of the aqueous system may be an aid to protection [1]. A guide to preservatives is

Table 20.1 Typical Bactericides and Their Manufacturers

Structure	Brand Name	Manufacturer
Monocyclic oxazolidine mix, 78%	BIOBAN CS 1135	Angus Corp.
Potassium *N*-hydroxymethyl-*N*-methyl-dithiodicarbamate and sodium 2-mercaptobenzothiazole	BUSAN 52	Buckmann Labs
Alkylamine hydrochlorides	COSAN 635W	Cosan Chem. Corp.
1,1′-(2-Butylene)bis(3,5,7-triaza-1-azoniaadamantane chloride), 90%	COSAN 265	Cosan Chem. Corp.
Methanol[[[2-dihydro-5-methyl-3(2*H*)-oxazolyl-1-methylethoxy] methoxy]methoxy], 50%	COSAN 145	Cosan Chem. Corp.
1-(3-Chloroallyl)-3,5,7,triaza-1-azoniaadamantane chloride	DOWCIL 100	Dow Chemical
p-Chloro-*m*-xylenol	OTTASEPT	Ferro
6-Acetoxy-2,4-triethyl-*s*-traizene, 95%	GIVGARD DXN	Givaudan
Formaldehyde donor	BIOCHEK 240	Merck & Co.
Benzyl bromoacetate, 90%	MERBAC 35	Merck & Co.
1,2-Dibromo-2,4-dicyanobutane, 98%	TEKTAMER 38	Merck & Co.
1,2-Dibromo-2,4-dicyanobutane, 25% (in water)	TEKTAMER 38 A.D.	Merck & Co.
Zinc pyridinethione-*N*-oxide and polybrominated salicylanilide	OMACIDE 645	Olin Chemical
1,2-Benzoisothiazolin-3-one, 35%	PROXEL CRL	ICI Chemical
1,2-Benzoisothiazolin-3-one, 17%	PROXEL GXL	ICI Chemical
5-Chloro-2-methyl-4-isothiazolin-3-one (8.6%) mixed with 2-methyl-4-isothiazoline-3-one (2.6%)	KATHON LX	Rohm & Haas
2-(Hydroxymethylamino)ethanol	TROYSAN 174	Troy Chemical
	COSAN 91	Cosan Chem. Corp.
2-(Hydroxymethylamino)-2-methylpropanol	TROYSAN 192	Troy Chemical
2-(Hydroxymethylamino)ethanol and bis(tributyltin)oxide	TROYSAN 364	Troy Chemical
30% Phenylmercuric acetate	TROYSAN PMA-30	Troy Chemical
Chloromethoxypropylmercuric acetate	TROYSAN CMP	Troy Chemical
Hexahydro-1,3,5-tri(hydroxyethyl)-*s*-triazine	BIOBAN GK	Angus Chemical
	VANCIDE TH	R.T. Vanderbilt
Glutaraldehyde, 50%	UCARCIDE Antimicrobial 750	Union Carbide
Undecylenic acid	—	Atochem Inc.
Mixed bicyclic oxazolidines, 50%	NUOSEPT 95	Nuodex Inc.
Mixed bicyclic oxazolidines, 78%	NUOSEPT 65	Nuodex Inc.
	COSAN 101	Cosan Chem. Corp.
Bactericide and fungicide, not identified	ULTRA-FRESH DM50	PMC Specialties Group Inc.

available from the Paint Research Association in England [2]. Some workers make the point that cobiocides (a combination of multiple biocides) may be needed [3]. Jakubowski and others described a preventive material for latex emulsions [4]. The Chicago Society for Paint Technology evaluated a variety of nonmercurial preservatives for aqueous coatings, looking also for enzymes inhibition [5]. The adhesive formulators may have even more need of biocides, as their formulations may contain a variety of starches or gums that can act as bacterial nourishment [6].

The dosage should be minimal. One of the facts of bacterial life is that they do mutate in time to find better ways to survive in their environment. Thus the bactericide you use will become their nourishment. Increasing dosage at that point is not crucial, and you should change the *type* rather than the amount.

Another recommendation is that storage, mixing, and application facilities must be thoroughly cleaned when the change is made in the formulation bacteristat. I remember visiting a facility some years ago where the formulator had problems, which we eventually discovered resided in his receiving tank which he had *never* cleaned. We found growth (algae? fungi?) that looked like Spanish moss.

It is not enough, sometimes, just to kill the bacteria. They survive by using enzymes to break down their food. Killing the bacteria does not eliminate the enzymes which can continue to degrade the cellulosic thickener or the surfactant vulnerable to attack. The message is clearly to start with a meticulously clean system and keep it this way. One company has a Japanese patent on enzyme inhibitors for coatings [7]. Winters et al. isolated three different cellulases (enzymes that degrade cellulosic additives) and characterized them for pH optima, substrate specificity, temperature stability, and other properties [8].

If you are the user of the coatings or latexes, why should the biocides incorporated be of concern to you? Strictly speaking, you want a coating or latex which you can apply directly. If, however, the formulation you receive is not used immediately but is stored in a dark, damp and warm place. Bacteria, molds, and even algae grow in it [47].

You then have two concerns: ensure that the microorganisms do not alter your coating or latex formulation while passing through receiving and processing, and that the coated articles are not attacked in the users' applications. It is important to have a good relationship with your formulation supplier. The coating or latex supplier will, in the main, have taken care of most of the potential problems. But there are points where *you* have to take the responsibility.

The physical layout of your facility is partly responsible for the exposure of the latex of coating formulation to bioactive organisms. The key points to remember in minimizing those exposures are

1. Keep the formulation covered, to exclude dust and wind blown bacteria. Plastic sheeting should cover your roller coaters and the pans that feed them, even when not in use.
2. All the temporary hoses and tubes used for transfer of the coating, from supplier delivery through the process to the coating application, must be rinsed and *hung* with the ends pointed downward. Hanging the tubes and hoses without loops eliminates growing colonies one might expect in the bottom of a loop.
3. Periodic cleanup of all tanks and vessels will minimize the seeding of biota from one batch to the next. For cleaning soap-and-water scrub with drying is essential. Woods [42] recommended steam, or formalin or bleach when steam is not available. Rinse well after the detergent, formalin, and bleach treatments. Keep the clean vessel covered during drying.
4. Vented storage tanks should be fitted with a water trap to keep air from coming into the tank without being washed. This technique is common to keep the latex or formulation from skinning over by drying, but its importance as a bacterial shield should be emphasized. Since the water trap will wash the bacteria from the air, it could be a source of bacteria if not cleaned.

These precautions are the minimum required. Many latex suppliers have reprints and special bulletins which can help you to design an easy-to-clean system for storage of the waterborne fluids to minimize bacterial contamination [9].

Testing for biological contamination can be done at your facility, with some care. This does not include the microscopic examination to identify the bacteria, which requires a specialist. However, there are quick and useful lab tests. You need an oven set at 30°C, small (1 cm^3 or smaller) disposable syringes, graduated in tenths, sterile culture dishes with an agar bacterial growth medium and sterile swabs. The latter may be bought in sterile packages from supply houses, as may the syringes. Woods described some detail in setting up your own biolab [43]. More rigorously, you may follow ASTM D2574-73.

Analysis of biocidal materials, for quality analysis purposes on incoming shipments, is not onerous, except that there may be precautions needed because of toxicity. The key tool would be infrared analysis (KBr pellet dispersion for solids and a thin film between two salt plates for liquids), to ensure uniformity. The analysis for concentration, or level of biocidal activity may be more appropriately done by more sophisticated labs, such as those of the supplier or an independent contract-testing lab.

One last caution on bactericides is necessary. Since the activated sludge process for sewage treatment uses bacteria, any bactericide spills or disposals to the sewer will upset this process. One must be very careful not to contaminate sewage with bactericidal residues.

II. FUNGICIDES

A substantial amount of work has gone into the identification and characterization of fungi and mildew growths on polymer-derived or coated surfaces. Mildew is a form of plant life that feeds on organic material. Fungi, more readily identified, are also forms of plant life that decompose organic materials and convert them into the elements needed for green plant growth.

Symposia on mildew attack on painted surfaces have been published [10], an Australian paper offers a good review [11]. The Federation of Societies of Coatings Technology (FSCT) consortium approach was described by Yeager [12]. The National Paint and Coatings Association (NPCA) published a booklet describing mildew growth on housing in shaded and unshaded areas [13], which is a refinement of the original procedure by G. G. Sward in 1952. A report of nonmercurial mildewicides describes the variability in testing situations between the U.S. Navy National Bureau of Standard and labs, while concluding that two of seventeen EPA-approved compounds were superior [41].

Criteria for exterior exposure testing of paint films for microbiological attack have been established [14]. Broome and Lowrey [15] described the dependence of film preservation by mercurials on weathering phenomena. The Columbus-Dayton-Indianapolis-Cincinnati (CDIC) Society for Paint Technology observed that mildew exposure studies should include the observation of protected as well as exposed areas [16]. Bravery et al. described a humidity-chamber collaborative study [17]. Smith [18] recommends heat stress, along with UV and leaching tests, for screening effectiveness of mildewicides. Zabel and Horner described an accelerated procedure for growing a mildew on painted panels [19] with the conclusion that reflectance was a better measure of film damage than visual ratings. There is an ASTM microscopic examination method for determining whether a surface degradation is due to dirt or microbial growth [20].

Many materials have been used to combat mildew and fungus, (Table 20.2) especially by the paint industry [21]. Machmer described tests wherein fluid protection against bacteria proved different from dry film protection against mildew [22]. Hoffmann examined government housing, offering a variety of exposure conditions with a broad spectrum of organic and inorganic preservatives [23]. Dupont outlines product selection criteria for nonmercurial mildewicides [24]. A thiazolylbenzimidazole was patented for making silicones mold and mildew resistant [25]. Ludwig recommend careful selection of ingredients, as no available system meets all requirements [26]. Woods briefly reviewed the antimicrobials for waterborne coatings [27]. Pendleton et al. [44] described a comparison of mildewcides in an alkyd and in a latex, with different recommendations for each.

The Environmental Protection Agency (EPA) has been regulating materials that are prohibited in coatings formulations as preservatives. For instance, pen-

Table 20.2 Typical Fungicides and Their Manufacturers

Structure	Brand Name	Supplier Manufacturer
Diiodomethyl *p*-tolyl sulfone	AMICAL 50	Abbott Labs
Diiodomethyl *p*-chlorophenyl sulone	AMICAL 77	Abbott Labs
Monocyclic oxazolidine mix 78%	BIOBAN CS 1135	Angus Corp.
Barium metaborate	BUSAN 11M1	Buckman Labs
Methylene bis(thiocyanate), 10%, 2-(thiocyanomethylthio)benzothiazole, 10%	BUSAN 1009	Buckman Labs
2-(thiocyanomethylthio)benzothiazole, 30%	BUSAN 1030	Buckman Labs
Potassium *N*-hydroxymethyl-*N*-methyl-dithiodicarbamate and sodium 2-mercaptobenzothiazole	BUSAN 52	Buckman Labs
Phenylmercuric acetate	COSAN PMA Series	Cosan Chem. Corp
Methanol[[[2-dihydro-5-methyl-3(2*H*)-oxazolyl-1-methylethoxyl]methoxy]methoxy], 50%	COSAN 145	Cosan Chem. Corp
1,1′-(2-Butylene)bis(3,5,7-triaza-1-azoniaadamantane chloride), 90%	COSAN 265	
Alkylamine hydrochlorides	COSAN 635W	
Tetrachloro-4-methylsulfonylpyridine	DOWCIL S-13	Dow Chemical
2,3,4-Trichloro-4-(propylsulfonyl)pyridine	DOWCIL A-40	Dow Chemical
Copper 8-quinolinolates	MICRO-CHEK	
	ISOTROL Series	Ferro Corp.
	NYTEK-GD	Maag Agric. Chem.
Tetrachloroisophthalonitrile	NOPCOCIDE N-96	Henkel
2,2′-Methylenebis(4-chlorophenol)	PRESERVATIVE 27-78	Maag Agric. Chem.
2-(4-Thiazolyl)benzimidazole	TK-100	Merck & Co.
Zinc naphthenate	M-GARD W550	Mooney & Co.
2-*n*-Octyl-4-isothiazolinone and zinc oxide	SKANE M-8 & ZnO	Rohm & Haas
2-*n*-Octyl-4-isothiazolinone	SKANE M-8	Rohm & Haas
bis(tributlyltin)oxide	TROYSAN TBTO	Troy Chemical
	STAY-CLEAN	Environ-Chem. Inc.
3-Iodo-2-propynylbutyl carbamate	TROYSAN POLY-PHASE AF-1	Troy Chemical
Phenylmercuric acetate, 18%	TROYSAN PMA-30	Troy Chemical
trans-1,2-Bis(*n*-propylsulfonyl)ethene, 95%	VANSIDE PA	R. T. Vanderbilt

tachlorophenol and creosote were prohibited for use in consumer coatings [28]. Mercurials and organotin materials are also regulated. Among the recommended alternatives for wood preservation is a carbamate [29].

Many reports show zinc oxide as having beneficial effects in fighting mildew. Werthan reported that it increased the effectiveness of mercurial mildewicides [30]. Mark described acrylic- and vinyl-based emulsion systems improved by additions of zinc oxide [31]. Madson described zinc oxide additions to acrylic and vinyl formulations [32]. Kronstein and Zipf [46] evaluated copper, zinc, lead, and tin oxides as fungicides in aqueous urethane- or vinyl acetate-based polymers, showing binder interaction with the metal oxides.

Pittman et al. have been working on tying the mildewicide species to a polymer backbone to reduce the toxicological exposure of the material to all but the species attacking the polymer [33]. Although the Paint Research Institute (Roon Foundation) is sponsoring further work with this concept through the FSCT Technical Committees, formulations have not been stable thus far. Wake reported work with similar emulsions containing pentachlorophenyl acrylate or methacrylate copolymers having antifungal activity [34]. I am mystified as to how the pentachlorophenol would be biologically active without hydrolysis and ingestion by the attacking species, although the Pittman article [32] shows photographs of the formulation with pentachlorophenol leached into the surrounding agar; the copolymerized pentachlorophenol ester monomer showed no such leaching. Pentachlorophenol (and creosote) cannot be offered to consumer markets, and protective measures are required for using the former [45].

The plastics world has much concern for the protection of materials from fungi. The PVC with plasticizer may be one of the prime substrates vulnerable to such attack. Allbee [35] calls the additives the main source of the problem, as pure synthetic polymers are seldom attacked, but the stabilizers, lubricants, fillers, and especially the plasticizers are essentially sources of nourishment. Annual listings of suppliers of biocides are available [36].

III. MARINE-COATING BIOCIDES

The attacking species on hull coatings for ships and boats include algae, tubeworms, and barnacles. These alter the hull shape and smoothness, increasing drag and fuel consumption. Some of the materials used for protection include copper oxides, organotin compounds, organomercurials, and organoleads [37]. The materials are formulated into slightly soluble binder species, from which they leach to the surface. Rascio and Capran [38] describe extender pigments added to cuprous oxide leaching paints. Ghanem used Alexandria harbor to test a wide variety of antifouling marine coatings, including some organotin copolymers [39]. Sherman describes three types of models for leaching materials from a coating or substrate to effect a controlled release [40]. The press reports the death of shellfish in marina areas, and concern for the use of organotin materials on small boat hulls may lead to regulations restricting application to ocean-going vessels.

IV. FORMULATION OF BIOCIDE ADDITIONS

The addition of bactericides and fungicides to waterborne formulations may create problems by causing localized coagulation. You are losing not only the binder and pigment of that coagulum, but also some of the bacteristat, usually an expensive chemical. The key points to remember are to predissolve and disperse the additive in water with any needed colloidal stabilizer, and ensure that the pH of the added emulsion and dispersion matches that of the coating formulation. Solutions of the organic biocides may be made in water-miscible solvents as well.

For addition of the inorganic materials that may be reactive with colloidal components (or the latex itself), check compatibility on a small sample first. Check the stability for at least a month, as the inorganic material may dissolve slowly, and the reaction may be slow. Zinc oxide may be dissolved in an amine- or ammonia-neutralized systems, and thus be ineffective (except as a carboxylate neutralizer) upon drying. Some salts act as buffers, and shift the pH to their own equilibrium point (usually high). The barium metaborate systems may only be useful around pH 9. Zinc and barium salts contribute polyvalent ions to the total salt load and thus may contribute to Schulz-Hardy coagulations.

REFERENCES

1. Kansas City Society for Paint Technol., *J. Paint Technol., 46(589),* 37 (1974).
2. A. A. Smith, Paint Research Assoc., Teddington, England (Nov. 1980).
3. T. E. Rusch et al., *Mod. Paint Ctgs.,* p. 53 (Feb. 1978).
4. J. A. Jakubowski et al., *J. Ctgs. Technol., 54(685),* 39 (Feb. 1982).
5. Chicago Society for Paint Technol., *J. Paint Technol., 43(563),* 80 (Dec. 1971).
6. F. H. Sharpell, *Adhesive Age,* p. 23 (Apr. 1982).
7. Troy Chemical Corp., Jap. Pat. 78 57230 (24 May 1978), Cf: *CA 89,* 181345x.
8. H. Winters et al., *Developments in Industrial Microbiology,* Vol. 14, Chap. 35, Amer. Inst. Biol. Sci., Washington (1973).
9. J. J. Gambino et al., *Resin Review, 32(4),* 16 (1979).
10. *J. Ctgs. Technol., 50(639),* 36–65 (Apr. 1978).
11. A. K. Kempson, *Austral. OCCA* p. 5, (Nov. 1976).
12. C. C. Yeager, *J. Ctgs. Technol., 53(680),* 47 (Sept. 1981).
13. NPCA Technical Div. Scientific Circular No. 802, National Paint and Coatings Associations, Washington.
14. ASTM D3456-75.
15. T. T. Broome and E. J. Lowrey, *J. Paint Technol., 42(543),* 227 (Apr. 1970).

16. CDIC Soc. for Paint Technol., *J. Paint Technol., 43(554),* 76 (1971).
17. A. F. Bravery et al., *JOCCA, 67(1),* 2 (1984).
18. R. A. Smith, *J. Ctgs. Technol., 57(643),* 56 (June 1982).
19. R. A. Zabel and W. E. Horner, *J. Ctgs. Technol., 53(675),* 33 (Apr. 1981).
20. ASTM D3274-82.
21. P. A. Layman, *Chem. Eng. News,* pp. 10–11 (Apr. 12, 1982).
22. W. E. Machmer, *Development in Industrial Microbiology,* Vol. 20, Chap. 3, Soc. for Industr. Microbiol, (1979).
23. E. Hoffmann, *J. Paint Technol., 43(558),* 54 (July 1971).
24. J. A. DuPont, *Mod. Paint Ctgs.,* p. 38 (Nov. 1978).
25. K. Shimizu, US 4247442 (Jan. 27, 1981).
26. Ludwig, *J. Paint Technol., 46,* 31 (July 1974).
27. W. B. Woods, *J. Water Borne Ctgs., 7(3),* 8 (Aug. 1984).
28. *US Federal Register, 49(136),* July 13 (1984).
29. J. Hansen, *Mod. Paint Ctgs.* p. 23 (Nov. 1984).
30. S. Werthan, *Paint Varnish Prod.,* p. 37 (Oct. 1966).
31. S. Mark, *Paint Varnish Prod.* p. 32 (July 1981).
32. W. H. Madson, *Amer. Paint Ctgs.,* p. 27 (Jan. 6, 1975).
33. C. U. Pittman, Jr. et al., *J. Ctgs. Technol., 50(636),* 49 (1978).
34. L. V. Wake, "Synthesis of Fungicidal Vinyl and Acrylic Latices By Emulsion Polymerization," Materials Research Labs, Ascot Vale (Australia) July 1981, MRL-R-822; Cf: NTIS AD-A107 848/4.
35. N. Allbee, *Plast. Compounding,* p. 41 (Mar./Apr. 1982).
36. Plastics Compounding *1986/87 Redbook,* Harcourt Brace Jovanovich Publications, Denver.
37. Skinner, *Paint Varnish Prod., 44,* 43 (1974).
38. V. J. D. Rascio and J. J. Capran, *J. Ctgs. Technol., 50(637),* 65 (Feb. 1978).
39. N. A. Ghanem et al, *J. Ctgs. Technol., 53(675),* 57 (Apr. 1981).
40. L. R. Sherman, *J. Appl. Polym. Sci., 27,* 997 (1982).
41. *Amer. Paint Ctgs. J.,* p. 28 (Apr. 20, 1987).
42. W. B. Woods, *J. Water Borne Ctgs., 5(4),* 2 (Nov. 1982).
43. W. B. Woods, *Amer. Paint Coatings J.,* p. 37 (Jul. 6, 1987).
44. D. E. Pendleton et al., *Mod. Paint Ctgs.,* p. 30 (Aug. 1987), and p. 148 (Sept. 1987).
45. *Federal Register, 49(136),* July 13 (1984).
46. M. Kronstein and M. E. Zipf, *Mod. Paint Ctgs.,* p. 132 (Sept. 1987).
47. W. B. Woods, *Paint Ctgs. Indust.,* p. 25 (Sept./Oct. 1987).

21

Fillers, Pigments, and Reinforcing Agents

I. INTRODUCTION

Solid particles are added to latex formulations for a variety of reasons. Each type has its own need for analysis and testing. These particles have a variety of effects on the formulation, both in the fluid state and in the final dry film. We shall discuss analysis and testing needs, the effects to be expected upon addition of the particles, and the particle classes.

II. ANALYSIS AND TESTING

Incoming pigment shipments should be tested before acceptance. Even commodity pigments (like titania), exhibit differences in surface activity from batch to batch. That means that the dispersant formulation used on the last batch may not work on the next. The testing on incoming shipments will ensure that you know the material will perform satisfactorily or that you know how much you have to adjust your process. You may also realize that the material is of no use to you and send it back.

Since the fluid properties are critical in formulations, I always regard the tests that affect those properties as my first set of queries. One may prepare an aqueous slurry quickly from the incoming dry particulate, or it may be delivered as a slurry. Hence, the pH of the aqueous slurry and the conductance may be indicators of quality and compatibility with formulations. The density of the slurry and total solids content should also be tested. One wants to measure total solids content in the delivered slurry, as one generally pays for the material by the pound. Density, pH, and conductivity may also reflect contamination.

The testing program should depend on your needs and the specifications the product must meet. If the paint must meet a "grind" specification, a simple sieve analysis may be desirable (or Hegman grind gauge on the slurry), unless you plan to use a mill or grinder to reduce particle size. The end product need may require the incoming material to be tested for color, brightness, whiteness, and other properties.

A record of such checks will give an idea as to how consistent the material or supplier is, and may give limits to acceptability for formulations. In no instance should a material shortcoming be compensated for in formulation without explanation and subsequent checks.

Primo Levi told a tale [12] on this lack of formulation documentation. It began with a pigment of poor quality in the late 1940s that he formulated with added ammonium chloride to minimize the bad effect for the paint company he worked for. He no longer worked there some 20 years later and was amused to hear a scientist in that company complaining about the formulation. It took much experimentation to eliminate the now nonfunctional ammonium chloride. No one know how it got there or when or why. The pigment quality had improved over the years to the point where the ammonium chloride was not needed.

III. EFFECTS ON FORMULATION PROPERTIES

A. Fluid Properties

The added solid particles do not usually affect fluid properties beyond the contribution to volume fraction in Einstein equation for viscosity. Indeed, many such particles simply modify the rheology of the formulation. They are, in the main, particulates added to serve some function in the final dried film. The exception will be noted in the detailed discussion where they occur, mainly as the solids dissolve.

The volume fraction occupied by the solid particles is only part of the story. Other parts include particle size (average and distribution), and how the several different particle sizes pack and blend with the latex particles (in the fluid for rheology sake, or as the dry film impacting gloss, smoothness, and porosity). Particle-size blending to improve rheology is a standard procedure in the concrete trade, in making solid rocket propellants, and in paint manufacture. Its theoretical basis is outlined by Farris [14].

Particle shape *drastically* affects other properties. Rheology becomes not a function of particle volume, but of the "rotationally inscribed volume." This can be an advantage; I made an excellent temporary adhesive of high viscosity for paper, using a dispersion of needle-shaped Attapulgus clay with good wet strength which was lost upon drying. The "shaped" particles can make rheology more non-Newtonian (probably dilatant) as their inscribed volumes interfere more at high-shear flow.

Particulate densities are higher than those of the latex particles and they may settle out. This should be avoided, they are frequently capable of "hard settlement," requiring a high shear mixer to redisperse them. Such a mechanical action can coagulate the latex. Hence, a great effort is made to prepare stable dispersions of solid particulates. The books of Patton [1] and the publications of Carr [2] give practical advice. Research at the Institute for Pigments Research in Stuttgart will also provide useful information. Our discussions of the organic chemicals used as pigment dispersants may be helpful [3]. Indeed, truck load lots of aqueous titania dispersions contain only small amounts of a substituted ethanolamine as dispersant. Remember also the discussion of Stokes' Law (see Chapter 2), and the controlling parameters in that equation (particle size, density differential, and dispersion medium viscosity).

B. Dry Film Properties

The ratio of pigment volume to the binder polymer volume (called the pigment volume concentration or PVC) is especially important in paint. At the "critical pigment-volume concentration" (CPVC), the film properties change drastically in terms of porosity and strength, and others. Strictly speaking, the CPVC is the point at which the binder volume fraction becomes so low that air spaces are occluded within the film. The debates as to the "true" or "scientific" definition of the CPVC and the techniques for evaluating or calculating continue. But the basic concept is valid, and measurements aimed at its characterization should be viewed as semiqualitative and for guidance only (at least until someone shows that the precision of such measurements is reasonable and related to some basic geometrical consideration measured by an independent technique).

1. Physical Properties

The physical properties of the film include strength (tensile, shear, compression, or combination thereof), hardness, flexibility, extensibility, and others. These are affected by the solid particulates. In general, strength, elasticity, and flexibility decrease. Modulus and hardness are likely to increase. The increase/decrease relation is usually linear with the amount of particulate added. But at the CPVC, there is a change in slope.

2. Optical Properties

The optical properties of concern here are those connected with the presence of the particle as a physical body rather than the color imparted. For instance, some pigments are called "flatting pigments" or "flatting agents." I prefer the latter term as these materials may be too small in particle size to provide any pigmentation. A normally glossy film has an extremely smooth flat surface, but the addition of the flatting agent covers that surface with tiny peaks or bumps. The "flatted" or "matte" surface has some feature that scatters light only at the surface, disrupting the reflected image. The term "flatting agent" is somewhat inappropriate, but the gloss and distinctness of image are destroyed by it.

Gloss and "flatness" may be related in other aspects. Some formulators have difficulty with their formulations, as they do not give the desired gloss; the formulators fault the pigment for being too poorly or too well dispersed. That may be the case, but should be confirmed by microscopic examination of the film surface before reformulation. The flat or matte film appearance may be due to other reasons for light scattering at the surface. Microcracks, micropinholes, microwaves, gel particles, or other sources of flatness could be the source; addition of coalescent may solve the problem.

Glossy paper coatings are 90 + % clay pigment at the surface, and development of their gloss is based on the supercalendering. The supercalender lines up all the platy clay particle surfaces in the same plane, preventing light scattering by most of the surface.

Particle shape is interesting and useful in other formulations. Some people argue that platy particulates can add "leaf" to impermeability and other film properties. Certainly "leafing" metallic pigments can produce a spectacular "candy apple" appearance on a variety of substrates. A needle-shaped green pigment in an extruded sheet gave a different shade of green viewed in the machine direction as contrasted to the across-machine direction; the color varies with the axis of viewing. The needles are oriented by the extrusion process.

Opacity and hiding are another concern. Not all pigments hide the substrate. My first colored coating (a high-solids polyurea in the early 1960s) was a transparent deep blue, based on Ultramarine Blue. However, addition of a small amount of simple kaolin clay imparted hiding quality with little reduction of the coloration.

Consider the pyrogenic silica pigment grades used for flatting paints. Dispersions of these, when complete enough, are transparent or translucent. However, when the dispersion is a colloid of flocs, the fluid is opaque. If you have a transparent or translucent dispersion, and determine its total solids, the dried mass is opaque because it is one big floc with air spaces between the particles.

Opacity in a dispersed phase or film is based on light scattering. The Rayleigh equation tells us that the intensity of the scattered light is dependent upon the particle size and the difference between the refractive indices of the dispersed phase and the dispersion medium. Scattering is not visible if the refractive indices are the same or if the particle size is too small.

Not all white rocks are good hiding pigments. Many ground minerals are white, but lack hiding quality. Again, we have to refer to the difference between particles refractive index and the refractive index of the medium the particle is suspended in. For hiding, the particles have to be suspended within the film of the coating binder. Here again, the pigment volume concentration (PVC) affects the hiding. If the particles are closely packed, scattering efficiency may be low, but if there are too few particles (at low PVC), you may not obtain any hiding. Film thicknesses make a difference, as the thicker films have more scattering sites through their depth. Titania is a preferred hiding pigment because of its high refractive index. Because it is expensive, cheaper "white" pigments are added as "spacers" so that the titania particles are not too crowded for efficient scattering. Some titania suppliers have developed computer programs for optimizing hiding at minimum cost. That is not to say you cannot make a white coating with limestone, precipitated calcium carbonate, or gypsum. There are paper coatings based on these inexpensive pigment materials; some made at the paper mill.

Paper coatings also depend heavily on air space scattering for some hiding. This should not be a surprise, as the paper coating is frequently formulated at a PVC substantially higher than the CPVC. Air spaces are well known to effect hiding, as with the whiteness of snow resulting from the air spaces between the water crystals. PPG Industries made use of air spaces in paint formulations to eliminate the need to build a plant to make titania [11]. Some paper people call these air-containing coatings "bubble coatings." Again, the air spaces are of the correct size and sufficiently different in refractive index to be effective scattering centers.

The "plastic" pigments were developed as low density replacements for the expensive hiding pigments. At various times over the past 15 years, plastic particulates based on poly(styrene), acrylics, and urea–formaldehyde resins have been offered for sale. I suspect that they may not be cost effective in many formulations. The poly(styrene) pigments could withstand supercalendering in some paper coating applications and not in others.

A second point might be made here: not all pigments in the coating will be visible as particulates. Another paint I made in the 1960s was an interior latex white with *only* titania as the pigment. A friend suggested a small amount of well-dispersed Ultramarine Blue to make it even whiter, and it worked. We could not see the blue particles, except in one batch where the blue was not well dispersed. My "polka dot" paint was a joke in that lab, as we did not know about the Zola formulations [13] in those days.

Titania brings up an interesting point of physical appearance, namely chalking. The term "chalking" is based on the fact that certain paint films become powdery upon the surface as they age outdoors. The powder is the white pigment loosened as the binder around it degrades and disappears. A microscopic examination of the chalking film shows that the pigment particle is isolated in a pit, or on a blister of paint film, ready to be washed or brushed off.

The pigments that absorb UV light (certain titanium, zinc, or other types) are absorbing energy that has to be dispersed and some is reemitted or even scattered unchanged. However, some reemissions are shifted to visible wavelength (fluorescence), whereas others are not visible. Hence, when considering how the binder will survive in a film as a particular formulation, you also have to consider the alternative energy dissipations used by the formulation components. If the radiation reemitted by the pigment(s) attacks the binder, the formulation may not survive as long as desired. Alternatively, that may be a good method to make a temporary coating or film. Certainly the photodegradable poly(ethylene)-pigmented blacks gather more heat to hasten their degradation.

IV. SOLID PARTICULATES

We divide particulate additives into three categories:

1. pigments (mainly colorants);
2. fillers (low-cost space occupiers to reduce costs); and,
3. reinforcing agents which add strength.

The chemical nature of these materials differs from group to group (variations occur within the categories, as well), and our discussions will emphasize structural details which impact fluid or final film properties. Our intent is not to be comprehensive, but to make the reader aware of potential problems, and of sources for help when a problem arises.

A. Pigments

Pigments were originally insoluble particulate colorants (as opposed to dyes which are soluble). However, some pigments used in modern practice have color only incidentally, as their real function is different. We will divide this category into four parts to cover some such functions, and will welcome contributions that outline yet further categories. The four types we will discuss are colorants, corrosion inhibitors, and conductive and texturizing pigments.

1. Colorants

Pigments were originally color bodies added to a paint or ink to yield a final film of desired color. Most were ground colored rocks, though chemistry contributed useful new materials. Rather than enumerate the myriad of color pigment types and the grades of each, let us mention some lessons to be learned from specific examples.

Coloration is usually accomplished with blends of pigments. Experimentation will give the results indicating the need to grind the pigments individually or together; overgrinding is possible. I know of one Phthalocyanine Blue pigment system ground too fine for the textile printing ink for which it was intended, and the blue disappeared within the fibers. A less well ground (larger particle size) blue was successful. Carr [2] showed how color yield was affected by type and degree of grinding.

Organic pigments serve well in inks, paints, and the like. Some fade with exposure to sunlight. Others present difficulties in preparing stable dispersions, and yet others in their compatibility with the polymer matrix into which they are mixed. However, all have their uses. The surface energetics involved in making dispersions in a polymer for the final use or in the dispersion medium suggest that the additive approach may help solve the problem. Fowkes (at Lehigh University) makes the point that recognizing the acid-base character of resin and pigment may be an aid to making them compatible, and an appropriately chosen additive may be the bridge needed for compatibility.

2. Corrosion Inhibitors

This is a class of pigments of particular interest, as they are somewhat soluble. Many act electrochemically to reduce the corrosion potential. Some act as traps to gather up corrosion catalysts and others neutralize any acids present; the solubility can be a serious formulation problem.

Most pigments in this class are based on polyvalent cations. The Schulz-Hardy Rule warns that this may act against colloidal stability of the dispersion, the emulsion polymer, or the total formulation. One may calculate the dissolved polyvalent ion concentration from the solubility product for many of these pigments [5]. We have demonstrated the coagulation (precipitation) of dispersant-grade water-soluble carboxylated polymers by polyvalent cations on several occasions [4,5]. This could also be a problem for the user of carboxylated latexes or thickeners. Using this as a guide, we were able to formulate corrosion inhibitive pigments into a latex system quite successfully [6], but it required testing pigment and dispersant formulations (mostly Daniels' "flow point" [19] and settling rate); combining the above successes, from previous section and Daniels' flow point, with the latex for further tests; and combining the above successes with rheology modifiers for further tests. The supplier literature for the final pigment chosen showed a 5-Hegman grind fineness limit noted for waterborne formulations, but we were able to get a 7+ grind with a Cowles disperser in a few minutes with our formulation. The client was pleased with the results, but the formulations were no longer usable after a few years. Investigation at the latex supplier's plant showed that they had changed the colloidal stabilization of their polymerization formulation to increase productivity; the whole coating system had to be reformulated.

Other problems can occur. The alkalinity of the dissolved portion of the pigment (barium metaborate) may cause some problems for pH-sensitive formulations. Thermal aid to dissolving (e.g., the disperser pumps in Joule heating during the dispersion process) can distribute an excess polyvalent cations into the system (use ice in the dispersion process). Some zinc-based pigment formulations can be ruined by adventitious amine or ammonia in the mixture, carried in by the surfactant, dispersant, or latex.

Some knowledge of corrosion-inhibitive pigments is based on testing procedures that are supposed to be accelerated aging processes. Although we have pointed out many times that heat-aging processes are only valid for single chemical reaction processes with applicable Arrhenius equation mechanisms, many other processes have been attempted. These include salt fog cabinets, UV and humidity chambers and gas (nitrous oxide, sulfur dioxide) exposure chambers. The Steel Structure Painting Council does not believe that the salt fog results correlate to weathering phenomena [8], and we have shown that formulations wherein salt fog or a variety of UV chambers did not correlate with outdoor exposures on two very good waterborne corrosion inhibitive formulations [9]. This exerts a considerable time constraint on development programs aimed at showing effectiveness of corrosion protection for coated panels on the Golden Gate Bridge or other harsh environment. One would hope that electrochemical techniques would be developed to afford appropriate corrosion protection.

3. Conductives

Conductive pigments are mainly metal particles, though a few oxide-based pigments are also conductive (e.g., barium titanate). The metals most commonly used in conductive applications are silver, copper, nickel, and occasionally gold, aluminum and zinc. Carbon black dispersions are slightly less conductive, but serve in some applications. One is more likely to find the metal powders sold separately for dispersion or already formulated into the (mostly solvent-borne) conductive coating, ink, paste, or adhesive.

Conductive-coating applications vary widely. The main use is to protect electronic equipment from stray electrical interference. The conductive coating can be a safe ground path for static discharge that could damage a computer. It can also protect computers from electromagnetic interference (EMI), a sort of radiation action on electronics that causes signal changes (e.g., microwaves can disrupt a pacemaker). Conductive inks can be used as wires to carry an electrical signal. We even advised one company on how they could paint plastic onto an electrode for a disposable medical test device.

The metal particle pigments are seldom formulated into waterborne systems, as the metal dissolves (corrodes) in time with the generation of hydrogen gas. Can stability is a *real* problem. However, less expensive metals (zinc, aluminum) are usually involved; gold, silver, and copper systems can survive. One may check the aqueous phase of a dispersion or coating formulation containing a metallic pigment by enclosing the mixture in a vessel attached to a gas buret to measure the hydrogen evolved, if any. In the one zinc powder formulation we tried, we found no hydrogen evolved so we could package it without fear [9].

There are some differences in conductive pigments, aside from their formulation weaknesses and strengths. Table 21.1 compares densities and resistivities (the inverse of conductivities), and the substantial differences in prices that are better documented in the commodities markets. The price of many coating formulations is based on the current market price of the metal used.

Table 21.1 Conductive Pigment Materials[a]

Material	Density, g/cm^3	Resistance, ohm cm
Silver	10.5	1.6×10^{-6}
Gold	19.3	2.3×10^{-6}
Copper	8.9	1.8×10^{-6}
Platinum	21.5	2.15×10^{-5}
Nickel	8.9	1.0×10^{-5}
Best silver-filled inks and coatings	—	1×10^{-4}
Best silver-filled epoxy	—	1×10^{-3}
Graphite	~2.2	1.3×10^{-3}
Graphite- or carbon-filled coatings	—	10 to 10^2
Polystyrene and other dielectrics	—	10^{16}

[a]Ref. 10

The formulation of conductive coatings or inks to achieve maximum conductivity requires formulating above the CPVC. One does not want the binder to inhibit maximum contact between the conductive particles, and therefore the binder supplies spot bonding of these particles along the edges. Is this possible? Certainly, as the paper coatings give good precedents at low binder content compared to the pigmentation level (substantially above the CPVC). With good strength, conductive coatings could do as well with appropriate formulation. The CPVC concept will be a guide to cost effective pigment choice.

Other formulation devices will be needed. The binder should be hydrophobic to minimize adsorbed and absorbed water and reduce corrosion. The additives must be chosen to minimize corrosion catalyst ions (sodium, chloride, sulfate) and hydrophilic components (volatile surfactants). Conductivity may also be maximized by choice of the conductive-particle shape. Metals in the form of flakes or needles may contact several other particles, ensuring good conductivity.

Application of conductive paint can be sensitive; it is necessary to ensure that the first coat applied has the same pigment level as the last. Since the metals used have high densities, settling can be a problem. The paints are best applied by spraying. Stirring in the spray application feed unit is recommended to make sure that the same pigmentation level is applied throughout. Such sprayers are commercially available.

The oxide-based conductive pigments are small contributors to the market. They are *expensive* but useful. Table 21.2 shows materials and their suppliers. These materials tend to be white (or nearly so), and do not conduct as well as the metal-containing paints, *however,* they are not corrosive, do not generate hydrogen, or otherwise threaten formulation stability. One may even be able to formulate for transparency with correct binder choice by refractive index matching.

Table 21.2 Conductive Oxide Pigments

Material	Supplier	Phone Numbers
Tin antimonate	Magnesium Elektron	(201) 782-5800
Rutile suboxide	Ebonex Technologies	(415) 420-0400
Barium titanate	Cerac, Inc.	(414) 289-9800
	GFS Chemicals	(614) 881-5501

4. Texturizers

Textured coatings are applied for several reasons. Many people like the three-dimensional surface appearance, and some coatings stimulate a desirable stucco or plaster surface effect. Others like the fact that the dirt absorbed from hand contact is kept on the high spots, and easy to clean.

Texture coatings are applied architecturally (outdoor or indoor) and on industrial products. There are two methods of preparing textured-film coating. One is to employ a large-particle-size "pigment" to produce a lumpy surface; the other is to have the binder drying process form the texture from the polymer (the "wrinkle" finishes available in drying oils 50 years ago). Our emphasis here is on the pigmentation technique, though we know of latex-based texture coating formulations based on modification of the "wrinkle" finish concept.

Texturizing pigments are simply *large* particles of some sort. Sand is often used, but even larger particles can be employed, such as those obtained from vermiculite. The latter is a choice material with a lower density than most pigment materials, and a lesser settling problem. This problem is serious in these formulations, and is usually solved by a very high viscosity at low shear, preferably pseudoplastic, but thixotropic will do. Mixing these pigments into the formulation is usually the last step before packaging, although some texture coatings have the texturizing added on site.

Fibrous materials are also used as texturizing agents. Some are inorganic and some are simple plastic (chopped textile fibers?); wood pulp can be used. If you want to try a quickly dispersed wood pulp fiber, add 2 g of toilet paper into 98 g of water in a Waring blender, and turn the blender on till you see no more sheet particles. This source is among the purest of available pulps, with no added resins for wet strength.

Application of these texturized paints presents some problems. Spray application is best (spinning bell preferred over nozzle to avoid clogging), as the texture is randomly distributed over the surface. Roller or brush applications may drag the larger particles, leaving tracks. The sprayer should be fitted with a stirrer.

B. Fillers

The main objective in adding a filler to any formulation is to reduce costs. This can be carried to extremes, as exemplified by the cheap, but heavy, carpet containing mostly limestone filler in the backcoating adhesive. However, where quality is not affected, a filler is a reasonable additive.

Thus the prime consideration is the filler cost. That limits the materials to clay, limestone, starch, sand, or other inexpensive locally available material. Fillers may even provide a desirable side effect, such as the spacing between titania particles to optimize hiding.

Since the filler (especially those bought on a cost basis from local sources) is inexpensive, one may expect a quality variation from batch to batch or shipment to shipment. One must do quality assurance checks in anticipation of potential problems. An aqueous extraction or slurry check of pH and conductivity, and a sieve or other particle size analysis are the minimum recommendations.

The main consideration with fillers concerns the physical properties. Most fillers have a deleterious effect. One may expect film hardness, stiffness, and density to increase, with reduction in tensile strength (or other strengths). There is a trade-off of physical property for cost reduction. In one adhesive application, I plotted adhesive add-on variations as cost, against the tensile strength obtained, to identify applied cost. This helped us chose among suppliers and formulations.

C. Reinforcements

Just as the filler materials can impair strength, other materials can improve it. These are called reinforcements (or occasionally reinforcing pigments). But first, let us examine the sources of strength within a polymer film. The contributors to strength (which may or may not be present in any particular polymer form may be:

1. *Molecular Weight*. The chains of a polymer become entangled (after reaching a critical molecular weight), and require energy to untangle.
2. *Cure or Cross-Linking*. Covalent (or ionic, in some cases) bonds link intermolecularly.
3. *Associative Phenomena*. In general, formation of some intermolecular bond of low energy, applies here.
   a. Hydrogen Bonding. Intermolecular hydrogen bonds, in great number, develop as much strength as the covalent bonds of a real "cure." (Cellulose wood pulp fibers are as strong as steel in "zero-span" tensile strength tests. Nylon, known as a strong fiber, has a molecular weight of only thousands, but has strength because of its large number of hydrogen bonds.)
   b. Crystallinity. The "hard" crystalline segments in poly(ethylene) (PE) require additional energy to lose this crystallinity before the molecules

can straighten out to break. ("Bleach bottle" crystalline PE is different from the waxier amorphous "sqeeze bottle" PE.)

c. Liquid Crystallinity. Intermolecular alignments of large polycyclic aromatic groups are responsible for the strength of KEVLAR fibers and some coatings formulated by the NDSU group led by Jones [18].
d. Ionic Groups. The alkali metal carboxylates gather as microaggregates in some nonpolar polymers; and the strength of SURLYN is explained by this mechanism (see publications of A. Eisenberg and others).

In each of the above cases, the bulk polymer develops the strength. However, in some formulations, it is the addition of a solid particulate material that *increases* polymer strength (instead of the expected decrease).

The requirement for a "reinforcing" pigment or filler is that the solid particulate material interacts in some way with the polymer. Depending on the examples chosen, one may find the solid particulate is hydrogen bonding to, curing, or being incorporated within the polymer system. Certainly compounding Nylon or cellulose fibers within a carboxylated polymer should yield a hydrogen-bonding interaction.

Some systems yield reinforcement by a mechanism that is unclear or only suspected. Silica reinforces some silicone rubbers. Carbon blacks reinforce rubbers, and the form of the carbon black and type of rubber determines the degree of reinforcement. The mechanisms of these reinforcements are not yet proven, but we can find useful formulations.

Specialty pigments are treated with chemicals to improve their reinforcing characteristics. Mica, talc, clay, and calcium carbonate (all weakening fillers) are converted to reinforcements upon surface treatment to make the solid interact with the polymer. Such surface-treated solids are commercially packaged to be stirred into the formulation. Table 21.3 gives a list of some typical materials.

Table 21.3 Typical Commercial Treated Reinforcing Fillers[a]

Material	Treatment	Supplier
NYFLAKE Mica	Unknown	Nyco
WOOLASTOKUP Wollastonite	Silane	Nyco
Varied	Silane, Titanate	ESSDEE Fillers
Glass spheres	varied	Potters Indust.
ULTRALINK Kaolin	Silane	Engelhard
TRANSLINK Kaolin	Silane	Engelhard

[a]Similar types are probably available from your pigment and filler supplier.

One may surface treat the filler separately before formulation and mix the treated filler into the formulation, or may even add the treating agent to the formulation in order to be attached to the filler surface. Surface treatments are simple (e.g., castor oil based fatty acid treatment of limestone [16]) or complex (e.g., with maleic anhydride, peroxide, polyethylene, or clay [17]). Other surface treatment chemicals (silanes, titanates) have been reviewed in ref. [15]. Some researchers think covalent or hydrogen bonds from particles to polymer are needed, others feel that it is a simple matter of compatibilizing the polar solid particle surface with the nonpolar polymer matrix. Actually, as long as the formulator sees a benefit, the mechanism is not important, except as a guide for other formulations.

V. REPRISE

Solid particles are added to the formulation for a good reason. The key needed to make the whole formulation stable in fluid form and make the final product functional is likely to be surface chemistry.

REFERENCES

1. T. C. Patton, *Paint Flow and Pigment Dispersion,* John Wiley & Son, New York, (1979).
2. W. O. Carr, *JOCCA, 61,* 397 (1978).
3. R. D. Athey, Jr., *TAPPI, 58(9),* 66, and *58(10),* 55 (1975).
4. R. D. Athey Jr., *Ind. Eng. Chem. Prod. R & D, 14,* 310 (1975).
5. R. D. Athey Jr., *J. Water Borne Ctgs., 5(3),* 25 (Aug. 1982).
6. R. D. Athey Jr. and Englina Porowska, unpublished work at Mellon Institute, 1979-1980.
7. B. Appleman, 1988 FSCT meeting presentation.
8. R. D. Athey Jr. et al., *J. Ctgs. Technol., 57(726),* 71 (July 1985).
9. R. D. Athey Jr., unpublished work at Mellon Institute, 1978.
10. Adapted from J. C. Bolger and S. L. Morano, *Adhesives Age,* p. 17 (June 1984).
11. J. A. Seiner and H. L. Gerhart, XI FATIPEC Congress Prepr. pp. 127–139 (1972).
12. P. Levi, *The Periodic Table,* Chap. Chromium (1986).
13. J. C. Zola, US 3811904.
14. R. J. Farris, *Trans. Soc. Rheology, 12(2),* 116 (1975).
15. R. D. Athey Jr., *J. Water Borne Ctgs.,* p. 7 (Feb. 1985).
16. E. J. Augustyn and A. S. DeSouza, *Plastics Compounding,* p. 62 (July/Aug. 1979).
17. N. G. Gaylord et al., *ACS Org. Ctg. Plas. Chem. Div. Prepr., 40,* 456 (1979).
18. D.-S. Chen and F. N. Jones, *J. Ctgs. Technol., 60(755),* 39 (Jan. 1988).
19. F. K. Daniel and P. Goldman, *Ind. Eng. Chem. Anal. Ed., 18,* 26 (1946).

Author Index

Adelman, R.L., 101
Adeoye, I.O., 82, 85
Ahmed, S.M., 72, 78
Agbezuge, L., 129
Aleksandrova, 64, 72, 78
Alfrey, T., 95, 100
Allbee, N., 243, 245
Anacreaon, R.W., 187
Anderson, D.W., 77, 236
Andrade, 153, 166
Appleman, B., 259
Arai, M., 82, 85
Aronson, J.R., 177, 187
Ashton, H.E., 175, 186
Augustyn, E.J., 260
Austin, L.K., 169
Austin, T.M., 147
Azad, A.R.M., 83, 85

Babian, R., 85
Bagchi, P., 76
Baier, R.E., 177, 184, 187
Balik, C.M., 86
Bannerjee, M., 82, 85
Baran, A.A., 168
Barr, J.K., 188
Barton, A.F.M., 209, 212, 218, 219, 222
Bassett, D.R., 206
Bataille, P., 62, 68, 73, 77
Becher, P., 9, 17
Becker, R.O., 68
Benecke, S., 85
Berens, A.R., 65, 69
Berry, G.C., 149
Bienenstock, A., 185, 188
Billmeyer, F.W., 33
Bird, R.B., 128, 154, 170, 174
Blackley, D.C., 12, 17, 56, 76, 90, 92, 144, 176, 186
Bolger, J.C., 259
Bul'shakova, T.G., 169
Booth, J.W., 199
Bragole, R.A., 126, 129
Braidich, E.V., 99, 100
Brandrup, T., 100
Bravery, A.F., 245
Brendley, W.H., Jr., 152
Brennan, S., 185, 188
Broome, T.T., 241, 244
Brown, P.O., 186, 187

Bufkin, B.G., 97, 101
Burkhart, R.D., 67, 68
Burnette, G.M., 28, 33
Burrell, H., 209
Byrne, K.M., 126, 129

Calgari, S., 65, 68
Campbell, D.R., 33, 168
Campbell, T.W., 57
Carlson, G.M., 177, 187
Carmichael, J.E., 131
Carr, W.O., 249, 252, 259
Carrock, F.E., 77
Carroll, B., 187
Carroll, M.J.B., 113
Cazes, J., 182, 188
CDIC Society for Coating Technology, 241, 245
Ceska, G.W., 76, 77
Chang, M., 73, 77
Chatelaine, J., 68
Cherry, B.W., 126, 129
Chilcoat, R., 185, 188
Chilcoat, S., 33
Chicago Society for Coating Technology, 187
Chou, Y.J., 73, 77
Chow, T.S., 175, 186
Chatterjee, A.J., 168
Christenson, J.J., 188
Clark, D.T., 159, 185, 188
Clayfield, E.J., 14, 17
Clemons, R.J., 100, 101
Coco, J.H., 111, 113
Coder, W., 189
Coffman, A.M., 101
Collins, E.A., 129, 166
Colwell, R.E., 128, 170
Compton, A.C., 178, 186
Conrad, M.M., 33, 199, 236
Coover, H.W., Jr., 66, 68
Covington, E.R., 66, 68
Croll, S.G., 175, 186
Culbertson, B.M., 98, 100
Cunningham, G.P., 147, 169
Czandema, A.W., 187
Czerepinski, R., 221, 222

Daniels, W.E., 68
Danyushin, G.V., 72, 78
Daroowalla, S., 207
Datyner, A., 74, 77
Debye, P., 11, 145
Dees, O.S., 67, 68
DeGraff, A.W., 76
DePugh, C.C., 117
DeSouza, A.S., 260
Dicker, D.H., 128
Dilks, A., 186
Doak, K.W., 77
Dolinski, J., 77
Doronin, A.S., 97, 101
Duncan, W.P., 186
DuPont, J.A., 245
Dwight, D., 185, 188

Eckler, P.E., 105, 107
Egusa, S., 76, 77
Einstein, A., 131, 248
Eirich, F.R., 153, 170
Eisenberg, A., 258
Eissler, R.L., 175, 186
El-Aaser, M.S., 68, 142
Elango, R., 53, 56
Ellis, R.A., 188
Englebrecht, R.S., 168
Eremenke, B.V., 169
Erickson, D.E., 17
Ermilov, P.I. 101
Everett, D.H., 72, 78
Everhart, D.S., 185, 188
Ewart, R.H., 57

Fallwell, W.F., Jr., 101
Fan, L.T., 56
Farnsworth, F., 186
Farris, R.J., 131, 248, 260

Feller, R.L., 85
Ferry, J.D., 153, 170
Field, R.E., 100
Filer, T.D., 168
Fitch, R.M., 33, 82, 85
Flory, P.J., 25, 28, 33, 100, 179
Flournoy, P.A., 187
Fochtman, E.G., 166
Force, C.G., 78
Fossick, G.N., 63, 67, 147
Foster, D.H., 143, 168
Fouassier, J.-P, 82, 85
Fowkes, F.M., 177, 187, 210, 219, 222, 253
Fox, T.G., 100
Franczek, W.E., 135, 140, 167
Frazee, J.D., 177, 187
Freeman, W., 128
Freling, J., 170
Friedl, H.R., 66, 68
Friel, J.F., 188
Friis, N., 62, 68, 74, 76
Fujii, M., 62, 68

Gabbert, J.D., 67, 68
Gabrielli, G., 83, 85
Galbraikh, N., 73, 77
Gambino, J.J., 244
Ganapathy, V., 53, 56
Gardella, J.A., 185, 188
Gardon, J.L., 57, 74, 77
Gaslini, F., 141
Gaylord, N.G., 260
Gedcke, D.A., 188
Gehman, D.R., 85
Gerhart, H.L., 259
Gerlock, J., 236
Ghanem, N.A., 243, 245
Gibbs, D.S., 66, 68
Girgis, M.M., 92
Glancy, J.W., 204, 206
Glass, J.E., 207
Glasstone, S., 57
Glintz, F.P., 68
Gmerek, H., 147, 169
Golden Gate Society for Coating Technology, 129
Goldfarb, I.J., 181, 187
Goddall, A.R., 77
Goodwin, J.W., 77
Gorenkova, G.A., 72, 78, 167
Graf, R.T., 177, 187
Grave, J.R., 97, 101
Greene, B.W., 139, 167
Greener, Y., 153, 170
Greet, R.J., 153, 170
Grieser, R.H., 176, 188
Griffin, W.C., 57
Griffith, W.C., 57
Grindstaff, T.H., 129
Gritskova, I.A., 186
Grommers, E.P., 64, 68
Grubert, H., 68
Grubisic, Z., 188
Grudus, G.M., 100
Gultepe, M.E., 72, 78
Guthrie, D.H., 221, 222

Haken, J.K., 176, 188
Hall, J., 33, 67
Hamielec, A.E., 74, 76, 77
Hansen, C.M., 209, 212, 218–221
Hansen, F.K., 57
Hansen, J., 245
Harkins, W.D., 57
Harrington, H.W., 178, 186
Hart, M.A., 186
Harwood, H.J., 93
Hashimoto, S., 83, 85
Hawkett, B.S., 74, 77
Haxo, H.E., Jr., 218, 222
Hearn, J., 142
Heinz, B., 187
Heitcamp, A., 125, 129, 199
Helin, A.F., 67, 68
Heller, D.S., 131
Hendrickson, F.G., 188
Henry, P., 185, 188
Hercules, D.M., 188
Herglotz, H.K., 188

Hildebrand, J., 209
Hill, S., 33
Hochberger, H., 184, 188
Hochheiser, S., 84
Hoffmann, E., 241, 245
Holmes, C.M., 126, 129
Holsworth, R.M., 135
Homula, A., 168
Hopf, H., 68
Horner, W.E., 241
Huntsberger, J.R., 126, 129
Huynh, H.K., 187

Immergut, H., 100
Ireland, R.W. 206
Isakson, K.E., 178, 187
Isferdiyaroglu, A.N., 113
Ishida, H., 187
Ishtani, A., 177, 187

Jackman, V., 126, 130
Jacobus, N., 188
Jakubowski, J.A., 239
Jaquith, R., 119, 128
Jenkins, R., 185, 188
Jenner, E.L., 33
Jensen, L.H., 185, 188
Jidai, E., 72, 78
Johnsen, S., 188
Johnson, D.J., 176, 182, 186, 188
Jones, H.L., 187
Jones, F.N., 258, 260
Jordan, W., 132, 147

Kamel, A.A., 168
Kamlet, M.J., 210, 222
Kane, P.F., 188
Kangas, D.A., 141
Katayama, Y., 129
Kazakevich, T.S., 77
Keillor, C.M., 66, 68
Keller, D.J., 129
Kelley, L.E., 100
Kempson, A.K., 244
Kerker, M., 134, 167
Keusch, P., 74, 76
Khodzhaeva, I.D., 175, 186
Kim, K.Y., 128, 170
Kinsman, S., 135, 167
Kline, B.B., 101
Klungness, J.H., 129
Knauss, C.J., 82, 85
Knechtges, D.P., 101
Kobayashi, H., 182, 187
Koenig, J.L., 187
Kosmodemyanskii, L.V., 77
Kossmann, H.H., 204, 206
Krasotina, T.S., 101
Krathovil, J., 134
Krishan, T., 66, 68, 91, 92
Krockenberger, D., 147, 169
Kronberg, B., 167
Kronstein, M., 243, 245
Krotki, E., 88, 92
Kruyt, H.R., 7, 17
Kulhanek, W., 167
Kuma, S., 182, 187
Kuo, C.-Y., 29, 33, 187
Kuortti, J., 167
Kvasnikov, Yu.P., 101

Laganis, D., 101
Lahey, T.F., 218, 222
Lancaster, P.E., 187
Larrabee, G.B., 188
Lattimer, R.P., 147
Laurie, O., 206
Layman, P.A., 245
Legrand, D.G., 185, 188
Leibrand, R.J., 186
Leipold, D.P., 204, 206
LePoutre, P., 187
Levi, P., 248, 259
Levine, E., 56
Lewin, S.Z., 188
Lewis, 153

Lichti, G., 74, 77
Lightfoot, E.N., 128, 154, 170
Lin, C.-C., 75, 77
Lindemann, M.K., 101
Lindlow, W., 56
Lindsey, R.V., 33
Litt, M.H., 68
Liu, C.A., 186
Livigni, R.A., 188
Lloyd, J.W., 92
Lowrey, E.J., 241, 244
Los Angeles Society for Coating Technology, 128, 152
Lukhovistkii, V.I., 68
Lumb, E.C., 14, 17
Luongo, J.P., 177, 187
Lyklema, J., 169
Lyon, H.R., 66, 68
Lyons, J.W., 128, 170

Machmer, W.E., 241, 245
Maddii, A., 83, 85
Madson, W.H., 243, 245
Magill, J.H., 170
Makuchi, K., 76, 77
Malac, J., 186
Manning. W.C., 130
Mantell, G.V., 67, 68
Margaritova, M., 66, 68, 92
Mark, H.F., 129, 187
Mark, S., 245
Maron, S.H., 132
Marten, F.L., 74, 77
Masuda, S., 67
Matijevic, E., 139
Matlin, W.A., 101
Matsuda, T., 152
Matsumoto, T., 83, 85
Matsushita, T., 74, 77
Mayer, W.P., 33
Mayer-Mader, R., 80, 85
Mayne, J.E.O., 85, 101
Mayo, F.R., 25, 26, 33, 80
McBane, B., 259
McCann, G.D., 168
McCarthy, D.C., 177, 185
McCarthy, J.J., 187, 188
McCracken, J.R., 74, 77
McCutcheon, J.W., 57, 199
McDonald, P.J., 68
McKillip, W.P., 98, 100
Mcsherry, J.J., 67, 68
Medvedeva, E.S., 77
Meeks, A.C., 181, 187
Meeks, M.R., 65, 69
Megee, J., 57
Meincke, E.R., 4, 30, 63, 92, 228
Merlen, A., 82, 85
Merriman, P., 186
Mertens, R., 82, 85
Middleman, S., 153, 170
Mikofalvy, 101
Min, T.I., 77
Minora, N., 186
Mirabella, F.M., 33, 187
Mittal, L., 85
Mittleman, M.L., 186
Morano, S.L., 259
Moriyama, N., 56
Morris, C.E.M., 56, 80, 85
Mosher, W.A., 33, 176

Nahum, L.Z., 56
Nomura, M., 73, 76
Northwest Society for Coating Technology, 226
Nowak, R.M., 77
Nowlin, T.E., 178, 186
NPCA (National Paint & Coatings Association), 245
Nyhagen, L., 62, 68
Nyquist, E.B., 97, 101

Oakes, J., 206
Obrecht, W., 168
O'Driscoll, K.F., 68

Ogo, Y., 73, 77
Okamoto, S., 120, 128
Okaya, T., 143, 168
Okubo, M., 74, 78
Olayemi, J.Y., 82, 85
Oliver, J.P., 144, 168
Olszewski, W.V., 188
One, H., 168
O'Neill, L.A., 141, 168
Ono, H., 72, 78, 83, 85, 139
Orchard, S.E., 153, 170
Owens, D.K., 125, 129

Panfilov, A.A., 68
Panina, I.S., 73, 76
Pappas, P., 223
Parfitt, G.D., 168
Park, G.S., 113
Parker, E.E., 169
Patella, R.F., 62, 68
Pattacini, R.C., 177, 187
Patton, T.C., 152, 170, 249, 259
Paul, T.K., 82, 85, 174
Pausch, J.B., 147
Pavelich, W.A., 188
Paxton, T.R., 56
Pendleton, D.E., 241, 245
Penlidis, A., 35, 56
Penwell, R.C., 186
Peterson, H, 66, 68
Pieper, J.H.A., 139
Pierce, P.E., 151, 153, 170, 259
Piirma, I., 72–74, 76, 77
Pittman, C.U., 243, 245
Plitz, I.M., 113
Poehlein, G., 17, 73, 76
Porowska, E., 259
Poulsen, K.V., 188
Price, C.C., 84, 95, 100, 115, 116, 198, 255
Progelhoff, R.C., 175, 186
Provder, T.L., 17, 29, 33, 135, 183, 187
Pryakhina, E.A., 89, 92
Pryor, W.A., 113
Puchin, V.A., 75, 77
Puumalainen, P., 185, 188

Raaschou-Nielson, H.K., 188
Raave, A., 56
Rabold, G.P., 72, 78
Rantanen, R., 188
Rascio, V.J.D., 243, 245
Rector, F.D., 101
Reder, A.S., 167
Reilly, C.N., 188
Riddle, E.H., 85, 113
Roberts, M.W., 126, 129
Robertson, A.A., 187
Rosen, S.L., 19, 28, 33
Rosenberg, M.L., 218, 222
Ross, J.R.H., 126, 129
Rothenhaeuser, 80, 86
Rudin, A., 19, 28, 33
Rusch, T.E., 244
Rygle, K.J., 147

Saito, S., 203, 206
Sakai, M.H., 101
Sakota, K., 143, 168
Salkind, M., 85
Sammak, E.G., 100
Samour, C., 144
Sano, T., 73, 77
Saremi, A.H., 113
Satas, D., 129
Satryan, J., 33
Saunders, F.L., 72, 78
Scatchard, G., 206
Schernau, W., 187
Schilling, F.C., 113
Schissel, P., 187
Schurz, J., 184, 188
Schwartz, A.M., 126, 129
Scolere, J., 17
Sedakova, L.I., 186
Seiner, J.A., 29, 33, 259
Sennett, P., 144, 168
Serrini, G., 129
Serrini-Lanza, G., 129

Seymour, R.B., 71, 75, 76
Sharev, Y.P., 68
Sharpell, F.H., 244
Shastry, J.S., 35, 56
Shaw, K.G., 204, 206
Shaw, P.A., 128, 152, 170, 178, 186, 188
Shchepetil'nikov, B.V., 83, 85
Sheetz, D.P., 168
Sherman, L.R., 243, 245
Shilov, G.I., 65, 69
Shimizu, K., 245
Shute, J., 140, 167
Sickfeld, J., 182, 187
Simunkova, E., 186
Skelly, N.E., 186
Skrovanek, D.J., 178, 187
Small, H., 134
Smith, A.A., 241, 244
Smith, A.L., 168
Smith, C.G., 178, 186
Smith, D.R., 66, 68
Smith, J.E., 166
Smith, J.W., 168
Smith, R.A., 245
Smith, W.V., 57
Smith-Ewart, 73, 74
Snoeyink, V.L. 168
Snow, A.M., 92
Sockis, I., 169
Sokoleva, N.I., 68
Solomon, R.A., 186
Sorenson, W.R., 57, 62, 68, 76
Sparrow, G.R., 185, 188
Spasov, V.A., 101
Squires, 153
Starnes, W.H., 33, 113
Stearns, 90
Stefferes, 80
Stehle, P.F., 101
Steichen, R.J., 147, 169
Stenius, P., 133
Stewart, W.E., 128, 154, 170
Stinson, S., 188
Stoisits, R.F., 167
Stone-Masui, J.H., 76, 78
Stout, G.H., 185, 188
Stratton, R.A., 168
Stryker, H.K., 67, 68
Suh, N.P., 186
Sung, N.H., 186
Swanson, J.W., 144, 168

Tadros, T.F., 72, 78
Taft, R.W., 210, 222
Talat-Erben, M., 113
Tarasova, Z.N., 186
Tereikovski, V.W., 168
Throne, J.L., 142
Tokarzewska, M., 88, 92
Tokiwa, F., 56
Tompsett, A.J., 63, 67, 147
Totty, G., 57
Travis, K., 113
Tryson, G.R., 188
Tsiirovasiles, J., 230, 231
Tsuruta, T., 68
Turck, U., 73, 77
Tyskowicz, A.S., 230

Ugelstad, J., 57, 73, 75, 77, 90, 92
Uraneck, C.A., 33
Ustinov, Z.M., 186

Van den Esker, M.J.W., 139, 167
Van Der Hoff, B.M.E., 57
Vanderhoff, J.W., 24, 33, 57
Van Essen, W.J., 129, 139
Van Wazer, J.R., 128, 154, 170
Vasilenko, A.I., 83, 85
Vegter, G.C., 64, 68
Venalainen, H., 188
Verbanic, C., 185, 188
Vijayendran, B.R., 62, 68
Villacorta, G.M., 113
Vincent, B., 72, 78
Volkov, V.A., 72, 78

Volpe, R.P., 101
Vona, J.A., 56
Voyutski, S.S., 186

Wake, L.V., 245
Walker, J.L., 76, 77
Warson, H., 85
Wax, L., 241
Werthan, S., 243, 245
Wessling, R.A., 66, 68
Weston, N.E., 33
Wheeler, B.D., 188
Wieloch, F., 129
Wiley, R.M., 77
Wilkinson, G., 185, 188
Williams, D.J., 74, 77
Williams, R.C., 178, 186
Wilson, A.S., 72, 78
Winters, H., 239, 244
Witseipe, W., 31
Witt, E., 95, 100, 131
Wolf, F. 73, 77
Wonnell, J., 100, 117, 226
Woods, W.B., 237, 240, 241, 243, 245
Wyroba, A., 101

Yamaki, J.-I., 129
Yap, W.T., 124, 129
Yaseen, M., 175, 186
Yeager, C.C., 241, 244
Yeliseeva, V.I., 97, 99, 101
Yeliseyva, V.I., 83, 85
Yocum, R.H., 97, 101
York, D., 68, 128, 160, 170

Zabel, R.A., 241
Zelinger, J., 186
Zil'berman, E.N., 68
Zimmerman, C.A., 4, 94, 100
Zimmt, W.S., 84
Zipf, M.E., 243, 245
Zisman, W.A., 177, 187
Zola, J.C., 251, 259
Zollars, R.L., 63, 68
Zuikov, A.V., 83, 85
Zutty, N.L., 67, 68

Subject Index

ABS (Acrylonitrile Butadiene Styrene resin), 30, 75, 79, 80, 88, 89
Absorbency, 193
ACACEMA (acetoacetylethyl methacrylate), 99, 100
Acetal, 124
Acetaldehyde, 63, 173, 213
Acetate, 15, 25, 28, 30, 39, 40, 53, 56, 61, 62, 63, 64, 65, 67, 68, 69, 83, 84, 98, 119, 140, 145, 146, 173, 178, 183, 195, 213, 221, 238, 242, 243
  vinyl (*See* vinyl acetate)
Acetic, 61, 63, 123, 173, 213
Acetoacetylethyl methacrylate, 99, 100
Acetone, 49, 88, 123, 212
Acetonitrile, 214
Acetophenone, 216, 218
Acetoxy, 238
Acetylacetonates, 99
Acetylene, 91, 123
Acidimetrey, 142
Acidity, 94, 234
Acrylamide, 73, 95, 141, 146, 201, 207, 231
  N-methylol, 31, 97, 98
  poly, 146, 201
Acrylate, 14, 42, 56, 61, 62, 66, 74, 75, 81, 82, 84, 95, 98, 111, 137, 141, 148, 173, 174, 179, 183, 197, 201, 202, 205
  (aziridinyl)ethyl, 97
  2-ethylhexyl, 83
  hydroxyethyl, 99
  hydroxypropyl, 99
  pentachlorophenyl, 243
Acrylates, 62, 65, 66, 75, 79, 81, 82, 83, 197
Acrylic, 24, 31, 32, 46, 81, 83, 84, 85, 95, 99, 105, 106, 113, 133, 137, 141, 142, 147, 185, 194, 210, 218, 223, 224, 230, 235, 236, 243, 245
  acid, 75, 94, 98, 100, 111, 112
    dimer (*See* acryloxypropionic *or* hydroxypropionic)
    polymer, 12, 14, 17, 20, 80, 94, 147, 193, 201, 202
  copolymer, 26, 42, 67, 80
  monomer, 42, 62, 66

[Acrylic, continued]
polyfunctional, 62
Acrylics, 19, 39, 42, 45, 79, 82, 83, 84, 104, 111, 229, 251
Acrylonitrile, 30, 47, 66, 72, 75, 79, 80, 81, 88, 90, 94, 111, 139, 141, 147, 186, 201, 208, 234
Acryloxypropionic acid, 94
Acyl-substituted, 229
Adhere, 125
Adhering, 40, 126
Adhesion, 97, 99, 104, 106, 115, 125, 126, 127, 129, 195
Adhesive, 3, 21, 63, 81, 83, 90, 91, 92, 96, 99, 109, 129, 137, 147, 148, 149, 150, 152, 163, 175, 178, 189, 191, 197, 234, 239, 248, 254, 257
Adhesives, 17, 30, 32, 61, 62, 64, 80, 81, 84, 88, 91, 92, 98, 99, 107, 128, 129, 141, 148, 156, 163, 165, 176, 177, 201, 244, 259
Adipate, 210, 215
Adipates, 208
Adsorb, 11, 12, 13, 14, 15, 17, 55, 183, 198, 201, 204
Adsorbed, 10, 12, 14, 24, 55, 133, 140, 143, 144, 196, 255
Adsorbing, 13, 55, 205
Adsorbs, 11, 205
Adsorbtion, 11, 12, 13, 14, 15, 16, 55, 62, 72, 83, 191, 192, 195, 198, 203, 204, 205
Adsorbtion-desorption, 14
Adsorptions, 14, 55, 72
Agar, 240, 243
Agbezuge, 129
AgCl, 120
Agglomerate, 55
Agglomerated, 72
Agglomerates, 55, 135, 136, 140
Agglomeration, 89, 135, 136, 139
Agglomerations, 140, 226
Aging, 75, 79, 80, 84, 92, 104, 112, 145, 146, 198, 204, 223, 233, 234, 235, 254
Agitation, 43, 49, 51, 55
AIBN(azobisisobutyronitrile), 22, 23, 49, 110, 173
Air-drying, 105
Aircraft, 210
Airplane, 21
Alcohol, 15, 63, 72, 73, 75, 94, 98, 103, 110, 140, 195, 198, 202, 203, 207, 208, 216, 226
Alcoholate, 202
Alcohols, 112, 203, 208, 226
Aldehydes, 112, 124, 237
Alfrey-Price, 29, 69, 73, 81, 82, 89, 93, 94
Algacide, 184
Algae, 239, 243
Alginate, 203, 205
Alginates, 203
ALGOR, 17
Aliphatic, 56, 71, 208, 219
Alkali, 61, 63, 71, 72, 84, 91, 96, 202, 258
Alkaline, 90, 111, 112, 123, 229, 230
Alkalinity, 104, 205, 253
Alkyd, 104, 105, 153, 241
Alkyds, 71, 103, 104, 105, 229, 230
Alkyl, 64, 65, 83, 145, 185, 208, 238, 242
Allylic, 233
Alpha-methylstyrene, 27
Alphabet, 185
Alum, 139
Alumina, 205
Aluminum, 45, 106, 122, 139, 185, 205, 254
Ambient, 24, 32, 36, 43, 46, 47, 49, 53, 62, 67, 90, 106, 111, 174, 233
AMICAL, 242
Amide, 75, 97, 98, 141, 145, 195, 225, 230
Amides, 76, 93, 97, 106, 141, 145
Amine, 26, 84, 93, 96, 97, 103, 104, 106, 111, 113, 141, 143, 224, 225, 229, 238, 242, 244, 253

Amines, 26, 96, 103, 104, 106, 229, 230, 233, 237
Aminimide, 98, 99
Aminimides, 98
Amino, 76, 96, 226
Ammine, 228
Ammonia, 90, 96, 106, 229, 230, 244, 253
Ammonium, 26, 98, 139, 226, 248
Amorphous, 31, 258
Amphoteric, 192
Analyse, 13, 30, 176
Analysed, 14, 173
Analyser, 124
Analyses, 14, 29, 94, 96, 109, 110, 131, 141, 147, 148, 154, 171, 172, 173, 176, 177, 179, 184, 185, 186, 194, 240
Analysing, 96
Analysis, 6, 7, 14, 15, 21, 29, 40, 49, 65, 66, 72, 94, 96, 112, 115, 125, 136, 141, 147, 150, 165, 166, 171, 172, 174, 175, 176, 177, 180, 185, 186, 187, 198, 205, 236, 240, 247, 248, 257
Analyst, 143
Analysts, 109
Analytical, 3, 4, 175, 176
Anatase, 185
Angle, 126, 129, 160, 176, 180, 185
Angles, 126, 129
ANGUS, 238, 242
Anhydride, 67, 80, 94, 95, 141, 201, 258
Aniline, 218
Anion, 23, 99, 104, 111, 112, 197
Anionic, 22, 25, 73, 91, 142, 143, 144, 175, 192, 193, 194, 197, 209
Anionics, 91
Anions, 24, 144, 234
Anisotropic, 32
Anomolous, 211
Antagonistic, 192
Antenna, 120
Anti-fouling, 243
Anti-oxidants, 183
Anti-sag, 152
Antifoam, 51, 137, 196
Antifoams, 51, 195
Antifungal, 243
Antimicrobial, 238
Antimicrobials, 241
Antimonate, 256
Antioxidant, 233, 236
Antioxidants, 5, 194, 233, 235
Apparatus, 38, 53, 129, 142
Aqueous, 11, 16, 22, 24, 33, 64, 74, 75, 80, 81, 82, 89, 90, 91, 106, 112, 119, 120, 146, 183, 184, 197, 237, 239, 243, 247, 249, 254, 257
Armor, 32
Aromatic, 71, 208, 219, 233, 236, 258
Aromatic-in-chain, 172
Arrhenius, 146, 223, 254
Artists, 79
Aryl, 65
Ash, 198
Ashing, 172
Asphalt, 92, 165
ASTM, 94, 100, 124, 127, 128, 129, 130, 137, 138, 148, 157, 165, 172, 174, 185, 186, 188, 210, 236, 240, 241, 244, 245
Atactic, 20, 206, 222, 236, 259, 260
Atlas, 57, 187
ATOCHEM, 238
ATR, 177, 178
ATR-FTIR, 178
Attapulgus, 140, 248
Auto-catalyst, 230
Auto-catalytic, 234
Autoacceleration, 74
Autoaccelerations, 74
Autocatalysis, 92
Autocatalytic, 61, 145
Autoemulsification, 82
Automate, 123

Automated, 163
Automatic, 123, 141, 158, 160, 162
Automatically, 164
Automation, 123, 140, 154
Avicel, 135
Azene, 234, 235
Azeotrope, 75, 211
Azeotropic, 29
Aziridenes, 225
Aziridienes, 230
Aziridinyl, 97
Azo, 22, 97, 111, 143
Azo-initiators, 110
Azobisisobutyronitrile, 22, 110
Azoniaadamantane, 238, 242

Bacteria, 104, 146, 183, 184, 229, 237, 239, 240, 241
Bactericide, 237, 238, 239, 240, 244
Bacteristat, 237, 239, 244
Baffles, 47
Balston, 38
Baltimore, 79
Barex, 80
Barium, 12, 65, 83, 140, 196, 237, 242, 244, 253, 254, 256
Barnacles, 243
Baroid, 154, 163
Barrier, 27, 28, 66, 75, 80, 104, 152, 193, 195
Base, 4, 43, 45, 81, 99, 111, 117, 142, 143, 147, 164, 172, 196, 210, 225, 234, 253
Batch, 38, 39, 42, 54, 65, 96, 115, 146, 240, 247, 248, 251, 257
Bath, 24, 38, 39, 40, 41, 46, 49, 104, 137
Battery, 158
Bayview, 159, 161
Beaker, 40, 46, 52, 138
Beam, 140, 161, 178, 185
Bear, 84, 149
Beckman, 123, 142
Bell, 256
Benecke, 85
Benzaldehyde, 142, 147, 173, 217
Benzene, 31, 49, 71, 83, 110, 111, 173, 216, 217, 219
Benzil, 181
Benzimidazole, 241, 242
Benzoate, 110, 111, 173
Benzoguanidine, 225
Benzoic, 111
Benzoisothiazolin, 238
Benzonitrile, 215
Benzothiazole, 242
Benzoyl, 49, 110, 111, 173
Benzyl, 77, 81, 82, 208, 210, 216, 238
Bias, 132
Bicarbonate, 39
Bicyclic, 238
Bingham, 121, 122, 149, 152, 153, 209
Biolab, 240
Bio-Rad, 143
Bioactive, 210, 239
BIOBAN, 238, 242
BIOCHEK, 238
Biochemical, 175
Biocidal, 240
Biocide, 229, 240, 244
Biocides, 5, 237, 239, 243, 244
Biota, 240
Biphenyl, 110, 173
Bisulfate, 24, 111
Bisulfite, 24
Bisulphate, 111
Blender, 136, 138, 174, 256
Blending, 248
Blends, 80, 92, 133, 134, 219, 252
Blessing, 193, 195
Blind, 54
Bloom, 72, 209
Blush, 64
Bohlin, 155
Bond, 20, 22, 27, 87, 91, 97, 104, 112, 124, 138, 147, 226, 234, 235, 257
Bonded, 20, 144, 224, 233
Bonding, 31, 32, 63, 144, 153, 177, 255, 257, 258
Bonds, 22, 31, 64, 65, 234, 257, 258

Bornyl, 81
Brabender, 123, 155, 164
Brass, 26, 46, 52
Brice-Phoenix, 140
Brinkmann, 123
Bromide, 98, 226
Bromination, 147
Bromine, 147
Bromoacetate, 238
Bromobenzene, 218
Bromoaphthalene, 218
Bronze, 26, 46, 52
Brook, 158
Brookfield, 121, 122, 154, 156, 157, 160, 163, 164, 165, 170
Brown, 178, 186, 187, 194, 234, 235
Brownian, 10
Brunswick, 175
Bubble, 122, 126, 137, 157, 163, 251
Bubbled, 38, 42
Bubbler, 37, 38
Bubbles, 42, 174
Bubbling, 123
Buchi, 49
Buckmann, 238, 242
Bueche, 184
Buffer, 5, 63, 97, 120
Buffering, 84
Buffers, 63, 196, 244
Buret, 42, 254
Buretted, 45
Burrell, 157, 163, 209, 222
Burrell-Severs, 157
BUSAN, 238, 242
Bushing, 36, 47, 53
Bushings, 36, 43
Butadiene, 10, 24, 256, 26, 29, 30, 40, 42, 45, 46, 49, 53, 65, 66, 72, 73, 75, 76, 79, 80, 83, 84, 87, 88, 89, 90, 96, 97, 98, 111, 141, 147, 173, 179, 233, 235
  poly-, 31
Butanol, 208, 226
Butoxy, 22
Butyl, 56, 65, 66, 73, 74, 83, 98, 210, 216, 219, 242
Butylene, 238, 242
Butyraldehyde, 213
Butyrate, 213
Butyric, 213
Butyrolactone, 217, 219
By-product, 19, 49, 88, 109, 111, 112, 196, 224, 234
By-products, 23, 24, 49, 94, 109, 110, 111, 112, 178, 198, 224, 226, 229

Cadmium, 234
Calcium, 72, 82, 92, 98, 106, 123, 139, 141, 173, 196, 202, 225, 228, 251, 258
Cannon-Fenske, 160, 162
Caprolactone, 105
Carbamate, 242, 243
Carbide, 67, 105, 107, 123, 173, 238
Carbodiimides, 225, 229, 230
Carbonate, 98, 139, 196, 202, 217, 218, 234, 251, 258
Carbonates, 63, 92
Carbonyl, 64, 226, 236
CARBOPOL, 9, 17
Carboxy, 23, 97, 110, 141, 225
Carboxyl, 96, 142, 143
Carboxylate, 76, 84, 90, 91, 93, 94, 95, 96, 99, 104, 106, 111, 143, 203, 244
Carboxylated, 66, 75, 76, 90, 91, 94, 96, 99, 100, 106, 119, 132, 140, 172, 173, 194, 201, 202, 203, 204, 226, 253, 258
Carboxylates, 96, 143, 192, 225, 230, 258
Carboxylation, 10, 96, 104, 144, 203
Carboxylic, 28, 29, 62, 76, 91, 103, 106, 112, 141, 143, 194, 197, 226, 229, 230
Carboxyls, 103
Carboxymethylcellulose, 202
Carcinogenicity, 42, 47
Cargille, 157

Carnegie-Mellon, 149
Caro's acid, 111
Carotene, 65, 234
Carpet, 5, 137, 149, 163, 173
  backing, 88, 138, 148
  backcoating, 257
Carri-Med, 159
Casein, 89, 203, 204
Castor, 258
Catalysis, 61, 97, 98, 104, 105, 226, 229, 230
Catalyst, 24, 25, 63, 67, 98, 104, 111, 202, 226, 230, 231, 255
Catalysts, 85, 104, 116, 124, 196, 230, 231, 253
Catalyzed, 97, 116, 226, 229, 234
Cation, 100, 104, 139, 172, 195, 203
Cationic, 22, 26, 56, 61, 91, 111, 113, 135, 142, 192, 209, 237
Cations, 76, 106, 139, 172, 189, 197, 202, 203, 206, 225, 228, 230, 253
Caulks, 152
Caustic, 234
CCC, 29, 139, 140
Cellulases, 239
Cellulose, 15, 19, 26, 39, 63, 64, 202, 203, 204, 235, 257, 258
Cellulosic, 30, 67, 112, 124, 202, 203, 204, 205, 229, 239
Cement, 8, 148, 149, 156, 158, 164, 185, 198
Centrifugal, 174, 175
Centrifuge, 132, 134, 180
Cerac, 256
Ceramics, 185
Cesium, 134
Cetyl alcohol, 75
Chain-Growth, 20
Chain transfer, 26, 27, 28, 31, 42, 45, 56, 73, 74, 80, 89, 90, 93, 110, 111, 112, 179, 233
Chalk, 251
Chalking, 251
Cheese, 8, 40, 54, 141, 153, 198
Chelant, 24
Chelate, 26
Chelating, 112
Chematrix, 120, 128
Chloride, 11, 12, 13, 41, 42, 46, 47, 61, 62, 64, 65, 66, 67, 69, 77, 80, 83, 91, 92, 98, 139, 140, 141, 147, 173, 177, 196, 202, 210, 213, 215, 216, 221, 226, 234, 238, 242, 248, 255
Chlorides, 112, 129, 140
Chlorinated, 72, 111, 219, 234
Chlorine, 67
Chloro, 87, 238
Chloroacetic, 202
Chloroallyl, 238
Chlorobenzene, 217
Chloroform, 216
Chloromethoxypropyl mercuric acetate, 238
Chlorophenol, 242
Chloroprene, 66, 87, 89, 91, 92, 141, 179
Chlorostyrene, 71
Chromate, 134
Chromatix, 182
Chromatography, 63, 96, 109, 132, 134, 143, 180, 182, 183, 186, 188, 194, 226
  gas, 63, 73, 112, 124, 147, 148, 172, 173, 176, 205
Chromium, 119, 230, 259
Cis, 24, 88, 90
Cis-poly(butadiene), 10, 31
Citrate, 41
Citric, 63
Clay, 64, 133, 135, 140, 144, 152, 177, 186, 203, 248, 250, 257, 258
CMC (carboxymethyl cellulose), 16, 72, 132, 202
CMP, 238
COABS, 242
Coacervation, 99
Coacervates, 219
Coagulant, 45, 64, 123, 140, 173, 197
Coagulants, 11, 135, 140, 145, 173

Coagulate, 4, 133, 139, 140, 146, 163, 172, 202, 203, 204, 249
Coagulated, 4, 138, 140, 141, 196, 197
Coagulates, 9, 136, 140
Coagulating, 106, 123, 196
Coagulation, 16, 39, 43, 46, 52, 63, 72, 83, 99, 123, 127, 135, 136, 138–140, 142, 143, 146, 173, 174, 191–193, 196–198, 205, 244, 253
Coagulum, 40, 127, 128, 135–137, 139, 142, 146, 198, 244
Coalesce, 83, 193, 209
Coalescense, 100, 175, 220, 229
Coalescent, 123, 174, 211, 212, 224, 250
Coalescents, 104, 207, 210, 211
Coating, can, 65, 84, 109, 147, 225, 230
  paper, 64, 65, 76, 84, 88, 99, 107, 109, 116, 122, 131, 141, 144, 148, 149, 164, 165, 177, 185, 250, 251, 255
Coalescing, 80, 218
Coaxial, 157
Cobalt, 99
Cobinders, 202, 204
Cobiocides, 239
Coefficient, 29, 52, 142, 171
Cole-Parmer, 142
Collagen, 204
Collodion, 142
Color, 39, 40, 59, 65, 79, 94, 97, 106, 112, 146, 171, 174, 223 234, 235, 248, 249, 250, 252
Colorants, 252, 253
Coloration, 75, 234, 250, 252
Colored, 80, 233, 250, 252, 253
Colorimeter, 235
Colorless, 65
Columbia, 161
Comonomer, 28, 29, 65, 67, 75, 132, 139, 141, 143
Comonomers, 11, 62, 65, 73, 75, 76, 99, 124, 139, 143
Compounding, 222
Computer, 6, 11, 14, 29, 33, 65, 132, 135, 155, 157, 160, 161, 177, 178, 182, 183, 187, 212, 251, 254, 256
Concrete, 8, 141, 239, 248
Condensation, 19, 49, 53, 97, 103, 104, 179, 181, 208, 224
Condense, 41, 88
Condensed, 40, 88
Condenser, 37, 38, 39, 49, 50, 53, 62
Conductance, 57, 247
Conductimetry, 141, 142
Conductive, 252, 254, 255, 256, 257
Conductivity, 142, 143, 144, 172, 198, 248, 255, 257
Conductometric, 141, 142, 173
Configuration, 20, 31, 164
Conjugated, 89, 234
Consistimeter, 155, 165
Consler, 174
Consolute, 146
Contact-Angle, 129, 130
Contaminants, 46, 49, 50
Contaminate, 229, 240
Contaminated, 46, 202, 240
Contaminating, 120
Convimeter, 155, 164
Copolymer, 21, 24, 26, 27, 28, 29, 31, 56, 63, 65, 75, 76, 81, 83, 88, 93, 94, 96, 97, 111, 140, 141, 194, 201, 212, 224
  acrylate, 62
  butyl/vinyl acetate, 62
    vinylidene chloride, 66
    ethyl, 66
  butadiene/acrylonitrile, 66, 208, 209
  ethylene/vinyl acetate, 30, 62–64, 67, 98, 195, 221
    propylene, 221
    vinyl chloride, 67, 98
  isobutylene/maleic, 193
  styrene/acrylonitrile, 72 (*See also* SAN.)
Copolymers, 71
Copolymerizable, 14

Copolymerization, 10, 21, 27, 28, 29, 30, 40, 65, 66, 73, 75, 83, 93, 96, 97
Copolymerizations, 94
Copolymerize, 71
Copolymerized, 28, 30, 62, 63, 139, 243
Copolymerizes, 66
Copolymerizing, 65, 143
Copolymers, 14, 15, 21, 29, 30, 42, 45, 47, 60, 62, 63, 64, 65, 67, 72, 73, 75, 76, 79, 81, 83, 92, 93, 94, 97, 98, 100, 110, 141, 179, 183, 194, 195, 201, 205, 235, 243
Copper, 26, 46, 52, 99, 242, 243, 254, 255
Core-shell, 74, 83
Corroding, 52, 139
Corrosion, 24, 52, 59, 66, 67, 80, 111, 112, 113, 117, 119, 121, 162, 174, 195, 196, 202, 226, 236, 252, 253, 254, 255
COSAN, 238, 242
Cosmetics, 9, 17
Cosolvency, 220
Cosolvent, 123, 174, 211, 212
Cosolvents, 104, 198, 207, 210, 211, 219
Cotton, 195
Couette, 155, 156, 158, 159, 160, 161, 163, 164
Coulometric, 124
Coulter, 128, 135, 140
Counterion, 113
Covalent, 31, 104, 257, 258
Covimat, 161
Cowles, 195, 253
CPVC, 249, 251, 255
Craters, 178
Crawl, 126
Crawling, 106
Crazing, 210
CRC, 186, 218, 222
Creaming, 8, 9, 10, 127
Creel, 42
Creosote, 243
Crescent, 124
Cresol, 208
Crosslink, 104, 106, 224, 225, 229, 235, 257
Crosslinkable, 60
Crosslinked, 17, 20, 204, 209, 212, 219
Crosslinker, 97, 229
Crosslinking, 20, 31, 32, 33, 60, 62, 83, 89, 146, 179, 195
Crotonic, 62, 94, 95
Crown, 41, 42, 45
CRTs, 183
Cryogenic, 53
Cryoscopy, 180
Crystal, 178, 185
Crystalline, 258
Crystallinity, 31, 258
Crystallite, 31
Crystallization, 66
Crystallize, 31
Crystallizing, 31
Crystals, 251
Cullet, 46
Cumene, 24
Cuprous, 243
Curability, 87, 98
Curable, 64, 75, 98, 99, 119
Curative, 97, 104, 224, 225, 226
Curatives, 5, 94, 139, 223, 225, 226, 227, 229
Cure, 3, 24, 27, 31, 32, 63, 66, 67, 75, 76, 80, 93, 94, 97, 98, 99, 100, 104, 105, 106, 116, 119, 137, 141, 143, 145, 177, 196, 223, 224, 225, 226, 230, 231, 236, 257
Cured, 20, 31, 75, 84, 103, 116, 178, 195, 230, 236
Cures, 87, 97, 98, 100, 104, 106, 117, 223, 224, 226, 230
Curing, 31, 62, 66, 75, 84, 87, 91, 93, 97, 98, 99, 100, 104, 197, 207, 223, 224, 226, 231, 258

Curtius, 106
Cushing, 156
Cyclohexanol, 215
Cyclohexanone, 216
Cyclohexene, 47, 49, 111, 147
Cyclohexyl, 81, 82, 215
Cyclopentadiene, 87

DBE, 215
DC, 85, 173, 185, 244
DE, 187
Deaerate, 42
Deaeration, 38
Deactivation, 28
Dean-Stark, 49, 124
Decalin, 216
Definition, 17, 20, 73, 119, 131, 150, 165, 166, 211, 249
Definitions, 16, 19, 20, 166, 211
Defoam, 137
Defoamer, 51, 137, 172, 196
Defoamers, 5, 51, 137, 195
Degradable, 146
Degradation, 60, 61, 65, 85, 92, 147, 176, 177, 178, 183, 188, 202, 203, 204, 236, 241, 252
Degradative, 60, 178
Degrade, 75, 123, 147, 235, 239
Degrades, 60, 65, 251
Dehydrated, 229
Dehydrohalogenate, 111, 234
Dehydrohalogenation, 65, 67, 92, 124, 234
Deionized, 38, 120, 196
Dem, 160
Demethylolation, 98
Den, 139, 188
Denier, 129
DeNouy, 125, 126
Densities, 209, 249, 254, 255
Density, 9, 10, 38, 55, 134, 137, 174, 185, 247, 248, 249, 251, 255, 256, 257
Dents, 122
Deoxygenate, 38
Deposition, 11, 52, 86, 148, 183
Derivatives, 87, 97, 103, 202, 212, 219
Desolubilized, 202
Desorb, 13, 14
Desorbing, 55
Desorption, 11, 14, 16, 74, 75, 191, 195, 198, 205
Destabilization, 197
Destabilize, 133, 134, 139
Destabilizers, 197, 230
Detector, 134, 154, 162, 164, 182, 183
Detergent, 149, 240
Detergents, 57, 199
Determination, 14, 39, 93, 122–124, 127–129, 133–135, 137–139, 141, 143, 144, 147, 149, 175, 176, 181–184, 187, 188, 198, 210, 212, 219, 250
Determinations, 40, 43, 46, 57, 76, 125, 126, 132, 136, 140, 151, 162, 174, 179, 221, 235
Determine, 13, 14, 65, 73, 88, 121, 122, 124, 127, 129, 132, 133, 139, 149, 163, 164, 178, 179, 184, 185, 203, 208, 219, 220
Determined, 93, 140, 147, 150, 181
Determining, 86, 94, 120, 145, 152, 241
Detriment, 25, 64
Detrimental, 149
Device, 39, 42, 47, 121, 124, 125, 126, 134, 135, 137, 144, 145, 154, 155, 156, 157, 158, 159, 160, 161, 163, 164, 165, 172, 175, 177, 178, 254
Devices, 47, 120, 121, 122, 123, 124, 126, 134, 140, 148, 149, 156, 161, 162, 163, 164 , 165, 177, 178, 181
Dextrin, 15
Dextrins, 63, 195
Di-butyl phthalate, 64, 215
Di-isopropylxanthogendisulfide, 91
Di-n-butyl sulfosuccinate, 193
Di-styrene, 112

Diabetic, 43
Diacetone, 98
Diacrylate, 83
Dialkyl, 209
Dialysis, 94, 142
Diamines, 100, 223
Diammonium, 63
Diamond, 143
Diapers, 226
Dibromide, 218
Dibromo, 238
Dicarboxylic, 94
Dichloride, 112, 216
Dichloroethylene, 66, 215
Dicker, 120, 128, 142
Dicyanobutane, 238
Dielectric, 11, 124, 143, 255
Diels-Alder, 49, 80, 88, 111, 147, 173
Diene, 67, 75, 80, 87–89, 92–94, 106
Diene-acrylics, 81
Diester, 94, 215
Diethyl, 210, 213
Diethylamine, 213
Diffraction, 177, 185
Difunctional, 26, 27
Diiodomethyl-p-cholorphenyl, 242
Diiodomethyl-p-tolyl, 242
Diisocyanate, 105
Dilatancy, 150, 164
Dilatant, 121, 152, 249
Dilatometric, 74
Diluents, 84
Dimer, 88, 94, 106, 111, 147
Dimerized, 91
Dimethacrylate, 31
Dimethyl, 87, 173, 187, 215, 217
Dimethylaminoethyl methacrylate, 95, 97
Dimethylaminopropyl methacrylate, 95
Dimethyldiallylammonium chloride, 140
Dimethylformamide, 216
Dimethylolpropionic, 105
Dimethylsulfoxide, 216
DIN, 160
Dioctyl-sulfosuccinate, 193
  phthalate, 210, 215
Diol, 106
Diolefin, 76
Diols, 105
Dionex, 143
Dioxide, 110, 236, 254
Dioxide-pyridine, 124
Diphenylmethane, 105
Dipole, 145, 211
Dipped, 3, 197
Dipping, 120
Dipropylamine, 213
Discolor, 75, 79, 198, 204
Discoloration, 76, 80, 84, 87, 92, 97, 104, 111, 112, 175, 235
Discolorations, 46, 62, 87, 234
Discolored, 116
Discolors, 60
Dispersancy, 203
Dispersant, 14, 15, 16, 65, 67, 106, 193–195, 247, 249, 253
Dispersants, 5, 12, 55, 62, 64, 98, 133, 142, 192, 196, 197, 249
Disperse, 67, 104, 194, 236, 244
Dispersed, 7–9, 16, 59, 90, 104, 124, 131, 144, 186, 195, 209, 250, 251, 256
Disperser, 253
Dispersers, 137
Disperses, 193
Dispersible, 103, 104, 172, 226
Dispersion, 8, 9, 11, 13–15, 20, 55, 65, 75, 103, 104, 106, 125, 134, 138, 142–145, 149, 165, 195, 198, 211, 212, 229, 234, 240, 244, 248–250, 253, 254, 259
Dispersions, 17, 21, 52, 80, 85, 87, 104, 105, 125, 138, 150, 156, 195, 196, 211, 229, 230, 234, 249, 250, 253, 254
Dispersive, 185
Disposable, 40, 67, 122, 240, 254
Disposal, 49, 237
Disposals, 240
Disposed, 50
Disposibles, 226

Disproportionate, 94
Disproportionation, 25, 26, 27
Distill, 90
Distillate, 49, 51, 109
Distillation, 48, 49, 53, 87, 90, 124
Distilled, 38, 39, 88, 120
Distugil, 92
Disulfide, 56, 112, 218
Ditertiarybutyl peroxide, 22
Dithiodicarbamate, 238, 242
Dithiodiglycolic, 112
Dimethyl dithioglycolate, 173, 187
Diurethane, 204
Divalent, 196
Divinyl, 31, 83
DMEA (dimethylethanolamine), 105
DMF (dimethylformamide), 75
DMSO (dimethylsulfoxide), 219
Docecylbenzenesulfonate, 193
Dodecyl, 72, 73
Dow, 67, 71, 76, 77, 98, 134, 136, 143, 238, 242
DOWCIL, 238, 242
Drawdown, 174
Dry-to-touch, 105
Drying, 32, 33, 40, 71, 72, 76, 84, 91, 98, 100, 105, 106, 109, 119, 120, 123, 126, 127, 137, 144, 148, 153, 172, 175, 197, 209–211, 226, 229, 230, 240, 244, 248, 256
DTIC, 187
DuPont, 30, 31, 67, 91, 92, 123, 126, 182, 218, 241, 245
DXN, 238
Dynatrol, 122, 154
Dyne-sec, 166
Dynes, 126, 166

Eastman, 100
Ebonex, 256
Ebulliometry, 180
ECO, 142
EDAX, 185
Effluent, 182, 183
Electrochemical, 3, 13, 120, 254
Electrochemically, 253
Electrode, 11, 13, 120, 142, 143, 145, 254
Electrodeless, 142
Electrodeposition, 142, 144
Electrodes, 11, 120, 128, 142, 145
Electrodialysis, 142
Electrolyte, 72
Electromagnetic, 254
Electromotive, 12, 14, 15
Electron, 96, 132, 135, 175, 185
Electroneutrality, 143
Electronic, 154, 156, 157, 159, 162, 165, 254
Electronically, 155, 157
Electronics, 135, 254
Electrophoretic, 8, 10, 11, 17, 72, 76, 83, 143, 183, 198
Electrophoresis, 144
Electrostat, 17
Electrostatic, 11, 192, 211
Electrostatically, 14
Electroviscous, 142
Elmer's Glue, 64
Elvace, 30
Elvax, 30
Elzone, 135
EMA, 67
Embrittlement, 75, 175
EMF (electromotive force), 13
EMI (electromagnetic interference), 254
Emission, 145, 185
Emitted, 185
Emulsification, 106, 208
Emulsified, 30, 56, 73, 91, 106, 141, 175
Emulsifier, 63, 73, 83, 89, 90, 91, 94, 95, 104, 111, 142, 201, 208
Emulsifiers, 57, 67, 72, 73, 83, 91, 111–113, 144, 192, 199
Emulsifies, 90, 193
Emulsify, 24, 63, 91, 103, 208
Emulsion, 3, 4, 5, 7–10, 12, 16, 17, 19–33, 35, 36, 38, 40, 46, 51, 55–57, 61–68, 71–76, 79, 80, 82, 83, 84, 87–92,

[Emulsion, continued]
96–99, 104, 106, 109, 111, 113, 115–117, 119, 125, 126, 129, 134, 135, 138–140, 142, 143, 145, 171, 172, 176, 192, 194–196, 198, 202, 209, 226, 229, 234, 239, 243–245, 253
Enamel, 98, 99
Encapsulation, 74
End-group, 91, 93, 105, 208
End-product, 116
Englehard, Kaolin, 258
Englewood, 68
Enthalpy, 181
Entropically, 15
Enviro-chem, 242
Environment, 112, 159, 187, 239, 241, 254
Environmental, 241
Enzyme, 146, 202, 239
EP, 85
EPA, 86, 173, 218, 222, 237, 241
Epichlorohydrin, 141, 221
Epoxidation, 111
Epoxy, 47, 93, 99, 100, 111, 207, 225, 227, 230, 255
Equipment, 36, 42, 46, 49, 76, 119, 124, 129, 131, 132, 134, 135, 142, 144, 184, 191, 194, 254
Error, 29, 39, 43, 120, 121, 130, 170, 182
ESA, 143
ESCA, 185
ESSDEE, 259
Ester, 24, 28, 29, 42, 80, 81, 94, 96, 99, 103, 104, 106, 111, 145, 185, 201, 208, 209, 223, 230, 243
Esterified, 84
Esterify, 208
Esters, 26, 31, 64, 65, 72, 76, 81–83, 85, 94, 100, 113, 205, 211
Estimate, 12, 49, 124, 163, 182, 219
Estimated, 45, 134
Estimates, 40, 212
Estimation, 45, 129
Ethanol, 238
Ethanolamine, 103, 215, 249
Ethene, 242
Ether, 26, 73, 88, 95, 97, 98, 144, 201, 213, 226, 229, 230
Etherification, 226
Ethers, 61, 98, 110, 202, 203, 211, 226
Ethyl, 65, 66, 73, 74, 81–84, 94, 97, 174, 202, 213
Ethylene, 30, 31, 47, 53, 60–65, 67, 69, 89, 98, 99, 105, 178, 193, 201, 205, 211, 212, 215–219, 258
Ethylenediaminetetraacetic, 39
Ethylenes, 252
Ethyleneimine, 140
Ethylhexanol, 208
Ethylhexyl, 81, 82, 83
Eur-Control, 158, 164
Europe, 64
European Coatings Journal, 33, 100, 129, 187, 222
EVA poly(ethylene-co-vinylacetate), 30, 62, 63, 235
Evaporate, 175, 229
Evaporation, 38, 49, 54, 172, 181, 209, 211
Evaporators, 49, 53
Evolution, 46
Evolve, 226
Evolved, 21, 178, 254
Exchange, 53, 74, 96, 99, 142, 143, 172, 208
Excited, 233
Exclusion, 134, 182
Exotherm, 28, 40, 53
Experiment, 14, 26, 42, 140, 165
Experimental, 11, 12, 47, 60, 144, 151, 189, 219, 236
Experimentalist, 4, 126
Experimentally, 184
Experimentation, 11, 116, 127, 145, 248, 252

Experimented, 138
Experimenter, 39, 136, 139
Experiments, 38, 40, 52, 116, 133, 140, 145, 181, 197, 207
Expert, 64, 186, 189
Expertise, 4
Experts, 176, 212
Explosion, 88
Explosions, 43
Explosive, 88
Exponents, 25
Extract, 147, 172, 208
Extractables, 229
Extracted, 230
Extraction, 257
Extracts, 109
Exudate, 193, 194
Exude, 229
EXXATE, 213, 218
EXXON, 218
Eye, 5, 47, 125

F-statistic, 212
Fabric, 31, 32, 81, 88, 98, 99, 117, 194, 233
Fabrics, 137, 209
Fade, 253
Fahrenheit, 41, 137
FANN, 155, 165
Faraday, 78, 206
Farinograph, 155
FATIPEC, 188, 259
Fatty acid, 15, 56, 64, 72, 91, 95, 103, 104, 106, 113, 204, 208, 258
FBI, 185
FDA, 64, 109, 225, 230
Fear, 134, 226, 254
Fell, 30, 147
Felt, 88, 96, 173, 258
Ferranti-Shirley, 158
FERRO, 238, 242
Ferrous, 24, 74
Fiber, 67, 98, 116, 117, 126, 135, 140, 142, 207, 235, 256–258
Fiber-binder, 115
Fibers, 20, 21, 32, 80, 116, 125, 126, 129, 130, 142, 195, 235, 252, 256–258
Fill, 7, 42, 45, 46, 116, 209
Filled, 23, 42, 145, 147, 172, 184, 255
Filler, 4, 5, 140, 146, 149, 154, 197, 257–259
Fillers, 104, 135, 139, 194, 198, 234, 243, 247, 251, 252, 257–259
Filling, 42, 54
Fillings, 184
Film, 53, 56, 60, 63, 64, 66, 72, 75, 83, 84, 98, 104, 109, 123, 125, 141, 153, 161, 173–179, 185, 189, 196, 197, 204, 207, 210, 211, 221, 240, 241–252, 256, 257
Film-Former, 187
Films, 83, 106, 125, 141, 174–178, 184, 185, 211, 230, 235, 241, 251
Filter, 40, 54, 128, 179, 183, 235
Filtered, 127, 184
Filtration, 38, 40, 54, 179, 198
Fineness, 253
Fines, 135, 140
Finger, 45
Fingerprints, 187
Fingers, 45
Finish, 29, 105, 256
Finished, 51, 105
Finishes, 105, 256
Finishing, 105, 129
Fittings, 36, 38, 181
Flake, 255
Flame, 5, 65, 67, 139, 194, 196, 208, 209, 234
Flammable, 211
Flange, 36, 47
Flanged, 47, 49
Flare, 53
Flask, 36, 38, 39, 40, 49, 88
Flat, 36, 49, 52, 175, 249, 250

Flatten, 178
Flavored, 147, 189
Flaws, 210
Flexibility, 32, 33, 59, 60, 65, 67, 88, 104, 106, 195, 224, 230, 233, 249
Flexible, 33, 175
Flexibilization, 75
Flexibilizers, 75
Flexural, 30
Float, 208
Floats, 192
Floc, 46, 49, 250
Flocculants, 89, 97
Flocculate, 10, 163
Flocculated, 186
Flocculating, 140
Flocculation, 40, 43, 49, 135
Flooded, 157
Flory-Fox, 93
Flow, 19, 20, 120, 121, 122, 123, 128, 134, 148–154, 158, 160–166, 170, 184, 194, 249, 250, 253, 259
Flowing, 20, 142, 162
Flowtester, 161
Flowtime, 162
Flowtimers, 160
Fluff, 80
Fluid, 11–13, 20, 38, 39, 43, 45, 47, 60, 104, 112, 122, 124–126, 135, 141, 145, 148–151, 153, 160, 162–166, 171, 174, 175, 194, 196, 209, 237, 241, 247, 248, 250, 252, 259
Fluids, 36, 49, 121, 152, 154, 159, 161–164, 175, 240
Fluorescense, 176, 185, 188, 252
Fluoresces, 60
Fluorocarbon, 175
FMLs, 218, 222
Foam, 8, 50, 51, 106, 126, 137, 138, 191, 193, 198
Foamability, 137, 138
Foamed, 46, 137, 138
Foaminess, 104, 198
Foaming, 40, 50, 51, 124, 137, 138, 202, 205
Ford, 152, 154, 158, 160–162, 236
Formaldehyde, 24, 75, 76, 97–100, 141, 146, 225, 226, 235–238, 251
Formalin, 240
Formally, 212
Formamide, 215
Formic, 213
Formylation, 226
Fortran, 33, 182
Fourier, 177
Foxboro, 142
Fractionation, 219
Fractions, 87, 143, 172, 194
Fragment, 176, 185
Freely, 33
Frees, 103
Freeze, 49, 76
Freezer, 174, 175
Freezing, 54
Freon, 213
Friction, 52, 171
Friend, 249, 251
Frit, 137
Frothing, 137, 148, 149
Frozen, 33
FSCT, 33, 67, 100, 109, 128, 152, 187, 236, 241, 243, 259
FTIR, 177, 178
Fuel, 243
Fugitive, 231
Fumaric, 75, 94
Fumes, 147
Functionalities, 100
Functionality, 23, 26, 62, 66, 99, 141, 143, 145, 183, 195, 208
Functionalized, 100
Functions, 24, 62, 83, 94, 189, 191, 203, 204, 211, 252
Fungi, 229, 239, 241, 243
Fungicidal, 245
Fungicide, 238

Fungicides, 241–244
Funnel, 37–39, 42, 91
Furan, 215
Furfural, 217

GA, 99, 158
Gas chromatography (*See also* Chromatography, gas)
Gasket, 42, 43, 45, 46, 47
Gasoline, 60, 79
Gauge, 47, 248
GC, 147, 176, 178
GE, 235
Gel, 26, 45, 80, 134, 148, 176, 179, 180, 182, 188, 197, 202, 250
Gel-like, 15
Gelatin, 195
Gellant, 197
Gellation, 97, 209
Gelling, 191, 197
Gels, 17, 80, 179, 184
Geon, 80
GIVAUDAN, 238
GIVGARD, 238
Glacial, 100
Glass, 13, 32, 36, 37, 42, 43, 46, 47, 52, 60, 79, 88, 92, 98, 120, 138, 142, 175, 203, 204, 206, 207, 259
Glass transition (or Tg), 71, 72, 81, 82, 83, 93, 96, 97, 99, 116, 174, 209, 224
Glassware, 125, 162, 184
Glaswerk, 160, 184
GLC, 198
Gloss, 196, 229, 230, 248, 250
Glossy, 211, 249, 250
Glutaraldehyde, 238
Glycerin, 36
Glycidyl, 95, 99, 141
Glycol, 31, 43, 83, 99, 105, 189, 211, 215, 219
Glycols, 84
GMA (glycidylmethacrylate), 99
GMBH, 98, 160
Gold, 254, 255
Golden, 100, 124, 129, 210, 219, 254
Goodrich, 125, 129
GPC, 180, 182, 183, 184, 188
GPCs, 182
Graft, 21, 65, 75, 82
Graftablility, 80
Graftable, 80
Grafted, 26, 63, 71, 195
Grafting, 21, 63, 74, 75, 83, 89, 99
Grafts, 30, 74
Granular, 174
Granules, 20, 178
Graphite, 255
Grating, 177, 178
Grave, 101
Gravel, 8
Gravimetric, 122, 124, 126, 127, 178
Gravimetrically, 179
Gravitational, 9
Gravity, 148, 149, 152, 163
Grazing, 185
Greases, 152
Greasy, 207
Green, 76, 241, 250
Grew, 4
Grind, 104, 174, 191, 248, 252, 253
Grinding, 149, 174, 175, 236, 252
GRS, 52
Guar, 203
Gum, 183, 209
Gums, 203, 239
GXL, 238
Gypsum, 251

H-P, 182
Haake, 122, 158, 164
Haas, 79, 80, 82, 84, 85, 97, 142, 143, 238, 242, 244
Hair, 17
Halide, 226
Halted, 224
Hammer, 46
Hazardous, 50, 210
Hazy, 64, 79, 106

HCl, 67, 173, 234
HDC (hydrodynamic chromatograph), 134, 182
Headspace, 147, 148
Health, 53
Heat, 36, 38, 40, 43, 45, 51, 52, 53, 60, 62, 63, 75, 76, 97, 105, 106, 142, 148, 175, 181, 197, 209, 224, 233, 234, 235, 236, 241, 252, 254
Heated, 38, 106, 123, 155, 161, 209, 235
Heater, 164
Heaters, 38, 47
Heating, 20, 38, 40, 43, 47, 51, 52, 80, 87, 123, 136, 138, 141, 144, 145, 164, 165, 197, 209, 233, 234, 235, 253
HEC (hydroxyethyl cellulose), 26
Hegman, 248, 253
Helical, 164
Helipath, 156
Helmholtz, 11, 24, 29, 142, 143, 192, 212
Hemi-acetal, 124
Henkel, 242
Heptane, 213
Hercules, 122, 146, 185, 187, 188, 204
Herculon, 126
Heterocycles, 237
Heterogeneous, 4, 163, 164, 166
Heterogeniety, 4
Heterophase, 3, 7, 17, 120, 163
HEVEA, 31
Hewlett-Packard, 181, 188
Hexachlorocyclopentadiene, 87
Hexahydro, 238
Hexane, 213
Hg, 123
HI, 124
Hindered, 233
Hindrance, 15
HLB, 57
Holder, 36, 163, 164
Holidays, 60
Homolytic, 22, 23
Homolytically, 110
Homopolymer, 29, 62, 63, 64, 66, 71, 80, 84, 93
Homopolymerization, 63, 66, 80
Homopolymerizations, 62
Homopolymerize, 230
Homopolymerized, 62, 64
Homopolymers, 27, 61, 71, 83, 88, 141, 201
Honey, 153
Hood, 40, 42, 46, 127, 240
Hook, 46
Hookean, 32
Hose, 36
Hoses, 26, 163, 240
Houwink, 184
HPLC, 198, 210
HSO, 111
HTHP, 155
HTHS, 157
Hull, 243
Hulls, 243, 244
Humidified, 54
Humidity, 208, 224, 236, 241, 254
HunterLab, 235
Hurt, 174, 175, 236
HYCAR, 80
Hydration, 96, 155, 164
Hydrocarbon, 10, 29, 60, 63, 72, 75, 80, 87, 88, 89, 90, 91, 92, 183, 185
Hydrocarbons, 75, 92
Hydrochlorides, 238, 242
Hydrocolloids, 106, 195
Hydrodynamic, 131, 132, 162, 183, 184, 201, 203, 205, 226
  chromatography (HDC), 134, 182
Hydrogen, 13, 23, 61, 63, 91, 92, 97, 104, 110, 119, 144, 177, 233, 254, 256, 257, 258
Hydrolyses, 105
Hydrolysis, 61, 63, 80, 84, 97, 104, 112, 145, 205, 234, 235, 243
Hydrolytic, 60, 61, 72, 106, 203, 224, 229
Hydrolyzable, 145
Hydrolyzates, 24

Hydrolyzed, 80, 201
Hydroperoxide, 24, 62, 64, 233
Hydrophilic, 16, 60, 62, 64, 80, 93, 96, 124, 142, 191–194, 204, 207, 255
Hydrophilicity, 63, 72, 195
Hydrophobe, 16, 72, 147, 192, 204
Hydrophobic, 16, 49, 65, 66, 71, 72, 74, 94, 96, 104, 106, 107, 125, 142, 191, 192, 194, 201, 204, 255
Hydrophobicity, 59, 64, 65, 67, 75, 229
Hydrophobized, 201, 204
Hydroquinone, 26
Hydrosols, 81, 82
Hydrostatic, 79, 180
Hydrotrope, 16, 65, 192–194
Hydroxide, 249
Hydroxy, 23, 24, 29, 84, 93, 97, 99, 141, 195, 225
Hydroxyethyl, 15, 26, 63, 95, 99, 141, 202, 204, 238
  acrylate, 76
Hydroxyethylcellulose, 146 (*See also* HEC.)
Hydroxyl, 105, 144
Hydroxylamine, 26
Hydroxylated, 99, 195
Hydroxylic, 140, 141, 202, 203, 206, 226
Hydroxymethylamino, 238
Hydroxypropionic, acrylic ester, 111
Hydroxypropyl, 95, 202
Hygroscopic, 205
Hypodermic, 38
Hysteresis, 151

I-Butyl, 81, 82
I-Propyl, 81, 82
IBM, 158
IBN (isobutyronitrile), 110
IBN-thioglycolate, 110
Ice, 40, 46, 49, 88, 174, 253
ICI, 68, 122, 237, 238
Igniting, 38
Imass, 126
Imbibe, 55, 234, 211
IMC, 105
Imino, 234, 235
Immersion, 38
Impact, 30, 75, 92, 104, 179, 185, 247, 248, 252
Impellers, 193
Impermeability, 250
Impermeable, 3, 60
Impurities, 94
Incompatibility, 17, 83
Incompatible, 72
Inert, 13
Infrared, 88, 123, 141, 175, 176, 177, 178, 187, 198, 240
Inhibit, 15, 33, 46, 92, 104, 194, 224, 255
Inhibited, 53, 146, 224
Inhibition, 20, 80, 135, 164, 229, 239
Inhibitive, 67, 253, 254
Inhibitor, 91, 94, 252
Inhibitors, 90, 112, 231, 239
Inhibits, 33, 110, 193
Inhomogeniety, 185
Initiate, 76, 196, 248
Initiated, 25, 27, 73, 74, 82, 110, 234
Initiation, 12, 22, 23, 24, 55, 72, 74, 75, 76, 82, 83, 84, 97, 111, 112, 124
  re-, 233
  photo-, 82
Initiator, 12, 20–23, 25, 27, 38, 39, 43, 45, 56, 52, 55, 56, 62, 63, 65, 72, 73, 76, 79, 90, 91, 96, 110, 111, 143, 146
Initiators, 4, 23, 24, 55, 62, 82, 89, 93, 110–112, 142, 143, 196
Injecting, 140, 173
Injection, 25, 37, 43, 45–47, 49, 51, 53, 73, 140, 147, 173, 177, 183, 186
Injector, 37, 173
Ink, 6, 8, 112, 125, 148, 164, 189, 209, 252, 254
Inks, 176, 253–255

Inlet, 176
Innocuous, 94, 197
Inorganic, 4, 172, 241, 244, 256
Insect, 240
Insensitive, 185
Insensitivity, 134
Insolubility, 55, 94
Insoluble, 14, 20, 26, 55, 56, 80, 82, 96, 146, 252
Inspected, 84
Inspection, 42, 54, 81
Inspectors, 194
Instability, 106
Instron, 126, 130
Insulation, 26
Intermolecularly, 257
Interaction, 28, 60, 97, 197, 203, 206, 209, 210, 212, 243, 258
Interactions, 32, 140, 210, 219
Interactive, 17
InterChemical, 222
Interface, 16, 17, 23, 56, 63, 78, 85, 129, 145, 191
Interfaces, 125
Interfacial, 83, 129, 177
Interfere, 124, 178, 249
Interfered, 142
Interference, 124, 254
Intermolecular, 153, 257, 258
Interparticle, 139
Intramolecular, 177
Intrinsic, 180, 184, 203
Inventor, 189
Inverse, 56, 254
Inversion, 90
Invert, 42
Iodine, 113, 124, 147
Iodine-pyridine, 124
Iodo, 242
Iodoform, 91
Ion, 12–14, 63, 96, 99, 119, 139, 141–143, 172, 185, 234, 249, 253
Ionic, 55, 96, 138, 184, 194, 196, 201, 202, 206, 231, 257, 258
Ionization, 11, 16, 197
Ionized, 16
Ions, 10, 13, 14, 15, 52, 55, 72, 74, 75, 120, 139, 143, 196, 244, 255
IR, 66, 84, 109, 140, 141, 172, 173, 174, 177, 178, 179, 185, 236, 252
IRB, 187
Iron, 24, 52, 112, 117, 124, 129, 139
Irradiation, 123
Irreproducible, 80
Irreversibly, 72
Irvine-Park, 157, 159
ISA, 159
Isoamyl, 213
Isobornyl, 81
Isobutyl, 193, 213
Isobutylene, 66
Isobutyronitrile, 22, 110, 173
Isocyanate, 93, 98, 99, 105, 106, 204, 229, 236
  blocked, 224
Isocyanatoethyl methacrylate, 98
Isolate, 49, 64, 94, 112, 143, 183, 203, 205
Isolated, 21, 26, 49, 80, 94, 109, 127, 172, 173, 239, 251
Isolating, 194, 212
Isolation, 46
Isols, 156
Isomer, 61, 88
Isomers, 20, 21, 88
Isophorone, 215
Isoprene, 10, 42, 75, 87, 88, 89, 90, 179
Isopropanol, 25, 123, 173, 203
Isopropyl, 173
Isotactic, 21
Isotherm, 143
Isothermal, 73
Isothiazoline, 238
Isothiazolinone, 242
ISOTROL, 242
ISS, 185
Itaconic, 75, 94, 95, 141

Itaconate, 94
IUPAC, 141
IV, 17, 180, 182, 184

JACS, 33, 57
Jena, 160, 184
Joule, 136, 165, 253
Joyce-Loebl, 134

Kaolin, 64, 140, 250, 259
Kaolin-latex, 149
Kathon, 238
Kayness, 126
KBr, 178, 240
Kelp, 203
Kem-Tone, 88
Kernco, 126
Ketene, 15
Ketene-ized, 204
Ketimine, 110
Ketone, 213
Ketones, 112, 124
KEVLAR, 258
Kinetic, 28, 74, 129
Kinetically, 19, 20
Kinetics, 20, 21, 22, 25, 27, 28, 57, 72, 73, 76, 91, 94, 112, 129, 223
Kingsport, 100
Kjeldahl, 141
Kobunshi, 77, 85
Koenig, 187
Komline-Sanderson, 144
KPG, 160
Kuraray, 90, 92

Lacquer, 71
Lamellar, 150
Laminate, 119
Laminated, 137, 148, 234
Laray, 161, 163
Laser, 144
Latex, 4–10, 12, 15, 17, 21, 24, 27, 29, 31, 40, 45, 46, 48, 49, 51–55, 59, 60, 64–66, 71–76, 79–83, 87–92, 94–97, 99, 100, 103–106, 109, 112, 113,
[Latex—continued]
116, 117, 119–127, 131–147, 149–151, 163, 172, 173, 175, 178, 189, 191–194, 196, 197, 202, 204, 206, 211, 218, 223, 226, 229, 234, 235, 237, 239, 240, 241, 244, 247–249, 251, 253, 256
Latexes, 6, 10, 21, 24, 31, 40, 43, 47–49, 54, 59, 63, 66, 67, 71, 72, 74–76, 80, 81, 82, 87, 88, 91, 92, 97–100, 105, 106, 112, 120, 127, 128, 131, 132, 135–137, 139–147, 172, 173, 175, 178, 186, 194, 198, 202–204, 212, 229, 230, 235, 239, 245, 253
Laundering, 59
Laurate, 73
Lauryl, 81, 198
  sulfate, 193
    sodium, 39, 75
    triethanolamine, 112, 192
LC, 160
Leach, 172, 243
Leachable, 65
Leached, 72, 243
Leaching, 36, 65, 241, 243
Lead, 56, 93, 120, 136, 139, 152, 186, 202, 233, 243
Leathery, 106
Lectron, 185
Leeds, 142
Lehigh, 10, 33, 63, 72, 74, 76, 83, 134, 142, 143, 145, 210, 253
Leneta, 106
Lhomargy, 159
Light scattering, 79, 83, 134, 135, 178, 179, 180, 185, 250, 251
Limestone, 106, 251, 257, 258
Limestone-latex, 149
Lion, 85
Lore, 139, 189, 203, 253
Lossen, 106
Lubricants, 243
Lubricated, 36, 207

Luer, 52
Luer-Lok, 43

M-cresol, 216
M-GARD, 242
MA, 130, 156, 157, 159
MAAG, 242
Magnesium, 65, 72, 92, 256
Magnetic, 36, 38, 51, 124, 176
Magnetically, 157
Maleate, 94, 197
Maleic, 63, 67, 80, 95, 201, 205, 258
  acid, 94
Mapco, 124
Marina, 244
Marine, 243
Mark, 125, 129, 183, 184, 187, 243, 245
Masterbatch, 39
Masterbatching, 38
Matec, 145
Matrecon, 210
MBTS (mercaptobenzothiazolylsulfide), 223
Mechrolab, 181, 182
MEHO (methyl/ether of hydroquinone), 45, 90
MEK (methyl/ether/ketone), 98, 105, 230
Meker, 40, 126, 127, 175
Melamine, 75, 99, 103, 178, 225, 236
Mellitates, 208
Mellon Institute, 126, 187, 259
Membrane, 13, 15, 120, 172, 180
MERBAC, 238
Mercaptan, 25, 26, 42, 56, 73, 112
Mercaptide, 23, 112
Mercaptobenzothiazole, 238, 242
Merck, 238, 242
Mercurial, 241, 243
Mercuric, 238, 242
Mercury, 174, 175, 241 (*See also* Hg.)
Mesitylene, 216
Metabisulfite, 82
Metaborate, 237, 242, 244, 253
Metal, 24, 36, 46, 47, 52, 62, 66, 75, 104, 105, 107, 119, 139, 196, 230, 234, 243, 254, 255, 258
Metallic, 250
Metallized, 106
Metals, 195, 202, 254
Metastability, 127
Metastable, 55, 128, 193
Meter, 13, 120, 128, 142, 154, 220, 235
Metering, 29
Meters, 124
Methacrylamide, 95, 97, 141
Methacrylate, 21, 24, 27, 66, 79, 82, 83, 95, 97, 141, 173, 185, 243
  2-isocyanatoethyl, 98
  hydroxyethyl, 99
  hydroxypropyl, 99
  pentachlorophenyl, 243
Methacrylates, 62, 79, 81, 82
Methacrylic, 24, 84, 85, 95, 99, 141
  acid, 75, 80, 94, 96, 98, 100
    polymer, 63
Methacrylonitrile, 110, 234
Methanol, 45, 74, 110, 123, 124, 173, 205, 212, 213, 226, 238, 242
Methoxy, 238, 242
Methyl, 21, 24, 27, 42, 61, 64, 66, 81, 82, 83, 87, 95, 98, 112, 145, 173, 185, 201, 202, 203, 205, 213, 238, 242
  cellulose, 39, 72
Methylal, 213
Methylene, 97, 210, 216, 226, 242
Methylethoxy, 238, 242
Methylisobutylcarbinol, 213
Methylol, 71, 93, 97, 98, 195, 226
Methylolamide, 63, 97, 98, 225, 226
Methylolated, 75, 141, 225, 226
Methylolation, 97, 98, 226
Methylomelamine, 97, 226
Methylopropanol, 238
Methylsulfonylpyridine, 242

Metuchen, 159
MF (melamine/formaldehyde), 195, 224, 230
MFFT (minimum film forming temperature), 174, 178, 211
Mica, 258, 259
Micellar, 85
Micelle, 16, 23, 55, 57, 72, 82, 90, 95, 132, 193, 194, 203
Michael, 94, 111, 112
Micro-chek, 242
Microaggregates, 258
Microbiological, 241
Microbiology, 244, 245
Microcomputer, 183
Microcracks, 185, 250
Microelectrophoresis, 144
Microemulsion, 9, 75
Microfine, 67
Microflocs, 54, 128
Microgels, 183
MicroMeritics, 134, 144
Micropinholes, 250
Micropore, 179
Microporous, 183
Microscope, 7, 132, 135, 144, 191, 240
Microscopic, 104, 237, 241, 250, 251
Microscopy, 96, 132, 135, 175, 178, 185
Microthene, 67
Microwave, 123, 125, 254
Microwaves, 250
Mie, 74
Mildew, 241, 243
Mildewicide, 104, 241, 243
Mill, 6, 90, 91, 149, 174, 175, 189, 223, 248, 251
Mineral, 98, 250
MIR, 177
Mirabella, 29, 33, 178, 187
Miscibility, 104, 220
Miscible, 134, 211, 212, 244
Mitech, 159
Mobay, 105
Modulus, 32, 249
Moisture, 61, 72, 75, 80, 104, 105, 107, 117, 175, 185, 193
Molecular Weight, 14, 15, 17, 19–21, 25–32, 56, 62, 71, 74, 82, 90, 91, 111, 113, 132, 134, 179–184, 204, 208, 219, 224, 257
Mono-alcohol, 204
Mono-esters, 99
Monocyclicoxazolidine, 242
Monodisperse, 72, 74, 83
Monomethyl, 26
Monosulfide, 112
Monsanto, 67
Monterey, 79
MOONEY, 242
MSDS, 141, 147
Mud-wrestlers, 15
Musk, 16
Myristic, 83

N-Butyl, 81, 82
N-diethyl, 26
N-dimethylacetamide, 215
N-dimethylacrylamide, 97
N-dimethylethanolamine, 105
N-Hexyl, 81, 82
N-hydroxymethyl-N-methyl, 238, 242
N-methyl, 216
N-methylol, 31, 95, 141, 231
N-methyloacrylamide, 97, 98
N-octyl, 242
N-Propyl, 81, 82
N-propylsulfonyl, 242
N-t-octylacrylamide, 97
N-vinylpyrrolidone, 63
NACE, 17
NaCl, 13 (*See also* Salt)
Nametre, 122, 159
Naphthenate, 242
Naval, 210
Navy, 241
NBRs (nitrite butadiene rubbers), 19
NBS, 241

Nd, 33, 57, 78, 188
NDSU, 258
Neoprene, 42, 91, 221
Nernst, 12, 13, 55
Newtonian, 121, 122, 150, 152, 159, 163
Nickel, 254, 255
NIOSH, 91
Nitrate, 173
Nitrile, 42, 76, 79, 80, 81, 82, 87, 88, 96, 98, 111, 127, 221, 234, 242
Nitrobenzene, 216, 218
Nitroethane, 215
Nitrogen, 22, 31, 38, 42, 46, 88, 97, 104, 110, 111, 141, 185, 234
Nitrogenous, 96
Nitropropane, 215
Nitrous, 254
Nitroxide, 72
NL, 154
NMR (nuclear magnetic roesonance), 109, 141, 173, 176
Nomograph, 153
Nonaqueous, 198
Noncarboxylated, 172
Noncharged, 15
Noncommercial, 60
Noncontact, 142
Nondestructively, 124
Nonfriable, 26
Nonfunctional, 91, 248
Noninitiator, 38
Nonionic, 15, 16, 62, 63, 72, 73, 91, 94, 96, 135, 144, 172, 175, 193, 197, 198, 201, 204, 205, 208
Nonlinear, 72
Nonmercurial, 239, 241
Nonmicelle, 65
NonNewtonian, 122, 149, 151, 152, 153, 159, 163, 165, 249
Nonoptical, 163
Nonoxygenated, 38
Nonpolar, 60, 66, 80, 126, 129, 258
Nonradicalized, 27
Nonsolvent, 20, 123, 173, 212, 219
Nonstandard, 73
Nonsurface, 12, 15
Nonsurfactant, 195
Nonswelling, 66
Nontechnical, 15
Nontransfer, 26
Nonvolatile, 40, 122, 123
Nonwoven, 31, 64, 66, 67, 76, 81, 84, 88, 98, 99, 116, 117, 119, 135, 137, 138, 141, 145, 147, 189, 233, 234, 235
NOPCOCIDE, 242
Norcross, 159, 163
Northrup, 142
Nozzle, 127, 137, 256
Nozzles, 116, 120
NPCA, 241, 244
NTIS, 129, 245
Nuclear, 124, 176
Nuodex, 237, 238
Nuosept, 238
Nusonics, 124
Nyco, 259
NYFLAKE, 259
Nylon, 19, 195, 257, 258
NYTEK-GD, 242

O-dichlorobenzene, 217
Oakes, 206
Oakland, 210
Octadecyl, 16
Octanol, 215
Octyl sulfate, 193
Octylphenoxypoly(ethylene oxide), 193
Off-specification, 115, 116
OH, 129, 159, 161
Ohm, 255
OK, 157
Oleate, 16, 208
Oleic, 90, 208
Oligomer, 75, 96, 97, 183
Oligomers, 19, 21, 24, 29, 83, 103, 142, 172, 208
Olin, 238
OMACIDE, 238

Organomercurials, 243
Organosilane, 117
Organosol, 64
Organotin, 243, 244
Orlon, 235
OSHA, 91, 237
Osmometers, 188
Osmometry, 15, 180, 187
Osmotic, 12, 14, 15, 172, 180
OTTASEPT, 238
Owl, 88
Oxalic, 226
Oxazolidine, 238
Oxazolyl, 238, 242
Oxen, 16
Oxidants, 146
Oxidation, 104, 112, 178, 230, 233, 234, 235
Oxide, 89, 97, 105, 106, 111, 113, 123, 139, 193, 196, 202, 212, 221, 236, 238, 242, 243, 244, 254, 255, 256
Oxides, 38, 92, 234
Oxidized, 123
Oxidizes, 124
Oxidizing, 123
Oxime, 98
Oxirane, 99
Oxyethylene, 144
Oxygen, 38, 61, 64, 66, 80, 88, 104, 111, 178, 186, 233
Oxygenated, 65, 67, 229
Oxygens, 144, 185
Ozone, 87, 92, 178

P-methylstyrene, 73
P-xylylene, 178
Paar, 160
Pacemaker, 254
Palmerton, 17, 76, 186
Papermaking, 144
Para-chloro-meta-xylenol, 238
Para-toluenesulfonic, 226
Parent, 71
Park, 86, 113, 158
Parr, 47
Paste, 254
Pastel, 106
Pastes, 234
Patent, 53, 63–67, 73, 74, 76, 83, 89–91, 94, 142, 144, 239
Patents, 4, 65, 83, 99, 175, 259
PC, 158
PE, 258
Pen-Kem, 144
Penta-chlorostyrenes, 71
Pentachlorophenol, 243
Pentadiene, 87
Pentanol, 213
Pentoxide, 123
Perkin-Elmer, 147, 182, 187
Permeation, 60, 75, 92, 99, 106, 134, 174–176, 180, 182, 188, 210
Peroxide, 22–24, 49, 62, 63, 75, 97, 110, 111, 172, 173, 236, 258
Peroxides, 110, 112, 113
Peroxydisulfate, 12, 23, 24, 62, 63, 72–74, 82, 84, 111, 112, 143, 146, 196
Phenol, 98, 243
Phenolic, 47, 230
Phenols, 208, 233, 237
Phenyl, 110, 111, 173, 238, 242
Phenylcyclohexene, 111, 173
Phenylmercuric, 242
Phosphate, 24, 63, 65, 106, 139, 172, 208, 209, 210, 217
Phosphorous, 123
Photocell, 162, 184
Photodegradable, 252
Photoelectron, 76
Photographs, 243
Photoinitiate, 82
Photometer, 134
Photosedimentometry, 132
Photovolt, 123
Phr, 90
Phthalate, 64, 65, 72, 208–210, 215, 216, 217
Phthalic, 103
Phthalocyanine, 252

Physica, 160
Physical, 3, 4, 7, 8, 10, 17, 30, 31, 60, 88, 137, 152, 165, 166, 171, 174, 177, 179, 187, 212, 239, 249, 251, 257
Physics, 7, 164
Pierced, 47
Pigment, 4, 5, 9, 12–15, 24, 33, 116, 117, 135, 140, 146, 149, 150, 163, 175, 177, 179, 185, 188, 191, 196, 204, 236, 244, 247–246, 258, 259
Pigment-binder, 115
Pigmentation, 178, 179, 236, 249, 255, 256
Pigmented, 184, 252
Pigments, 5, 9, 12, 31, 104, 116, 131, 135, 193, 194, 196, 198, 234, 236, 243, 247, 249–258
Pimelate, 210, 213
Pinholes, 60
Pipeline, 54, 152
Pipes, 127, 139
Piping, 46, 52, 54, 120, 163, 191
PKa, 95, 116, 197, 226
Plasma, 172, 178, 185, 186
PLASMOD, 172
Plaster, 256
Plasti-corder, 156
Plastic, 21, 30–33, 52, 60, 65, 66, 75, 121–123, 149, 152, 153, 164, 177, 178, 209, 236, 240, 251, 254, 256
Plasticization, 75, 229
Plasticize, 65, 72, 209–211
Plasticizer, 60, 64–66, 80, 83, 123, 147, 192, 207–210, 219, 224, 243
Plasticizers, 5, 20, 72, 109, 183, 198
Plasticizing, 226
Plastics, 71, 129, 210, 222, 243, 245, 260
Plastisol, 64, 156, 164, 175, 209
Platinum, 255
Platy, 250
PMA, 238, 242
PMC, 238
Poise, 166
Poisonous, 53
Polar, 29, 60, 75, 83, 104, 124, 147, 179, 193, 195, 258
Polarity, 29, 80, 183, 195, 208, 229
Poly-ene, 100
Poly-isocyanates, 106
Polyacid, 103, 208
Polyacrylates, 208
Polyalcohol, 103
Polybrominated, 238
Polycarbodiimides, 229
Polycarbonate, 172, 178, 186
Polycarboxylate, 201–203, 206
Polycarboxylated, 230
Polycarboxylic, 63
Polycondensation, 56, 57
Polycyclic, 258
Polyesters, 19, 105, 208
Polyether, 72, 105–107
Polyethylene, 172, 174, 185, 258
Polyfunctional, 62, 208
Polyions, 194
Polyisocyanate, 229
Polyisocyanates, 105, 204
Polymer-plasticizer, 210
Polymerizable, 205
Polymerization, 3, 4, 9, 12, 16, 20–32, 35, 36, 38–40, 42, 43, 45, 50, 52–57, 61–65, 67, 68, 72–76, 79, 80–85, 87–91, 93–100, 103, 104, 109–113, 115, 127, 132, 133, 144, 146, 173, 179, 192, 194–196, 208, 245, 253
Polymerizations, 41, 46, 66, 82, 148
Polymerized, 106, 147
Polymers, 4, 5, 7, 10, 14, 15, 19–23, 26, 27, 29–32, 35, 36, 46, 63, 65, 67, 71, 72, 75, 80, 83, 84 87–89, 91, 93, 98, 99, 103–106, 115–117, 127, 129, 135, 139, 140, 170–174, 176–179, 183, 185–187, 194, 195, 201–203, 205–209, 212,

[Polymers—continued]
219, 221, 225, 226, 229–231, 234, 243, 253, 258
Polyol, 105
Polyols, 84
Polyphosphate, 12, 14, 133
Polysciences, 136
Polystyrene, 74, 82, 110, 255
Polyurea, 250
Polyurethane, 105, 106, 204, 229, 230
Polyvalent, 76, 100, 104, 106, 139, 197, 202, 203, 206, 225, 228, 230, 244, 253
Pond, 210
Pores, 15, 60, 182
Porosities, 182
Porosity, 60, 174, 183, 248, 249
Porous, 74, 182, 194, 197
Postpolymerization, 189
Poststabilization, 205
Potassium, 238, 242
Potato, 66
Potatoes, 96
Potentiometric, 142
Potentiometry, 141
PPG, 251
PRA, 239
Praxis, 124
Preblended, 80
Precautions, 42, 44, 53, 240
Precipitate, 104, 202, 219
Precipitated, 205, 251
Precipitation, 104, 202, 203, 205, 252, 253
Precoating, 186
Predissolve, 244
Predissolved, 39
Preemulsification, 192, 208, 226
Preemulsified, 208, 229
Preemulsify, 208, 209
Prefilter, 127
Prefloc, 52, 127
Preplasticized, 66
Preservatives, 237, 239, 241, 242
Pressure, 30, 35, 36, 40–43, 45–47, 52, 53, 63, 65, 67, 73, 81,
[Pressure—continued]
88, 90, 98, 117, 126, 147–149, 155, 157, 158, 160–165, 172, 179–183, 188
Pressures, 62, 79, 180
Pressurize, 46, 90, 91
Pressurized, 45, 65, 154, 155
Preweighed, 45
PRI, 243
Profilometer, 153
Propellants, 248
Propionitrile, 214
Propylene, 32, 67, 99, 123, 125, 126, 138, 189, 212, 215, 218, 221
Propylsulfonyl, 242
Propynyl, 242
Protein, 5, 8, 124, 204, 205
Proteins, 90, 143, 183, 203, 204
Proxel, 238
Pseudo-gel, 12, 15
Pseudoplastic, 121, 152, 256
Pseudoplasticity, 150, 164
Putty, 153, 209
PVC, 19, 64, 65, 66, 75, 80, 147, 209, 234, 243, 249, 251
PVDC, 66
PVOH, 124
PWB, 68
Pyrex, 174
Pyridine, 95, 96, 217, 242
Pyridines, 219
Pyridinethione-N-oxide, 238
Pyridinium-hydroiodide, 124
Pyrogenic, 250
Pyrolysis, 176
Pyrrolidone, 216, 217

QA (quality assurance), 125, 127, 131, 133, 177, 180, 198, 240, 247, 257
QC (quality control), 122, 161, 185
Quadrille, 40
Quadruple, 10
Quaternary, 26, 113
Quinoline, 217, 219
Quinolinolates, 242

Rack, 42, 43, 44, 45
Radiation, 76, 84, 185, 235, 252, 254
Radical, 22–28, 63, 73–75, 88, 98, 110–112, 208, 233, 235
Radicals, 12, 45, 83, 104, 179
Radio-carbon, 15
Radio-tagged, 14
Radiometer, 142
Rain, 64, 84, 85, 193, 251
Rains, 7
Rainwater, 54
Rame'-hart, 126
RAPRA, 68, 76, 78, 86, 92, 101, 186
Rayleigh, 250
Reach, 43, 46, 90, 100, 104, 132, 147, 257
Reactor, 21, 30, 46, 47, 53, 54, 65, 90
Reactors, 43, 63
Recrystallization, 110
Recycle, 149
Recycled, 30, 46
Red, 65, 80, 123, 138, 141, 210, 234–236, 240
Red-orange, 67
Redbook, 222, 236, 245
Reddish, 234
Redisperse, 249
Redissolved, 173
Redox, 24, 111, 124
Reductants, 146
Reemissions, 251, 252
Reemphasize, 8
Reemulsification, 90
Reequilibration, 196, 198, 202
Reestablish, 198
Reflect, 29
Reflectance, 84, 141, 175, 177–179, 241
Reflected, 185, 250
Reflux, 37, 39, 40, 53, 62, 90
Reforms, 23
Reformulate, 115
Reformulated, 253
Reformulation, 250
Reformulations, 221
REFR, 69, 73, 81, 82, 89
Refractive, 60, 72, 124, 133, 134, 182, 211, 250, 251, 256
Refrigrator, 239
Regression, 212
Reichold, 30
Reilly, 185, 188
Reinforcement, 31, 32, 257, 258
Reinforcing, 247, 252, 257–259
Reinhold, 17, 85, 113, 129
Reinitiation, 233
Reinspect, 42
Reirradiated, 235
Rejection, 15
Rejuvenation, 120
Relabel, 46
Relaxation, 124
Relaxed, 178
Reodorants, 5
Repeatability, 125
Repeatable, 136
Repel, 192, 243
Repellant, 119
Repetition, 125
Replicate, 219
Replicates, 14
Reprocess, 49
Reproducibility, 125, 205
Reproducible, 36, 128, 147, 207
Reproducibly, 7
Repulsion, 11
Residual, 14, 46, 47, 49, 63, 65, 87, 92, 96, 124, 142, 146, 147, 171, 173
Residue, 109, 127, 128, 162, 173, 226, 240, 251
Resilience, 87, 88, 92
Resin, 33, 36, 38, 40, 49, 66, 67, 75, 80, 81, 89, 103, 104, 143, 153, 172, 179, 188, 207, 224, 230, 244, 253
Resinification, 233
Resins, 67, 75, 76, 79, 80, 88, 89, 99, 104, 141–143, 153, 178, 179, 195, 225, 251, 256
Resist, 59, 60, 255

Resistance, 15, 59, 60, 71, 75, 79, 83, 84, 92, 104, 117, 136, 143, 163, 174, 181, 182, 230
Resistant, 193, 241
Resisting, 62, 99
Resistivities, 254
Resonance, 124, 176
Retard, 5
Retardance, 67
Retardancy, 208
Retardant, 65, 67, 139, 194, 234
Retardants, 194, 196
Retarding, 209
Retards, 88
Retention, 144, 148
Reuse, 63
Reversion, 231
Rewettability, 193
Rewetting, 193
Rf, 172
RheoChan, 158
Rheogram, 149
Rheological, 172, 174, 202
Rheologist, 165
Rheology, 26, 80, 121, 148, 152, 153, 155, 158, 165, 170, 186, 201, 202, 203, 204, 223, 248, 253, 260
Rheomat, 161
Rheometer, 154, 155, 157, 159, 161
Rheometrics, 160
Rheometry, 122, 149
Rheopectic, 121
Rheopexy, 151
Rheotron, 156
Rheotronic, 161
RIA, 142
Ringstand, 46
Rinse, 127, 240
Rinsed, 46, 181, 240
Rocket, 248
Rocks, 248, 249, 250, 252, 253
Rodent, 240
Rohm, 79, 80, 82, 84, 85, 97, 98, 142, 143, 238, 242, 244
Romicon, 142
Ronbun, 85
Ronigen-Pettinger, 54
Roon, 230, 243
Rosin, 91
Rotor, 155, 156
Rotovisco, 122, 158
Rough, 121
Round, 36, 49, 128
Rub Test, 136
Rubber, 8, 10, 21, 26, 31–33, 36, 42, 43, 45, 49, 52, 57, 60, 66, 80, 87, 88, 90, 91, 106, 120, 121, 123, 126, 127, 136, 138, 147, 153, 165, 175, 178, 184, 209, 219, 221–223, 231, 234, 258
Rupture, 47, 52
Rutile, 185, 256

S-Butyl, 81, 82
S-triazine, 238
Saddle, 158
Sadtler, 187
Salicylanilide, 238
SALS, 180
Salt, 10–13, 55, 62, 67, 72, 106, 124, 134, 135, 139, 140, 144, 172–174, 178, 184, 189, 194–196, 198, 202, 203, 206, 231, 240, 244, 254
Salting, 196, 197, 202
Salts, 5, 11, 12, 23, 24, 26, 55, 62, 64, 65, 72, 80, 99, 112, 124, 126, 134, 135, 139, 140, 142–144, 172, 189, 193, 196, 197, 230,. 237, 244
Sampe, 187
Sampling, 37–40, 42–47, 53, 138
SAN (styrene/acrylonitrile), 72
Sand, 8, 137, 256, 257
Saturants, 81
Saturate, 132
Saturated, 27, 32, 74, 98, 117, 133, 181, 235
Saturating, 14
Saturation, 32, 67, 125, 131

Sawtoothing, 140
SAXS, 185
SB, 120
SBR (styrene/butadiene rubber(s), 19, 56, 87, 112, 147, 173, 226
Scatchard, 196, 202, 206
Scatter, 140
Scattered, 128, 185, 250, 251
Scattering, light (*See* Light scattering.)
Scatters, 250
Scavenger, 26
Schleicher, 142
Schott, 160, 184
Schuell, 142
Schulz-Hardy, 139, 143, 146, 196, 244, 253
Scratch, 104
Scratched, 42
Scratches, 43, 122
Screen, 40, 46, 117, 127, 128, 140
Screened, 113
Screening, 128, 241
Screens, 235
Screws, 46
Scrub, 240
Scrubbing, 59
Scum, 104
Seal, 54, 55
Sealed, 88
Sealer, 64
Sealing, 47
Sebacates, 208
Seed, 21, 74, 75, 96
Seeded, 65, 66, 74, 76
Seeding, 75, 240
Seeps, 172
Seive, 127
Self-crosslinking, 66, 75, 97, 98, 116, 124, 137, 146, 231
Self-curing, 100
Self-react, 111
Self-stabilizing, 143
Semipermeable, 15, 172
Semiqualitative, 249
Semisolid, 207
Semisoluble, 106
Semiwater-soluble, 75
Sensadyne, 126
Sensor, 43, 46, 158
Sensors, 163
Serum, 72, 144, 145
Settling, 8, 9, 10, 17, 55, 127, 249, 253, 255, 256
Shamrock, 143
Shearing, 136, 137, 162
Sheen, 160
Shellfish, 244
Sherwin-Williams, 88
Shield, 42, 43, 203, 240
Shielded, 45, 46
Shields, 33
Shortstop, 45, 62, 80
Shortstopped, 46, 80
Shortstopping, 26
Shot, 88, 145, 181, 186
Sieve, 248, 257
Silane, 31, 259
Silanes, 258
Silica, 14, 31, 172, 195, 250, 258
Silicone, 31, 36, 122, 175, 178, 219, 258
Silicones, 165, 241
Siliconized, 106
Silver, 63, 74, 196, 254, 255
SIMS, 185
Sinclair-Koppers, 74, 77
SKANE, 242
Slimicide, 184
Sludge, 240
Slurried, 20
Slurries, 154, 163
Slurry, 247, 248, 257
SME, 188
Smog, 84
Smoke, 8
Smoky, 127
Snow, 91, 92, 251
Soap, 55, 72, 74, 103, 104, 132, 133, 240

Soapless, 82
Soda, 41, 42
Sodium, 11, 12, 13, 24, 39, 72, 82, 139, 140, 141, 144, 172, 173, 185, 195, 202, 238, 242, 255
SOHIO, 80
Solidify, 209
Solids, 4, 10, 29, 39, 40, 43, 46, 48, 49, 53, 104, 122, 123, 125, 127, 129, 131–133, 136, 147, 171, 173, 174, 178, 195, 198, 205, 207, 208, 223, 236, 240, 247, 248, 250, 258
Sols, 12
Solubilities, 29, 203
Solubility, 29, 31, 60, 71, 75, 80, 82, 94, 97, 175, 183, 196, 206, 207, 213–219, 229, 253
Solubility Parameter, 208–210, 212, 219–222
Soluble, 10, 15, 17, 23, 24, 28–30, 38, 55, 56, 65, 67, 82, 89, 94, 96–99, 103, 105, 142, 143, 146, 172, 183, 191, 193–198, 201–203, 205, 206, 211, 219, 226, 229, 234, 243, 252, 253
Solute, 209
Solution, 9, 15, 20–22, 28, 39, 45, 51, 56, 61, 67, 71, 75, 80, 82, 90, 95, 106, 120, 125, 126, 139, 140, 147, 164–166, 171, 173, 179–182, 184, 186, 192, 204, 207, 219, 220, 229, 234
Solutions, 21, 24, 52, 139, 147, 165, 180, 181, 184, 203, 205, 209, 229, 234, 244
Solvency, 209
Solvent, 8, 15, 20, 21, 23, 25, 59, 60, 71, 75, 79, 84, 95, 98, 103, 105, 106, 110, 123, 134, 174, 175, 177, 180–184, 207, 209, 210, 212–219, 221, 226, 229, 230, 254
Solvents, 11, 20, 28, 71, 80, 123, 147, 153, 188, 209, 210, 212, 218–221, 244
Sonic, 124
Soot, 8
Soya, 204
SP, 84, 85
Spaghetti, 30, 33
SPAN, 68, 134, 257
Spanish, 239
Spectra, 141, 176, 177, 178, 179
Spectra-Tech, 187
Spectral, 178
Spectrapor, 142
Spectrograms, 187
Spectrograph, 178, 179
Spectrometers, 160
Spectronic, 133, 140
Spectrophotometer, 140
Spectrophotometrically, 14
Spectroscopy, 66, 76, 84, 173, 174, 175, 176, 177, 178, 185, 187, 198
Spectrum, 88, 142, 177, 178, 179, 182, 210, 235, 241
Spillage, 52
Spills, 240
SPR, 85
Spray, 45, 116, 120, 127, 137, 233, 255, 256
Sprayability, 129, 198
Sprayable, 104
Sprayed, 174
Sprayer, 257
Sprayers, 255
Spraying, 120, 125, 174
Spreading, 183
Stabilitization, 83
Stability, 7, 8, 9, 17, 29, 40, 51, 67, 72, 84, 93, 95, 96, 98, 99, 119, 133, 136, 139, 140, 143, 145, 146, 147, 195–198, 202, 204, 226, 239, 244, 253, 254, 256
Stabilization, 15, 24, 38, 55, 56, 72, 76, 90, 94, 106, 111, 112, 139, 171, 192–195, 198, 202, 204, 253
Stabilize, 11, 15, 24, 90, 104, 211

Stabilized, 15, 140, 175, 195, 196, 203, 209
Stabilizer, 16, 56, 73, 146, 192, 193, 198, 234, 236, 244
Stabilizers, 4, 5, 38, 40, 112, 174, 191, 192, 194, 197, 198, 229, 233–236, 243
Stabilizes, 63
Stabilizing, 16, 17
Stable, 9, 10, 12, 17, 20, 43, 63, 94, 106, 111, 136, 145, 146, 192, 194, 197, 198, 205, 223, 233, 243, 249, 253, 259
Stack, 53, 134
Staining, 87, 194
Stainless, 36, 37, 40, 43, 47, 52, 127
Standard, 13, 22, 25, 36, 40, 42–44, 49, 50, 52, 65, 72, 74, 91, 94, 103, 106, 119, 120–126, 134, 136–138, 141–144, 147, 148, 159, 162, 174, 177, 179, 181, 182, 185, 191, 192, 194, 196, 198, 223, 234–236, 248
Standardization, 136
Standardize, 137, 148
Standardized, 120
Standards, 73, 121, 127, 134, 136, 137, 141, 147, 157, 165, 183, 236
Starch, 5, 15, 64, 73, 99, 144, 155, 164, 195, 257
Starch-latex, 164
Starches, 30, 63, 90, 202, 239
Starnes, 33, 113
Static, 254
Statistical, 6, 116, 189, 210, 212
Statistically, 52, 65, 117, 179, 197
Statistics, 179, 212
Stay-Clean, 242
Steam, 38, 43, 49, 97, 240
Stearate, 16, 119
Stearic, 83, 90
Stearyl, 82
Steel, 9, 36, 37, 40, 43, 47, 52, 80, 98, 111, 127, 148, 174, 254, 257
Step-Growth, 20
Stereo-regular, 31
Stereochemistry, 31
Stereoisomerization, 20
Stereoisomers, 20
Stereoregular, 65
Steric, 15, 16, 193, 195
Sterile, 240
Stir, 51, 105, 143, 192, 208, 258
Stirred, 65, 73, 205
Stirrer, 16, 24, 36, 37, 38, 43, 47, 51, 52, 53, 54, 155, 193, 257
Stirring, 35, 36, 38, 39, 51, 55, 136, 140, 143, 149, 192, 195, 208, 255
Stoichiometric, 12
Stokes, 9, 55, 134, 249
Stopwatch, 139, 182, 184
Storage, 9, 40, 46, 54, 55, 85, 111, 112, 116, 119, 145, 146, 148, 203, 204, 209, 210, 226, 233, 239, 240
Store, 42, 46, 52, 119, 145
Stored, 54, 94, 120, 146, 149, 153, 239
Storing, 111, 120
Stormer, 160
Story, 116, 128, 248
Stoving, 99
Strain, 32
Strains, 175
Stratton, 144
Strength, 11, 30, 31, 32, 55, 59, 60, 62, 75, 83, 84, 92, 98, 104, 106, 117, 129, 137, 143, 171, 174, 179, 184, 195–197, 202, 203, 206, 207, 212, 231, 233, 248, 249, 252, 255–258
Strengths, 249, 254, 257
Stress, 31–33, 47, 52, 136, 148–52, 153, 155, 156, 159, 163, 166, 241
Stresses, 30, 175, 176
Stretching, 32, 177, 178
Stringiness, 153
Strip, 49, 64, 140, 158, 161, 183

Stripped, 88, 90, 94, 103
Stripping, 47–51, 53, 63, 65, 97, 106, 109, 111, 112, 172
Strips, 175
Structural, 19, 21, 25, 61, 179, 212, 236, 252
Structurally, 20, 27
Structure, 19, 21, 23, 30, 53, 59, 60, 65, 66, 68, 83, 87, 88, 104, 117, 125, 177, 178, 185, 188, 192, 195, 197, 198, 201, 204, 212, 229, 234, 237, 238, 242, 254
Structures, 84, 179, 189, 208, 209, 233, 235
Stucco, 256
Styrene, 10, 24, 25, 28–30, 40, 45, 47, 49, 53, 62, 71–76, 79, 83, 88, 90, 94, 96, 98, 106, 111, 112, 127, 139, 141, 142, 145, 147, 173, 179, 181, 182, 186, 212, 230, 251
Styrene-acrylics, 81
Styrenes, 39, 42
Submicroscopic, 7, 17
Suboxide, 256
Substituent, 236
Substituents, 208
Substituted, 71, 76, 97, 175, 195, 249
Substrate, 30, 97, 116, 120, 175, 191, 193, 197, 198, 235, 237, 239, 241, 243, 250, 253
Substrate-binder, 115
Substrates, 74, 99, 106, 175, 194, 210, 243, 250
Subsurface, 137
Succinate, 193
Sucrose, 124
Suffocation, 53
Sugar, 30, 189
Sugars, 82, 90
Sulfamate, 139
Sulfate, 12, 14, 24, 39, 64, 72, 75, 76, 93, 112, 143, 144, 172, 193, 196, 197, 198, 255
Sulfates, 192
Sulfite, 82
Sulfole, 56
Sulfonate, 14, 82, 91, 93, 193, 196, 197
Sulfonated, 82
Sulfonates, 192
Sulfone, 217, 242
Sulfoxalate, 24
Sulfoxide, 217
Sulfur, 81, 76, 87, 91, 111, 124, 223, 226, 254
Sulphate, 23, 24
Sun, 236
Sunlight, 253
Supernatant, 14
Surface, 10–17, 20, 24, 26, 29, 31, 37, 38, 52, 54, 55, 57, 72, 74, 78, 82, 94, 96–99, 103, 121, 123, 125, 126, 129, 132, 133, 138, 142–145, 162, 164, 172, 175, 178, 179, 183–186, 191–199, 202, 205, 208, 210, 226, 229, 230, 236, 241, 243, 247, 249, 250, 251, 253, 256, 258, 259
Surfaces, 76, 84, 122, 124, 188
Surfactant, 16, 25, 26, 51, 57, 62, 64, 65, 72–75, 82, 83, 90, 91, 95–97, 104, 106, 126, 127, 132, 133, 135, 136, 139, 143, 144, 171, 172, 192–197, 202–205, 207–209, 229, 239, 253
Surfactant-free, 82, 89
Surfactants, 5, 11, 55, 56, 67, 89, 113, 125, 138, 140, 142, 147, 174, 191, 198, 237, 255
SURLYN, 258
Sward, 241
Swedlow, 210
Sweeco, 54
Sybron, 123
Symbols, 166
Symmetrical, 27
Symmetry, 66
Synchrolectic, 156, 163, 164

Syndiotactic, 21
Syntheses, 36, 61, 88, 94, 97, 109, 110, 124, 147, 148, 183, 188, 230
Synthesis, 15, 19, 20, 32, 71, 73, 91, 105, 179, 194, 202, 233, 245
Synthesiser, 59
Synthetic, 64, 79, 91, 196, 205, 243
Synthetics, 63, 77, 236
Syringe, 38, 39, 42, 43, 44, 45, 46, 52, 181, 240
System, 7, 11, 20, 49, 50, 52, 53, 55–57, 62, 75, 80, 83, 89, 90, 95, 100, 120, 122–124, 126, 128, 134, 137, 142, 144, 145, 147, 150, 153, 157, 163, 171, 174, 175, 181, 183, 184, 194–198, 203, 208, 210, 211, 224, 231, 233, 237, 239–241, 252, 253, 258

T-butanol, 67, 111
T-Butyl, 81, 82
T-butylstyrene, 73
Tack, 8
Tackifiers, 91
Tackiness, 175
Tacky, 20, 91, 147, 174, 175, 211
Tactile, 209
Talc, 175, 258
Talcs, 117
Tamed, 53
Tank, 54, 55, 65, 73, 116, 145, 148, 152, 191, 239, 240, 257
Tanks, 54, 112, 120, 139, 240
Tape, 47, 81, 98
Taper, 36
TAPPI, 12, 17, 57, 77, 109, 117, 128, 129, 137, 138, 187, 188, 189, 195, 202, 236, 259
Tapping, 43
Target, 48
Taring, 123
Tartaric, 63
Taste, 64, 109
Tastes, 47
TBTO, 242
Teach, 5, 100
Teflon, 36, 42, 174, 185
Tekmar, 161
Tekmar's, 126
Tektamer, 238
Telechelic, 23, 26, 27, 91, 93, 110, 112, 187, 208
Telomers, 19
Temperature, 24, 27, 32, 35, 39–41, 43, 46, 47, 49, 60, 62, 64, 65, 67, 71, 72, 88, 90, 91, 98, 103, 117, 124, 136, 144, 146–149, 156–158, 160, 162–166, 172, 174, 181, 184, 203, 207, 208, 211, 221, 223, 224, 230, 231, 239, 253
Temperatures, 81, 93, 106, 116, 159, 197, 209
Tensile, 30, 98, 115, 117, 126, 153, 171, 174, 175, 231, 249, 257
Tensiometer, 125, 126
Tension, 17, 57, 74, 83, 103, 125, 126, 129, 132, 192–194, 198, 199, 202, 205
Terminated, 20
Terminating, 28
Termination, 22, 25–27, 55, 63, 73, 74
Terminology, 20, 29, 75, 79
Terpenes, 66
Testing, 6, 7, 14, 15, 21, 30, 32, 40, 94, 115, 117, 119, 121, 122, 124, 125, 128, 137, 153, 159, 160, 161, 171, 174, 175, 197, 98, 205, 208, 235, 236, 240, 241, 247, 248, 253
Tests, 65, 127, 131, 136, 233, 257
Tetrabromide, 73
Tetrachloride, 56, 73, 112, 216
Tetrachlororoethylene, 65
Tetrachlororoisophthalo, 242
Tetrafunctional, 112
Tetrahydrofuran, 215
Tetralin, 217, 218
Tetramethylsuccinonitrile, 23, 49, 110, 173

Textile, 6, 76, 79, 99, 112, 126, 130, 140, 156, 189, 198, 252, 256
Textiles, 32, 64, 65, 66, 67, 81, 84, 97, 98, 235, 236, 257
Texture, 105, 252, 256
Tg, 32, 60, 61, 64, 66, 69, 71–73, 81–83, 88, 89, 93–97, 99, 174, 209, 224
Thermal, 19, 20, 62, 123, 145, 146, 149, 178, 223, 253
Thermally, 22, 27, 80, 99
Thermistor, 181, 182
Thermoanalysis, 187
Thermochemical, 3
Thermodynamic, 83, 125, 129, 223
Thermodynamics, 165, 223
Thermometer, 36, 37, 39, 47
Thermoplastic, 20, 21, 30, 31, 106, 178, 224
Thermoplasticity, 26
Thermoset, 20, 99, 224
Thermostat, 42–46, 184
Thermostatic, 47
Thermostatted, 181
Thiazole, 242, 237
Thiazolyl, 241
Thickener, 52, 67, 140, 197, 239
Thickeners, 5, 63, 80, 94, 123, 146, 172, 197, 201–205, 253
Thickening, 121, 150
Thinner, 49, 53, 178, 219
Thinning, 121, 150
Thio-organic, 234
Thiocyanomethylthio, 242
Thioethers, 233
Thioglycolate, 112, 173
Thioglycolic, 23, 112
Thixotropic, 121, 152, 164, 256
Thixotropy, 151
Tin, 175, 241, 243, 256
Tinge, 234
Tire, 4, 30, 31, 53, 63, 88, 92, 96, 99, 189, 192, 207, 209
Tires, 26, 31, 175
Tissue, 256
Titanate, 26, 31, 228, 254, 256, 259
Titanates, 63, 258
Titania, 14, 98, 185, 186, 247, 249, 251, 257
Titanium, 236, 251
Titratable, 103
Titrate, 96, 132
Titrating, 132, 142, 147
Titration, 72, 123, 129, 132, 141, 142, 173
Titrations, 124, 133
Titrator, 141
Titrimetry, 94, 142
TK, 242
TMI, 154
TMSN (tetramethyl/succinonitrile), 23, 110
TN, 100
Toluene, 71, 216
Tone, 105, 107
Topcoat, 236
Torque, 36, 52, 155, 156, 158, 163, 164
Toughness, 75, 106
Towels, 147
Toxic, 23, 91, 99, 113, 237
Toxicity, 50, 65, 88, 99, 109, 211, 219, 240
Toxicological, 65, 103, 243
Toxicology, 91
Traction, 88
Trademark, 17
Tragacanth, 203
Trans, 24, 88, 206, 242, 260
Transducer, 160, 163
Translink, 259
Transluscent, 80, 154, 211, 250
Transmission, 104, 107, 141, 174, 177, 178, 179
Transmittance, 140
Transmitted, 177, 178
Transmitter, 158
Transmitting, 52
Transparency, 59, 60, 220, 256
Transparent, 60, 71, 79, 134, 154, 159, 160, 209, 211, 250
Transport, 13, 40, 52, 128, 170

Triacrylates, 100
Trials, 9, 116, 142, 164, 196, 198
Triangle, 86
Triazene, 238
Triazoles, 237
Tributyltin, 238, 242
Trichlorethylene, 216
Trichlorethane, 215
Trichresyl, 210, 217
Tridecyl, 56
Triester, 209
Triethanolamine, 73, 112, 198
Trifluoromethanesulfonic, 226
Trimer, 106
Triol, 103
Tristearin, 181
Trivalent, 196
Trollers, 161
Troy, 238, 242, 244
Troysan, 238, 242
TSC, 122, 131, 139
Tuberculin, 43
Tubeworms, 243
Tulsa, 157
Turbidimetry, 132, 133
Turbidity, 39, 104, 132, 134, 172, 197, 211, 220

Ubbelohde, 160, 162
UC, 209, 210
UCARCIDE, 238
UF (Urea/formaldehyde), 195, 224, 230
Ugine, 68
Ultra, 56, 180
Ultra-fresh, 238
Ultracentrifugation, 135, 145
Ultrafiltration, 142
Ultralink, 259
Ultramarine, 250, 251
Ultrasound, 72
Un-crystallize, 258
Undecylenic, 238
Uniloc, 142
Union, 67, 105, 107, 160, 173, 238
Unlinking, 231
Unpigmented, 178
Unsaturated, 27, 62, 64, 112, 124, 230
Unsaturates, 147
Unsaturation, 92, 104
Unsaturations, 113
Unshielded, 43
Uraneck, 25, 33, 73
Urea, 75, 106, 225, 229, 251
Urethane, 105, 106, 175, 219, 224, 230, 243
USI, 67
UV (ultraviolet), 5, 60, 62, 75, 82, 134, 178, 182, 194, 233, 235, 236, 241, 251, 254
UV-Humidity, 236

Vacuum, 38, 46, 48, 49, 50, 51, 53, 90, 91, 106, 123, 175
Vancide, 238
Vanderbilt, 238, 242
Vanderhoff, 56, 57, 68, 74, 76, 83, 134, 142
Vanside, 242
Vapor, 49, 104, 107, 141, 147, 179, 180, 181, 187, 188
Variac, 38
Varnish, 71, 245
VC, 64, 65
VDC, 66
Vent, 46, 52
Vented, 47, 53, 240
Vermiculite, 256
Versatic, 64
Version, 55, 178
Versions, 201
Versus, 116, 137, 138, 165, 181, 183, 184, 226, 231
Vertical, 140, 164
Vessel, 16, 24, 26, 27, 35, 38, 52, 53, 90, 109, 240, 254
Vessels, 35, 40, 52, 53, 88, 240, 244
VI, 113
Vibration, 72, 122
Vibrational, 154, 159, 165
Villacorta, 113
Vinegar, 119, 146

Vinyl, 15, 25, 26, 28, 30, 39–42, 46, 47, 49, 53, 56, 61–69, 71, 72, 77, 80, 83, 84, 95, 96, 98, 105, 119, 140, 145–147, 173, 178, 183, 195, 201–203, 207, 221, 234, 243, 245
Vinyl acetate, 15, 25, 28, 39, 40, 53, 61–64, 69, 83, 84, 98, 119, 140, 145, 146, 173, 178, 183, 195, 221, 243
Vinyl acetate, poly-, 119, 178, 195
Vinyl-acrylics, 81
Vinylacetylene, 91
Vinyl alcohol, poly, 15, 63, 72, 140, 202, 207
Vinylanthracene, 14
Vinylbenzylchloride, 71
Vinylcyclohexene, 47, 49, 88, 111, 147, 173
Vinyl Chloride, 41, 42, 46, 47, 62, 64, 65, 67, 69, 80, 98, 147, 221, 234 (*See also* VC.)
  poly, 80, 221 (*See also* PVC.)
Vinyl ether, 61
  methyl, 201
Vinylidene chloride, 61, 65, 66, 67, 69, 98, 234 (*See also* VDC)
Vinyl pyridine, 95, 96
Vinylpyrrolidone, 73
Vinyls, 19, 61, 66, 69
Virtis, 49
Visco-amylo-graph, 155
Viscobalance, 159
Viscocorder, 156
Viscoelastic, 153, 164, 165, 170
Viscolab, 160
Viscometer, 121, 122, 148, 154–165, 184
Viscometric, 149, 182
Viscometry, 120, 121, 125, 128, 146, 149, 150, 154, 155, 157, 159, 160, 165, 180, 184
Viscontrol, 159
Viscosities, 36, 153, 155, 162, 163, 206
Viscosity, 9, 11, 12, 15, 17, 20, 36, 40, 52, 74, 88, 89, 116, 120–122, 125, 126, 128, 131, 142–144, 146–166, 170, 171, 175, 179, 180, 184, 191, 192, 194, 197, 201–205, 209, 211, 220, 224, 248, 249, 256
Viscotek, 161
ViscoTester, 158
Viscous, 11, 15, 39, 127, 148, 149, 160, 172, 193, 208
Visible, 4, 67, 139, 140, 177, 235, 250–252
Visiting, 239
Viskositat, 206
Visual, 241
VM, 161
VOC (volatile organic compounds), 221, 226, 229
Volatile, 65, 72, 123, 147, 208, 255
Volatility, 65, 237
Voltage, 11
Volumetric, 38, 42
Volumetrically, 49, 124
VOR, 155, 160
Vortex, 52, 188
Vorti-Sieve, 54
Voss, 100, 152, 219, 230
VPO (vapor phase osmometer), 180, 181, 182, 187
VTA, 157
Vulcanization, 87, 223
Vulnerable, 224
VWR Scientific, 123, 126
Vyosokomol, 69, 85, 92

Wacker, 69
Waring, 136, 138, 174, 256
Waste, 50, 179, 210, 218, 222
Watch, 40, 50, 52, 123, 139, 143, 144
Watching, 140, 235
Water, 8–11, 13, 14, 16, 21, 23, 24, 28, 29, 33, 38–43, 46, 48, 49, 52, 54–56, 59, 60, 63–68, 75, 82, 83, 88–90, 92, 94, 96–99, 103–106, 117,

[Water—continued]
119, 120, 122–124, 126, 127, 129, 132, 134–136, 139, 140, 142–144, 146, 147, 155, 157, 164, 172, 173, 175, 177, 179, 184, 185, 191–199, 201–208, 211, 212, 214, 223–226, 229, 230, 234, 236, 238, 240, 241, 243–245, 251, 253–256, 259, 260
Wavelength, 177, 185, 235, 250, 251
Wax, 72, 73, 119, 138, 183
Waxier, 258
WB, 129
Weathering, 72, 75, 84, 92, 178, 236, 241, 251, 254
Web, 115–117, 125, 129, 131, 137, 138, 144, 146, 198
Webb, 177, 187
Webmakers, 116
Webs, 125, 126, 137, 233, 235
Welded, 47, 53
Wells-Brookfield, 157
Wescan, 142, 162, 181
Wetability, 177
Wetness, 104
White, 40, 64, 165, 250, 251, 255
Whiten, 236
Whiteness, 248, 251
Whitening, 250
Wicking, 148
Wiley, 73, 76, 77, 128, 170, 174, 188, 259
Wiley-Interscience, 33, 68, 100
Wire, 26, 98, 99, 125, 126, 174, 175
Wires, 254
Witco, 56
Wollastokup, 259
Wollastonite, 259
Wood, 129, 195, 256, 257
Woodpulp, 207
Wool, 98

X-axis, 212
X-ray, 76, 176, 185, 188
X-rays, 185
XRF, 185, 186
Xylene, 215, 216

YSI, 142

Zahn, 158, 160, 161, 162
Zero, 13, 15, 40, 41, 55, 160, 180, 181, 184, 257
Zeta-Meter, 144
Zeta potential, 11, 13, 143, 144, 145
Zinc, 75, 92, 106, 139, 141, 196, 202, 225, 228, 236, 238, 242–244, 251, 253, 254
Zinc Oxide (ZnO), 139, 202, 236, 242, 243
Zirconium, 75, 225, 228